S260 Geology
Science: Level 2

The

Block 4
Surface processes

Prepared for the Course Team by Evelyn Brown, Angelo Coe (Block Chair), Peter Skelton, with a concluding Section by Chris Wilson

The S260 Core Course Team

David Rothery *(Course Team Chairman and Author)*

Glynda Easterbrook *(Course Manager and Author)*

Iain Gilmour *(Multimedia Development Coordinator and Author)*

Angela Coe *(Block 4 Chair and Author)*

Other members of the Course Team

Gerry Bearman *(Editor)*

Roger Beck *(Reader)*

Andrew Bell *(Author)*

Steve Best *(Graphic Artist)*

Evelyn Brown *(Author)*

Sarah Crompton *(Designer)*

Janet Dryden *(Secretary)*

Neil Edwards *(Multimedia)*

Nigel Harris *(Author)*

David Jackson *(BBC)*

Jenny Nockles *(Design Assistant)*

Pam Owen *(Graphic Artist)*

David Palmer *(Author & Multimedia)*

Rita Quill *(Course Secretary)*

Jon Rosewell *(Multimedia)*

Dick Sharp *(Editor)*

Peter Skelton *(Author)*

Denise Swann *(Secretary)*

Tag Taylor *(Design Co-ordinator)*

Andy Tindle *(Multimedia)*

Fiona Thomson *(Multimedia)*

Mike Widdowson *(Author)*

Chris Wilson *(Author*

John Taylor *(Graphic Artist)*

This publication forms part of an Open University course S260 *Geology*. The complete list of texts which make up this course can be found at the back. Details of this and other Open University courses can be obtained from the Student Registration and Enquiry Service, The Open University, PO Box 197, Milton Keynes MK7 6BJ, United Kingdom: tel. +44 (0)845 300 60 90, email general-enquiries@open.ac.uk

Alternatively, you may visit the Open University website at http://www.open.ac.uk where you can learn more about the wide range of courses and packs offered at all levels by The Open University.

To purchase a selection of Open University course materials visit http://www.ouw.co.uk, or contact Open University Worldwide, Michael Young Building, Walton Hall, Milton Keynes MK7 6AA, United Kingdom for a brochure. tel. +44 (0)1908 858793; fax +44 (0)1908 858787; email ouw-customer-services@open.ac.uk

The Open University, Walton Hall, Milton Keynes, MK7 6AA

First published 1999. Reprinted with corrections 2002. Second edition 2007

Edited, designed and typeset by The Open University.

Printed in Europe by the Alden Group, Oxfordshire

ISBN 978 0 7492 1873 7

2.1

BLOCK 4 SURFACE PROCESSES

CONTENTS

1 **Introduction to Block 4** 5
- 1.1 The observation → process → environment concept 5
- 1.2 Objectives for Section 1 8

2 **Sedimentary processes and change: looking for clues** 9
- 2.1 Objectives for Section 2 10

3 **From rocks to sedimentary materials: the weathering process** 10
- 3.1 An introduction to weathering 10
- 3.2 Physical weathering: from rock exposure to fragments 10
- 3.3 Chemical weathering: decomposing and dissolving 11
  - 3.3.1 The susceptibility of silicate minerals to chemical weathering 13
  - 3.3.2 Rocks and their rates of chemical weathering 14
- 3.4 Concluding comments 15
- 3.5 Summary of Section 3 15
- 3.6 Objectives for Section 3 16

4 **Sediments on the move** 17
- 4.1 Transport by fluids 18
  - 4.1.1 How sediments move in fluids 18
  - 4.1.2 Starting sediment moving in fluids 21
  - 4.1.3 Abrasion during transport 24
- 4.2 Sediment deposition in fluids 25
  - 4.2.1 Sedimentary fabrics and textural maturity 27
  - 4.2.2 Layering in suspension deposits 29
  - 4.2.3 Layering in bedload deposits 30
  - 4.2.4 Bedforms produced by water currents 30
  - 4.2.5 Wave-produced bedforms 37
  - 4.2.6 Aeolian bedforms 42
- 4.3 Transport and deposition by sediment gravity flows 45
  - 4.3.1 Grain flows 45
  - 4.3.2 Debris flows and mud flows 46
  - 4.3.3 Density currents and turbidity currents 46
- 4.4 Glacial transport 49
- 4.5 Summary of Section 4 49
- 4.6 Objectives for Section 4 50

5 **Sediments from solution** 52
- 5.1 Biological processes and sediment formation 52
- 5.2 Chemical precipitation and sediment formation 54
- 5.3 Bedforms in carbonate sediments 56
- 5.4 Summary of Section 5 57
- 5.5 Objectives for Section 5 57

6 **From sediments to rocks** 58
- 6.1 Compaction 58
- 6.2 Cementation in sandstones 59
  - 6.2.1 Silica cement 59
  - 6.2.2 Carbonate cement 60
  - 6.2.3 Iron oxide cement 60
- 6.3 Cementation in limestones 60
- 6.4 Nodules 61
- 6.5 Naming and classifying sedimentary rocks 62
  - 6.5.1 Classifying siliciclastic rocks by grain size 62
  - 6.5.2 Classifying siliciclastic rocks by mineralogical composition 63
  - 6.5.3 Classifying limestones 63
- 6.6 Concluding comments 64
- 6.7 Summary of Section 6 65
- 6.8 Objectives for Section 6 65

7 **Introduction to fossils** 66
- 7.1 Scientific value of fossils 67
- 7.2 Objectives for Section 7 68

8 **Interpreting fossils** 69
- 8.1 Objective for Section 8 71

9 **Fossil classification and palaeobiology** 72
- 9.1 Classifying body fossils 72
  - 9.1.1 Specimens with one skeletal element 73
  - 9.1.2 Specimens with two skeletal elements 74
  - 9.1.3 Specimens with many skeletal elements 74
- 9.2 Making sense of body fossils 77
  - 9.2.1 Ecology of marine animals and their fossils 77
- 9.3 Interpreting fossils as living organisms 80
  - 9.3.1 The bivalve *Circomphalus* (Fossil Specimen I) 81
  - 9.3.2 The echinoids *Pseudodiadema* and *Micraster* (Fossil Specimens C and J) 85
  - 9.3.3 Gastropods (Fossil Specimens A and K) 90
  - 9.3.4 Ammonoids (Fossil Specimen D) 93
  - 9.3.5 Brachiopods (Fossil Specimens F and L) 96
  - 9.3.6 Crinoids (Fossil Specimen H) 97
  - 9.3.7 Corals (Fossil Specimens E and M) 99
  - 9.3.8 Trilobites (Fossil Specimen G) 100
- 9.4 Relationships and classification 102
  - 9.4.1 Recognizing bodyplans and phyla 103
- 9.5 Body fossils of other organisms 105

9.6 Trace fossils 106
9.6.1 Mining buried food in quiet environments 108
9.6.2 Feeding in current-swept environments 110
9.6.3 Traces and water depth 114
9.7 Summary of Section 9 115
9.8 Objectives for Section 9 117

10 Fossilization 118
10.1 From death to final burial 118
10.1.1 Death 118
10.1.2 Decay and disintegration 120
10.1.3 Fossilization potential 121
10.1.4 Biological attack 122
10.1.5 Physical effects 123
10.1.6 Chemical attack 124
10.1.7 Burial and final burial: pre-fossilization 124
10.2 Diagenesis following final burial 125
10.2.1 Moulds 125
10.2.2 Casts and skeletal alteration 126
10.2.3 Exceptional preservation of soft tissues 127
10.2.4 Sedimentary compaction 128
10.3 Metamorphism of fossils 128
10.4 Summary of Section 10 128
10.5 Objectives for Section 10 129

11 Fossils as clues to past environments 130
11.1 Clues to physical conditions 132
11.2 Palaeoecological relationships between fossil organisms 135
11.3 Summary of Section 11 137
11.4 Objective for Section 11 138

12 Introduction to sedimentary environments 139
12.1 Factors that control sediment deposition 140
12.2 Facies 142
12.2.1 Vertical successions of facies 142
12.2.2 Walther's Law 144
12.3 Vertical successions of facies and graphic logs 144
12.4 Facies and environmental models 148
12.5 Summary of Section 12 148
12.6 Objectives for Section 12 149

13 Alluvial environments 150
13.1 The river valley 152
13.2 Fluvial channel styles and their typical sedimentary successions 152
13.2.1 Meandering river systems 153
13.2.2 Braided river systems 157
13.3 Alluvial fans 160
13.4 Summary of Section 13 160
13.5 Objectives for Section 13 161

14 Deserts 162
14.1 Wind in the desert 162
14.2 Water in the desert 164
14.3 Summary of Section 14 165
14.4 Objectives for Section 14 165

15 Siliciclastic coastal and continental shelf environments 166
15.1 Subdivision of the coastal and continental shelf environment 166
15.2 Interaction of fluvial, tidal and wave processes 167
15.3 Strandplains 172
15.4 Barrier islands 175
15.5 Deltas 177
15.5.1 Controls on delta building 177
15.5.2 Delta architecture and deltaic successions 178
15.6 Summary of Section 15 181
15.7 Objectives for Section 15 182

16 Shallow-marine carbonate environments 183
16.1 Controls and processes affecting carbonate deposition 184
16.2 Similarities and differences between shallow-marine carbonates and siliciclastics 186
16.3 Carbonate platforms 187
16.4 Peritidal facies 187
16.4.1 Tidal flats 188
16.4.2 Lagoons 191
16.4.3 Carbonate sandbodies 192
16.5 Carbonate build-ups (including reefs) and biostromes 192
16.6 Carbonate platform facies models 197
16.6.1 Carboniferous-age carbonate ramp 197
16.6.2 Modern-day South Florida 198
16.7 Summary of Section 16 198
16.8 Objectives for Section 16 201

17 Deep-sea environments 202
17.1 Processes and sedimentation in the deep sea 202
17.2 Sediment gravity flows in the deep sea 205
17.2.1 Sediment gravity flow processes 205
17.2.2 Turbidites and the Bouma Sequence 205
17.2.3 Debris flows and mud flows 209
17.2.4 Sediment gravity flow successions and facies models 209
17.3 Hemipelagic and pelagic sediments 212
17.3.1 Hemipelagic and pelagic depositional and erosional processes 212
17.3.2 Pelagic successions 212
17.4 Summary of Section 17 213
17.5 Objectives for Section 17 214

18 A matter of time 215

Answers to Questions 219

Acknowledgements 231

Index 233

1 INTRODUCTION TO BLOCK 4

This Block is about sediments, sedimentary rocks and fossils. It is comparable to Block 3 in that there is a lot about geological *processes*, in this case the sedimentary processes that act on the Earth's *surface*. In a similar fashion to Block 3, we will examine processes that we can see happening today and use our understanding of these to help interpret events that occurred in the past and which are recorded in the geological record. This Block will give you the opportunities to practise the skills of observation and reasoning introduced in Blocks 1 and 2 and to learn how to put these together to understand the Earth's surface processes and sedimentary environments. Spend time studying the Figures and Plates carefully in this Block because they contain a lot of valuable information.

To give you an overall flavour of this Block and to illustrate the principles and methodologies for studying sedimentary rocks and their fossils, the first Activity involves watching a video about the kind of features you can see along a modern coast. In it, we compare features of this modern coast with ancient sedimentary deposits that are also interpreted as having been deposited along a coast, but in this case over 300 Ma ago. In both the ancient and modern examples, the animals and plants or their remains, which together may be termed the **biota**, are also considered.

On Ordnance Survey maps, the coastline is defined by the mean high-tide level. However, the position of the coastline changes rapidly. If you look at old maps of Britain drawn several centuries ago, you will find that along the east coast of England, entire villages have disappeared as the coastline has retreated. For example, it is estimated that the Holderness coast of Humberside has retreated some 4–5 km since Roman times. Locally, rates of retreat may be very high. At Covehithe, on the East Anglian coast, the rate of retreat between 1925 and 1950 averaged 5 m yr^{-1}, and during 1953 a storm surge removed a 12 m depth of cliff in one day! Conversely, the rate of growth of new land at Dungeness foreland in Kent averaged 5 m yr^{-1} between 1600 and 1800.

If we look back over geological time-scales, we find that the coastline has changed even more dramatically because of both tectonic processes and sea-level change. In fact, for many periods of geological time, Britain was entirely covered by the sea. Thus, the coastline is a very dynamic environment that is constantly changing in response to the interaction of different sedimentary processes and organisms.

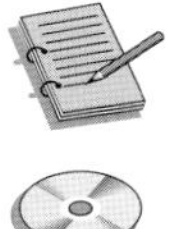

Activity 1.1

Now complete Activity 1.1 which involves watching the video sequence *Coastal processes* on DVD 3 and answering some questions about material shown in the sequence.

1.1 THE OBSERVATION → PROCESS → ENVIRONMENT CONCEPT

In previous Blocks, you were introduced to the concept that careful observations on rocks can be used to determine the processes that formed them and the possible environments in which these processes may have occurred. In effect you have the following logical thought sequence:

observation → process → environment

This thought sequence provides a conceptual framework in this Block and at Summer School, and we will continually refer back to it. It is important to realize that when studying sedimentary rocks and their fossils, many useful and diagnostic observations can be made on hand specimens and on exposures in the field without the need for sophisticated equipment. The processes of formation and deposition of sedimentary rocks can be readily deduced from putting together these observations; for instance, questions like 'was the sedimentary rock deposited by wind or water?' can be answered by looking at the texture of the grains in the rock, any sedimentary structures that formed at the time of deposition and any fossils the rock contains. Similarly, a question like 'was the direction of transport always the same or was it changing?' can be answered by making observations on the type of sedimentary structure and deducing how the structures formed. The correct answers to these kinds of questions allow us to hypothesize about the different possible environments in which the rocks may have been deposited. However, to determine which unique environment is represented, many observations about the rock and its field relationships are often required, though it is amazing how much one can determine from just a single hand specimen!

An example of how the observation → process → environment concept can be used is illustrated in the following question.

Question 1.1 For all the parts of this exercise, you will need to use RS 10 and the small card showing the grain-size scale. Part (a) involves completing Table 1.1 and parts (b) & (c) involve answering a number of short questions.

(a) *Observations*

You have already seen in Block 2 that we can make a series of observations about sedimentary rocks. These can be grouped into the useful checklist in Table 1.1 so that you know you have not missed anything. Remember that even 'negative' observations, i.e. recording that something is *not* present, are often as important as 'positive' ones. Do not forget to use your grain-size scale. If you are having difficulty deciding on the composition, then look at the thin section of this rock (TS W). Do not be daunted if you cannot see all the features (which are difficult to ascertain in fine-grained sedimentary deposits, for example). Just record as much as you can on the right-hand side of Table 1.1. Remember that basic features like colour and hardness give an indication of the composition.

(b) *Processes*

From these observations we can deduce the processes that were responsible for the deposition of the original sediment. Answering the following questions will guide you through this stage.

(i) Was RS 10 deposited where the energy of the transporting medium was relatively high or low, or where there was a mixture of high and low energy?

(*Hint:* In siliciclastic rocks, particle size generally indicates the energy conditions: clay and silt indicate fairly low energy, whereas larger particles indicate increasing energy because higher current speeds are required to move them and the clay and silt grains have been winnowed away. In carbonate rocks, the size of the clasts is often dependent on the organisms present rather than the energy conditions, because some produce large shells and skeletons and others small ones.)

(ii) Do the grains look as though they have (1) been moved around and sorted or been derived from a rock with well-sorted grains, or (2) are they poorly sorted because they have not been transported far or because they have been derived from a rock with a mixture of grain sizes?

(*Hint:* The sorting and grain morphology of RS 10 should form the basis of your answer to this question.)

Table 1.1 Checklist for sedimentary rock interpretation. For use with Question 1.1.

Property		Questions	Observations for RS 10
Composition	Grains	What is the composition of the most abundant grains?	
	Matrix	Is there any finer-grained fragmentary material infilling the spaces between larger grains? If so, what is it?	
	Cement	Is there any crystalline material precipitated around the edges of grains, or in the spaces between grains? If so, what is it?	
Texture	Crystalline or fragmentary	Is the rock crystalline or fragmentary?	
	Grain size	What is the most abundant grain size present (use your grain-size scale)?	
	Grain sorting	Are the grains all more or less of the same size (i.e. well-sorted) or different sizes (i.e. poorly sorted)?	
	Grain **morphology**:		
	shape or form	Are the grains long and thin or equidimensional?	
	roundness	Do the grains have rounded or angular corners?	
	sphericity	Are the grains like spheres?	
	Grain surface texture	Are any quartz grains present smooth and glassy, or are they frosted?	
	Grain fabric (packing)	Are the grains orientated in any preferred direction?	
		Are the grains closely packed together? Are the grains matrix- or grain-supported?	
Fossils		Can you see the remains of any biota or their movements?	
Sedimentary structures		Are there any obvious layers in the rocks?	

(iii) Was the sediment deposited in water or in air?

(*Hint:* Remember from Block 2 that the surface texture of the grains tells you this. If the grains are smooth and glassy, the sediment has been deposited in water because the grains are cushioned; if it was deposited in air, the grains are constantly colliding and knocking tiny chips off each other creating surface pits which give the grains a frosted appearance.)

(iv) What other evidence (or lack of it) is there to support your answer to part (iii)?

(v) Does the presence (or absence) of any sedimentary structures tell you anything about the processes?

(c) *Environment*

In what possible environments could this rock have been deposited?

Let us now return to thinking about the video that you watched in Activity 1.1. In this video, you saw the evidence for several different processes having operated, including waves, tides, wind and the action of animals and plants, and we examined what features they produce along a modern coastline. This knowledge then allowed us to make observations on the Carboniferous rocks and deduce what processes were responsible for depositing these rocks.

In this instance, it was relatively easy because we were making a direct comparison between the modern-day coast and a Carboniferous example to teach you the concepts. However, in the real world, when we look at a series of specimens and rock exposures, geologists do not have the magic power to instantly decide the environment of deposition. They have to make observations and determine the processes from first principles. With experience, this can be done quite quickly, but beware – even professional geologists can make mistakes through lack of careful observations!

Subsequent Sections of this Block are structured on the observation → process → environment concept. Sections 2–11 will consider observations and processes; i.e., how to make relevant observations about sediments, sedimentary rocks, sedimentary structures, fossils and about the mechanics, chemistry and biology of the different types of surface processes on the Earth. Sections 12–17 will demonstrate how observations and processes can be used to reconstruct the environments of deposition (e.g. deep sea, desert, etc.). Ancient environments reconstructed from evidence preserved in the rock record are termed **palaeoenvironments**.

Before you move on, attempt Activity 1.2 which introduces you to the Sedimentary Environments Panorama that you will use several times throughout your study of this Block.

Activity 1.2

The features and processes that occur along the coast are described on the Sedimentary Environments Panorama on DVD 3. This Activity will introduce you to the DVD package and reinforce your understanding of coastal processes and features.

1.2 Objectives for Section 1

Now you have completed this Section, you should be able to:

1.1 Describe, in very general terms, the relationships between sediments and sedimentary rocks; living biota and fossils; and sedimentary structures and bedforms in modern and ancient coastal environments.

1.2 List the key observations that can be made about sedimentary rocks that reveal the processes contributing to their formation.

1.3 Give examples from the video sequence *Coastal processes* on DVD 3 or RS 10 of how observations can be used to interpret sedimentary processes.

1.4 Summarize how a series of observations and interpretations of the processes for a succession of sedimentary deposits can contribute to its interpretation as a succession deposited along a coast.

2 SEDIMENTARY PROCESSES AND CHANGE: LOOKING FOR CLUES

The video sequence *Coastal processes*, which you watched in Activity 1.1, is a salutary reminder that beaches are more than sites of leisure, pleasure and general relaxation. They are also considerably more geologically active than the idyllic scenes depicted in holiday postcards would have us believe. Complex interactions between moving water, shifting sediments and the activities of living organisms are taking place continuously. As you saw in the video, these same processes leave behind clues that may be preserved in sedimentary rocks.

As geologists, we assume the role of detectives. We hunt for clues, decide what they can tell us about the processes that shaped sedimentary rocks, and then put them together to reconstruct the 'scene of the crime'. In essence, we use the principle of **consilience** (i.e. corroboration of an hypothesis by independent testimony, as in a legal trial), a principle frequently demonstrated by Sherlock Holmes, probably the greatest of all fictional detectives, who would also have made an excellent geologist! Happily, geologists are luckier than most detectives because much of what is found at the site of deposition is relevant evidence. An important skill is learning to recognize what counts, and what does not. Before we can begin to interpret the evidence, we have to look at sediments in the making to find out what leads to the particular assemblage of characteristics, or suite of clues, observed in the rocks. This assemblage provides a record of the history of the sediment from its source to its site of deposition.

Activity 2.1

You should now work through Activity 2.1 which introduces you to some of the evidence that might be used to interpret sediment deposition in a particular sedimentary environment. You will need to look at Plates 2.1–2.6 to complete this Activity.

The answers to Activity 2.1 illustrate not only how much variety there is among even a limited range of sedimentary deposits, but also how dynamic surface and sedimentary processes are. From the moment a rock is exposed at the Earth's surface and begins the weathering processes that lead to its disintegration, to the point when the sedimentary materials generated are deposited, buried and compacted to form new sedimentary rocks, change is occurring. This change takes place on the scale of exposures to mineral grains, and on time-scales ranging from millennia (or longer) to periods as short as decades, years or even only days. Some of these changes have already been hinted at through the observations we asked you to record in Activity 2.1. The next stage in the detective hunt is to ask *why* and *how* the changes occur. This is the stage at which you can begin to relate observation to *process*. Here are some of the questions you may have asked yourself as you worked through the Activity.

- Why do rock exposures disintegrate, leaving shattered fragments that accumulate on hill slopes?
- How were the quartz sand grains in Plate 2.6a and b extricated from the rocks from which they are derived?
- How do rock fragments like those in Plate 2.1 become rounded pebbles? And why are wind-blown sand grains so different from those deposited in water?
- Why are the pebbles in some deposits grain-supported whereas in others they are matrix-supported?

- How is weathered debris sorted out such that pebbles can end up in one coastal locality, whereas sand or mud is found in others? For that matter, where does mud come from in the first place?

You may already have an idea about the answers to some of these questions. If so, you will soon find out if you are correct. In the next three Sections, you will see how geologists set about answering them in a logical fashion, using the sorts of observations of features you made in Activity 2.1 to help understand the processes that produced these features, and others that you encountered in the video sequence *Coastal processes* (Activity 1.1).

2.1 OBJECTIVES FOR SECTION 2

Now you have completed this Section, you should be able to:

2.1 Describe some of the features that may characterize sedimentary deposits.

2.2 Compare some of the features of sedimentary deposits, and record their differences.

2.3 Suggest some of the changes that take place as sedimentary materials are moved from their site of origin to their site of deposition.

3 FROM ROCKS TO SEDIMENTARY MATERIALS: THE WEATHERING PROCESS

3.1 AN INTRODUCTION TO WEATHERING

We left Section 2 with a whole lot of questions that require answers. In this Section we will explain, among other things, why and how solid rock exposures disintegrate, and what the products are likely to be. In the first place, rocks disintegrate because they are exposed to the atmosphere at the Earth's surface, where they are attacked by water (in the form of frost, rain and river water), and invaded by plants and animals. Rock breakdown due to these agents comes under the general heading of weathering, a concept introduced in Block 2 Section 7.

Question 3.1 Two main types of weathering were described in Block 2. What are these, and what are the main products of each?

We can also distinguish a third form of weathering, *biological weathering*, which includes the roles played by plants, and even the microscopic organisms and larger animals living in soils. However, as this biological havoc is wreaked through a combination of physical and chemical processes, the outcomes of biological weathering are the same as those of physical or chemical weathering. An essential ingredient in all forms of weathering is water, to a greater or lesser extent.

3.2 PHYSICAL WEATHERING: FROM ROCK EXPOSURE TO FRAGMENTS

As its name suggests, physical weathering leads to rock breakdown by purely mechanical means, as a consequence of temperature changes or other causes of stress within rock bodies. Frost shattering is one example of physical weathering. Water may look innocuous when it collects in cracks and crevices within rocks, but it can be remarkably destructive if it freezes. This is because water expands when it freezes and so ice takes up more room than water in its liquid state, in fact 9.2% more room.

It may seem implausible that something as apparently fragile as ice could shatter something as durable as granite. Appearances, however, are deceptive. The pressure attained by water freezing in rock crevices can reach 14 MPa, whereas the tensile strength of granite (the pressure at which it will yield by fracturing) is only 4 MPa.

In highland regions, frost shattering can create spectacularly alien landscapes, spiked with jagged peaks and littered with rock fragments (Figure 3.1). It takes many cycles of freezing and thawing to shatter rocks as extensively as those shown in Figure 3.1, although the process was probably hastened by the well-cleaved and jointed nature of these particular rocks.

❑ Why should cleavage and jointing be important?

■ They provide convenient pathways for percolating water.

Figure 3.1 A frost-shattered landscape in the Glyder Range of Snowdonia.

Freeze–thaw processes are most likely to be active in climatic zones where temperatures fluctuate intermittently across the freezing point of water for significant periods of the year, and when rocks are well cleaved or jointed. This means that both climate and the state of the rocks themselves determine the effectiveness of frost shattering. This form of physical weathering is not an issue at lower latitudes where temperatures never fall below 0 °C, and at very high latitudes or altitudes where they rarely rise above 0 °C.

Plants also exploit cracks and crevices in rocks, penetrating them with their roots which expand as they grow longer until they are capable of exerting pressures every bit as great as freezing water.

3.3 Chemical weathering: decomposing and dissolving

Physical weathering opens the way for chemical weathering to proceed more easily. This is because small rock fragments present a greater overall surface area for chemical attack than would the same volume of rock forming part of an exposure.

The whole process of chemical weathering is enhanced by the fact that surface waters, be they rain or river waters, are not pure water. Among other substances they contain carbon dioxide (CO_2) dissolved from the atmosphere. The result is the formation of bicarbonate ions (HCO_3^-) and hydrogen ions (H^+) in solution. The release of these hydrogen ions renders the waters slightly acidic (Equation 3.1):

$$\underset{\text{carbon dioxide}}{CO_2(g)} + \underset{\text{water}}{H_2O(l)} \rightleftharpoons \underset{\text{hydrogen ions}}{H^+(aq)} + \underset{\text{bicarbonate ions}}{HCO_3^-(aq)} \tag{3.1}$$

The H^+ ions from acidic water can displace the metallic ions in a mineral, allowing them to be released into solution, or to be mobilized in other ways depending on water chemistry and climatic factors. The water in soils overlying rocks is often made more acidic as the result of plant activity. Plant roots liberate various organic acids, which means there are even more H^+ ions available to encourage chemical weathering of minerals and rock fragments in the soil. In lowland areas where there is a reasonably thick soil cover and plenty of vegetation, the predominant form of weathering is likely to be chemical weathering.

Whether a mineral ends up completely decomposed, or a new mineral is formed, depends upon the composition and crystal structure of the mineral being weathered. For silicate minerals, it is the way in which their fundamental SiO_4 groups are shared that is important (see Block 2 Section 4.2).

Question 3.2 What are the silicate structures of each of the main rock-forming silicate minerals: olivines, pyroxenes, amphiboles, micas, feldspars and quartz?

Olivines, pyroxenes and some amphibole minerals decompose completely and all their constituents are mobilized. Micas, feldspars and some other types of amphibole are only partly decomposed. These are all minerals in which Al substitutes for Si in some of the SiO_4 tetrahedra. The metallic ions (and some silica) are released into solution, and what is left of the aluminosilicate lattice forms a sheet-like structure, analogous to the structure of the micas, producing a new group of aluminosilicate minerals known as clay minerals. Clays are dominant constituents of muds and soils. Quartz is very pure SiO_2 and has no cleavage, so there are no pathways along which chemical attack can occur. Consequently, it is very resistant to chemical attack and remains largely undecomposed during chemical weathering to form a *residual mineral*, although a little silica may be mobilized in solution given sufficient time.

Mafic minerals (olivines, pyroxenes, amphiboles and biotite mica) contain iron in its reduced state as Fe^{2+}. When these minerals are chemically weathered, iron is released into solution in the same state.

❏ What do you think will happen to this Fe^{2+} when it encounters atmospheric oxygen, dissolved in water?

■ It will be oxidized rapidly to its more oxidized state, Fe^{3+}, to form ferric oxide.

Ferric oxide is highly insoluble and is often left behind as a brown or reddish-coloured material, coating rock surfaces at the site of weathering.

Question 3.3 If exposures of basalt, granite and pure quartzite each underwent extensive chemical weathering, explain which of the rock types would be likely to show a surface coating of iron oxide.

Some ferric oxide may be precipitated onto the surfaces of quartz grains during transport, or after deposition, giving them a yellow or light brown colour. Sandstones and mudstones with an intense red colour caused by ferric oxide coating the grains are known as **red beds** (e.g. see Plate 4.3), and are typical of formation in highly oxidizing environments.

There are exceptions to the scenario we have just presented. For example, when waterlogged swamps and marshes are cut off from the atmosphere, so that any oxygen used up by decomposing vegetation cannot be replenished, or when marine basins develop stagnant bottom waters due to lack of circulation or high organic matter productivity, *reduction* of iron can take place. In these circumstances the brassy-yellow iron (Fe^{2+}) sulfide mineral, pyrite (FeS_2) may form. So the presence of pyrite in a sedimentary rock is often a good indicator of **anoxic** conditions (deficient in oxygen) at some stage during or after the deposition of the original sediment.

3.3.1 The susceptibility of silicate minerals to chemical weathering

Silicate minerals vary considerably in their response to chemical weathering. You have already seen that quartz, for example, is highly resistant relative to other igneous silicate minerals. This is because the higher the Si : O ratio of a silicate mineral, the higher the degree of polymerization of silicate tetrahedra, and so the *less likely* it is to decompose.

Question 3.4 Table 3.1 lists the main groups of igneous silicate minerals, together with their Si : O ratios. Predict the order in which you would expect these silicate minerals to break down during chemical weathering (putting the most easily decomposed first).

Table 3.1 The Si : O ratios of the common igneous silicate minerals.

Silicate mineral group	Si : O ratio
olivine	1 : 4
pyroxene	1 : 3
amphibole	4 : 11 (1 : 2.8)
mica	2 : 5 (1 : 2.5)
Ca-feldspar	1 : 4
K- and Na-feldspars	3 : 8 (1 : 2.7)
quartz	1 : 2

If the predicted order seems familiar to you, it should do because it is very similar to the order in which these minerals would crystallize from a basaltic magma if fractional crystallization were to proceed to completion (Block 2 Plate 6.5). This should not surprise you because the polymerization of silicate tetrahedra increases as temperature of crystallization decreases. As quartz, with the chemical composition SiO_2, has the highest degree of polymerization, it is the most resistant to chemical weathering, and so is left as a residue when rocks disintegrate. A word of caution, however: the nature of the metallic ions can also influence the susceptibility of minerals to chemical weathering. This is true of the micas. Iron-rich biotite mica is more susceptible than muscovite, and so muscovite is more likely than biotite to occur in sedimentary rocks.

You should now understand why quartz is such an important mineral in sedimentary rocks whereas feldspar is a lot less common than it is in igneous rocks; why some sedimentary rocks may contain flakes or layers of mica; and why olivines virtually never occur in sedimentary rocks. It follows from all this that we can use the mineralogical composition of a sedimentary rock to infer something about the extent to which the original sediments have been chemically weathered. Bear in mind, though, that weathering doesn't stop once rock fragments and mineral grains are moved away from their parent rock. It may carry on through various stages of transport, and even after deposition and burial.

Question 3.5 Suppose you were examining a sandstone containing 70% quartz and 30% fresh K-feldspar grains, both derived from the same source. What might you infer about the amount of chemical weathering experienced by the source rock, and the resulting sediment during transport?

We refer to humans (and cheese) at an advanced stage of development as 'mature'. We use the same concept when describing sediments composed of silicate minerals, i.e. the siliciclastic sediments, and talk about their **compositional maturity**. Compositionally mature sediments contain almost nothing except quartz or clay minerals (the end products of chemical weathering). Conversely, compositionally immature sediments contain a good deal of undecomposed rock fragments, fresh feldspar and maybe other silicate minerals which we should expect to decompose rapidly. So the degree of compositional maturity is a qualitative measure of the extent to which a sediment has suffered chemical weathering, throughout *all* the stages in its history.

We have to bear this potentially complex history in mind when we try to use the compositional maturity of a sedimentary rock as a clue to unravelling its history. For example, if a rock is compositionally immature, this could mean it was derived from an igneous or metamorphic rock by almost nothing but physical weathering, and that any subsequent transport involved no water. Alternatively, transport by water might have occurred, but not for long before deposition, so that no significant decomposition of minerals took place. On the other hand, if a sediment is derived from a pre-existing sedimentary rock that is already compositionally mature, then the next generation of sediment will be compositionally mature, too, regardless of how weathering took place, or for how long the sediment may have been transported in water.

Activity 3.1

Now test your understanding of compositional maturity by attempting Activity 3.1.

(a)

(b)

Figure 3.2 Different rates of chemical weathering shown by the clarity of the inscriptions on gravestones. (a) Negligible chemical weathering of a slate gravestone, dated 1708. (b) Relatively rapid chemical weathering of a gravestone made of sandstone with a calcite cement, dated 1878, and 170 years younger than the slate gravestone.

3.3.2 Rocks and their rates of chemical weathering

It follows that if some silicate minerals are more easily decomposed than others, then rocks containing the less resistant minerals will be chemically weathered more rapidly than those containing the more resistant ones.

Question 3.6 Imagine that you have a block of granite and a block of gabbro. You allow these to undergo years of weathering under the same climatic conditions and with the same degree of exposure. Explain which of the rock types you would expect to weather faster chemically.

In addition to mineralogy, there are two other important factors that affect the rate at which rocks are chemically weathered: their texture and the local climatic conditions.

❑ How do the textural relationships between the mineral grains in igneous, metamorphic and sedimentary rocks vary?

■ In igneous and metamorphic rocks, the grains are intergrown (interlocking) to produce a crystalline texture. In most sedimentary rocks, the grains are discrete and separated (although held together in some way), to give a fragmental texture.

Chemical weathering begins along grain boundaries but it is much harder for water to penetrate when crystals interlock (Figure 3.2a) than when mineral grains are simply cemented together, especially if the cement does not fill all of the intergrain spaces, or is less resistant to chemical attack than the mineral grains, for example if the cement is made of calcite (Figure 3.2b). This difference in rates of weathering between crystalline and fragmental textures explains why in general geologically ancient igneous and metamorphic rocks may persist as highland areas, long after younger but less resistant sedimentary rocks have been weathered away.

You have already seen how climate affects the likelihood of frost shattering occurring (Section 3.2). Climate also affects the rate at which chemical weathering takes place as it is enhanced by warm, wet conditions.

❑ Can you think of two reasons for this?

■ First, chemical reactions proceed faster at higher temperatures. Secondly, vegetation is likely to be more abundant, leading to an increase in the amount of organic acids available.

3.4 Concluding comments

We hope we have answered satisfactorily some of the questions posed at the end of Section 2. At the same time, you should have become aware of the different types of sedimentary materials that are produced by weathering, as well as gaining an insight into how we can use the mineralogy of siliciclastic sedimentary rocks to unravel the history of chemical weathering of a sediment. However, our answers inevitably raise more questions, because this is what science is all about. Two of these are:

- If chemical weathering continues as sediment moves away from its parent rock, does physical breakdown carry on too?
- What happens to the soluble metal ions and silica, carried away in solution following chemical weathering?

For the answers to the first of these questions, read on into Section 4. The answer to the second will have to wait until Section 5.

Activity 3.2

You should now be able to answer some of the questions we asked at the end of Section 2. Activity 3.2 will let you see whether you can do this. This Activity also provides you with some practice in constructing written answers to questions.

3.5 Summary of section 3

- Weathering represents the outcomes of interactions between rocks and the atmosphere, water, plants and other organisms. Physical weathering simply fragments rocks whereas chemical weathering decomposes them.
- Physical weathering is due to mechanical stresses within rocks caused, for example, by water freezing and plant growth.
- Chemical weathering occurs when minerals react with water containing hydrogen ions. The products of the chemical weathering of silicate minerals depend upon their compositions and crystalline structures but may include metal ions and silica in solution, insoluble residual materials, and the production of new materials such as clay minerals and iron oxide.
- Oxidation involves the conversion of Fe^{2+} ions to Fe^{3+} ions and their precipitation as insoluble ferric oxide.
- The resistance of igneous silicate minerals to chemical weathering increases as their Si : O ratios increase, and the order in which they decompose is the same as that in which they would crystallize from a magma.
- The mineralogy of siliciclastic sedimentary rocks is used to determine their compositional maturity, which is a measure of how much chemical weathering they have experienced throughout their history.
- The rate of weathering of a rock depends on its mineralogy, texture, climatic conditions, the extent of vegetation cover and how well jointed or cleaved it is.

3.6 OBJECTIVES FOR SECTION 3

Now you have completed this Section, you should be able to

3.1 Describe an example of how physical weathering takes place, and be able to recognize frost-shattering from photographs.

3.2 Describe the roles of water, carbon dioxide and plants in weathering processes.

3.3 Explain how the chemical weathering of silicate minerals takes place and predict the likely chemical weathering products of a rock from its mineralogy, explaining the bases of your predictions.

3.4 Make deductions about the extent of chemical weathering of a siliciclastic sedimentary rock from its mineralogical composition and degree of compositional maturity.

3.5 Predict the likely rate of weathering of a rock, knowing factors such as its mineralogy, texture, the local climatic conditions, and the extent of vegetation cover.

Now try the following questions to test your understanding of Section 3. You will need RS 5 (quartzite) and RS 19 (gabbro) from your Home Kit to answer Question 3.9.

Question 3.7 Plates 2.1 and 4.4 show different landscapes.

(a) For each, decide whether physical or chemical weathering is the dominant process shaping the landscape, and explain how you have arrived at your conclusions.

(b) Which factors visible in these Plates might enhance the rate at which weathering is taking place?

Question 3.8 Table 3.2 is designed to show the end products of the chemical weathering of the main groups of igneous silicate minerals. The weathering products of olivine are already shown. Complete Table 3.2 to show the weathering products of the remaining groups of minerals.

Table 3.2 The end products of the chemical weathering of the main groups of igneous silicate minerals.

Mineral group	Products in solution	New materials	Residual minerals
olivines	metallic ions, silica	ferric oxide	none
pyroxenes			
amphiboles			
biotite mica			
muscovite mica			
feldspar			
quartz			

Question 3.9 Examine RS 19 (gabbro) and RS 5 (quartzite) from your Home Kit.

(a) Which minerals are present in each of these rocks?

(b) What would be the end products of the chemical weathering of each rock? Make it clear whether the products you describe are in solution, insoluble residues or new materials.

4 SEDIMENTS ON THE MOVE

Many of the processes described in this Section are illustrated on the Sedimentary Environments Panorama by video clips. Where we think you might find this helpful, we have put a DVD icon in the margin. You need only refer to the DVD here if you need further clarification.

As you have just seen, the products of weathering consist of rock fragments produced by physical weathering, residual quartz grains that have been released by chemical weathering (although resistant themselves), and other products of chemical weathering, new minerals (particularly clay minerals), and various ions in solution (Figure 4.1). Crystals of mica and feldspar (mainly K-feldspar) may also be released as residual products if chemical weathering is not very advanced.

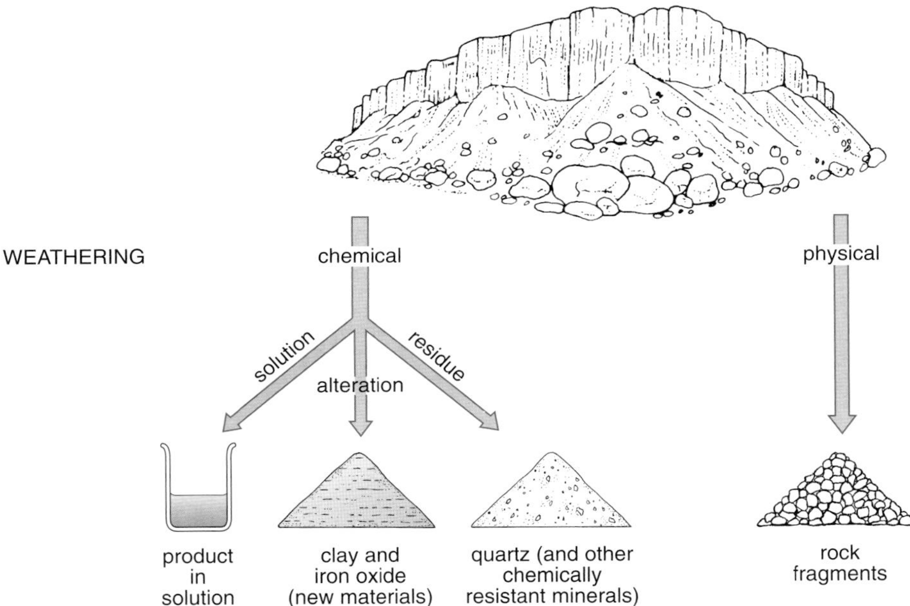

Figure 4.1 A summary of the products of weathering.

Generally speaking, once rock fragments and mineral grains have been released by a combination of physical and chemical weathering, they begin the journey to their eventual site of deposition. This journey is referred to as transport and a record of at least part of it may be contained in the features of the sedimentary rocks that are formed. The process of removal of weathered debris, which acts to lower the surface of the land, is called erosion. This word is borrowed from the Latin verb *erodere*, 'to gnaw', which is the same word that lends its name to rodents, a group of animals such as mice and rats which are obliged to keep their front teeth filed down by constant gnawing. Without significant erosion, the solid products of weathering simply accumulate ***in situ*** (i.e. in the place where they formed) to form the starting materials of soils.

There are two main ways in which sediments are transported: (i) as the result of fluid movement, and (ii) by gravity.

4.1 TRANSPORT BY FLUIDS

We begin by looking at the role of fluids in transporting sediments. Fluids are extremely important agents of sediment transport. If the term **fluid** conjures up for you a picture of something that is only a *liquid,* then this is too limited a picture because the term is used to describe *any* substance that will flow. In the surface environment, the most obvious medium that flows is water. However, there is something else that flows, too, which you can experience every day.

❑ Can you think what this is?

■ The answer is air, or wind (which is how we experience air movement).

Although water is a liquid and air is a mixture of gases, they are both fluids. From a sedimentary geologist's point of view, the only two things that distinguish them are their viscosities and densities.

❑ Can you recall from Block 3, and the discussion of magmas, what is meant by the viscosity of a fluid?

■ It is a measure of how freely a fluid can flow. The *lower* the viscosity, the more easily the fluid will flow.

You might think that water flows pretty freely but its viscosity is around 50 times greater than that of air at room temperature, nearly the same order of magnitude difference as there is between the viscosities of water and olive oil, for example. Water is also 1000 times more dense than air. All this means that although there are many similarities in how water and wind transport and deposit sediment, because both are fluids, there are also differences. So water-transported sediments and wind-blown **aeolian** (pronounced 'ay-owe-lee-an') sediments have features in common and features apart (*Aeolus* was the ancient Greek god of the winds). The distinguishing feature of transport by a fluid, be it water or wind, is that the sediment is carried along in some way by the fluid, and does not move unless the fluid moves.

We will now turn our attention to what happens when rock fragments and mineral grains encounter a moving fluid, using water as our example.

4.1.1 HOW SEDIMENTS MOVE IN FLUIDS

When a mass of sediment enters moving water such as a stream or river, then one of two things is likely to happen to it: either the material will sink to the bottom and stay there, or it will be carried along in some way by the water. Imagine that you are stirring a bucket of water and that, as you stir, you drop a handful of sand and pebbles into it.

❑ Which factors will determine whether a particular sand grain or pebble falls to the bottom of the bucket, motionless, or is swept round the bucket by the moving water?

■ How fast the water is moving (its *speed*), how big the particle is (its *grain size*) and the *density* of the particle. You may have thought of a fourth factor: the *shape* of the particle.

The faster the water is moving, the smaller the particle, the smaller the difference in density between the particle and the water, and the flatter or flakier the particle, the more likely it is to be swept round in the flow. As the bulk of sedimentary materials (rock fragments, quartz grains and clay minerals) do not vary greatly in density, we can ignore this factor. We will consider the significance of flat or flaky grains later.

How the sediment is swept along depends on the way in which the water itself is flowing. It may seem rather strange to you, but there are basically two ways in which water, and any other fluid you care to name, can flow: in a laminar fashion, or turbulently. During **laminar flow**, individual parcels of the fluid move in straight lines parallel to each other (Figure 4.2a); *laminar* simply means 'layered'. When **turbulent flow** occurs, the parcels of fluid move in a more chaotic fashion, sometimes moving upwards and sometimes downwards to create complex eddies, even though the *overall* flow is in one direction (Figure 4.2b). You will see the difference between these two types of flow by watching a lighted cigarette resting in an ashtray or by burning an incense stick in a room where there is no air movement. The smoke begins to rise in parallel streams (laminar flow) but as it drifts higher these break down into a pattern of swirls (turbulent flow). A similar effect can be seen in flowing water injected with dye (Figure 4.2c).

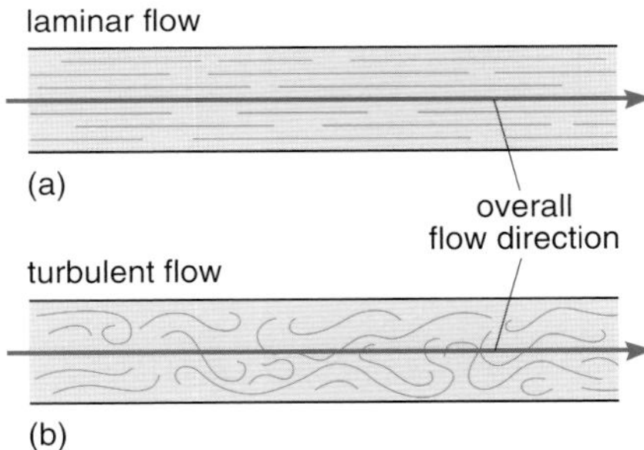

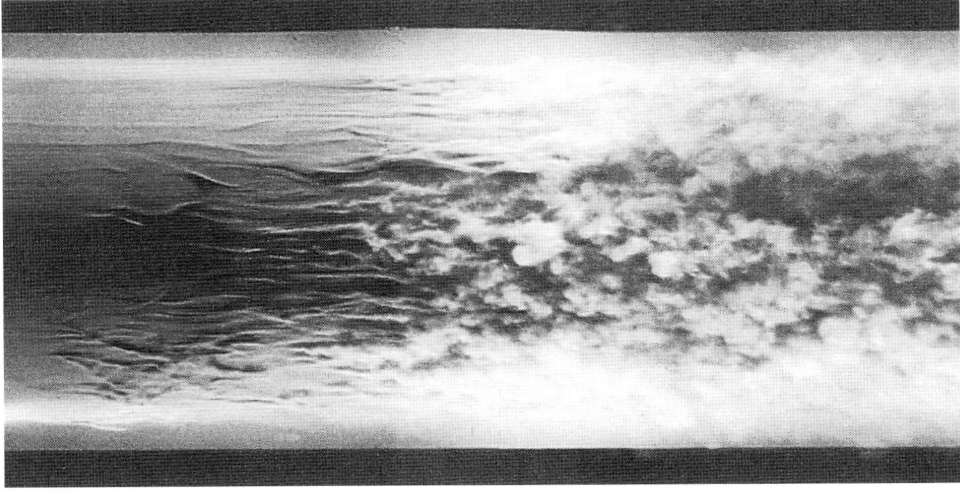

Figure 4.2 The movement of water during (a) laminar and (b) turbulent flow. (c) The transition from laminar to turbulent flow as water, injected with dye, flows over a plate.

There are two major factors that determine whether flow in a fluid is laminar or turbulent. You have already encountered them, but not in this context: the *speed* at which the fluid is moving and its *viscosity*. The likelihood of turbulent flow increases as speed increases and viscosity decreases.

❑ For a given speed, which of the following fluids is most likely, and which is the least likely, to exhibit turbulent flow: air; rhyolitic lava; water?

■ Air is the most likely to because its viscosity is lowest. Conversely, rhyolitic lava is the least likely to because its viscosity is highest.

Felsic magmas such as rhyolite are so viscous that all flow is usually laminar. Effectively all air flow is turbulent, and water needs to flow at only a few cm s^{-1} before it becomes turbulent.

Let's return now to our mass of sediment which has just entered a river, where flow will almost certainly be turbulent. Some of it will be swept along with the water. The largest particles in this sediment will roll or slide across the river bed

whereas somewhat smaller ones will bounce along, knocking into each other and ricocheting upwards (Figure 4.3). This bouncing mode of travel is referred to as **saltation** (from the Latin *saltare*, 'to leap'). Together the rolling, sliding and saltating grains make up what is called the **bedload** of the moving sediment, because they are continuously or intermittently in contact with the bed during transport. The finest particles are caught up in the turbulent eddies, carried in suspension, and never touch base. These form the **suspension load** of the moving sediment (Figure 4.3).

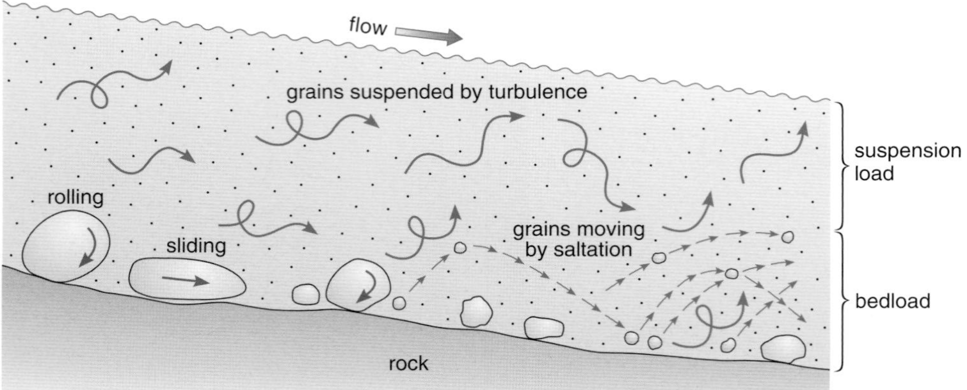

Figure 4.3 Different forms of particle movement in turbulent water.

We cannot fix rigid particle size limits for these different ways of moving because a sand grain that is content to bounce along the bed one day might well have been swept along in suspension, had the water flow been faster. However, we can say that clay particles and very small quartz grains less than about 0.1–0.2 mm in diameter are almost always carried in suspension whereas coarser-grained material will be carried in suspension only at high flow speeds. These observations will prove important when we come to look at what happens when the flow is no longer able to carry the grains, i.e. when **deposition** occurs.

As air is a fluid, wind is capable of moving sediment in much the same way as water. Cast your mind back to the 'thought experiment' involving the bucket of water and imagine that, instead, you drop the same handful of gravel and sand into an airstream moving over a flat surface. As before, some of the sediment will fall through the airstream and come to rest, motionless, on the surface below, but the remainder will be swept along in some way by the wind.

Question 4.1 Assuming that the air is moving at the same speed as the water was in your bucket, explain whether you would expect the sizes of the largest sediment particles swept along by the wind to be the same as those kept moving by the water.

One consequence of these differences between wind and water is that during normal aeolian transport, such as occurs in desert areas, bedload transport by saltation is very much the dominant process. Generally speaking, only grains less than about 0.1 mm will ever make it permanently into suspension, and pebbles are too big to slide or roll as bedload. The carrying capacity of air flows such as tornadoes (whirlwinds) is exceptional so we will not consider them further. As air offers much less frictional resistance to the saltating grains than water, because of its lower viscosity, they can leap higher above the underlying sediment than they would if cushioned by water, forming a moving cloud up to as much as a metre above the surface (Figure 4.4). They also hit the sand grains at rest beneath them at considerably higher speed when they descend. As a result, the assaulted grains are pushed forward as they are struck; not enough to throw them up into the air, but sufficient to cause a slow creep forward in the direction towards which the wind is blowing. An impacting sand grain can

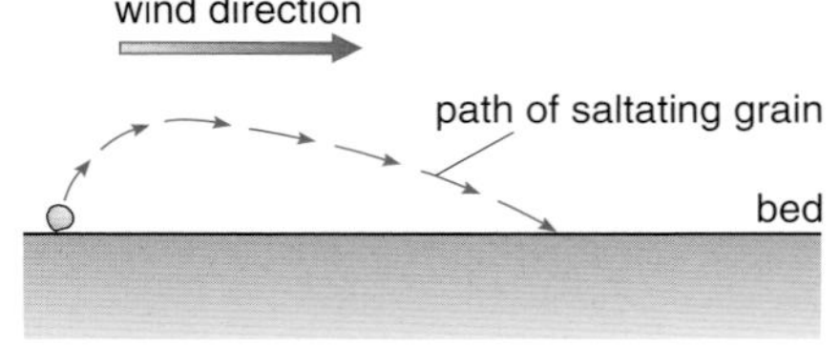

Figure 4.4 The saltation of a sand grain in air.

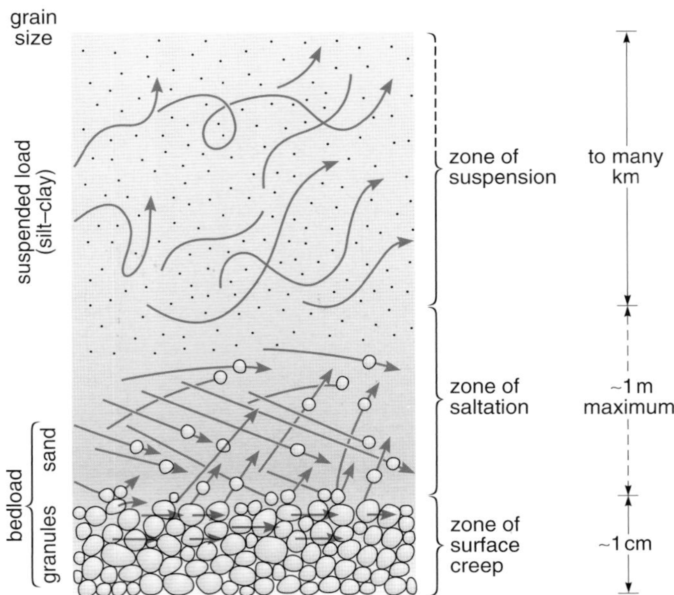

Figure 4.5 The movement of sand grains in suspension, by saltation and by surface creep during aeolian transport.

move another grain up to six times its own diameter through this process of **surface creep** (Figure 4.5). The wind, like water, lifts the finer-grained silts and clays directly into suspension.

These various modes of transport are not equally effective. Large grains rolling or sliding over a river bed, or creeping across a desert floor, are tardy in comparison with saltating grains. The saltating grains, in turn, cannot keep up with finer-grained debris being carried along in suspension.

❑ What will be the consequence of these different rates of movement when, for example, wind blows over a desert floor, weathered rock debris enters a river, or seawater flows over a beach?

■ Finer-grained material will be moved faster and further than coarser-grained material, so it will be separated out, or sorted.

You met the concept of sorting in Block 2 (Section 7.1). The more effectively siliciclastic sediment is separated out to give a narrow range of grain sizes in a sedimentary deposit, the better sorted it is (see *Bookmark*). We will explore a little more fully how sorting occurs in Sections 4.1.2 and 4.2.1.

4.1.2 STARTING SEDIMENT MOVING IN FLUIDS

We have seen what happens to material swept along by flowing water, in a river for example, and compared the outcomes of this with what happens when the same sediment is moved by the wind. But what about the sediment that comes to rest on the bed of a river? Does this represent some form of irreversible deposition? This is highly unlikely, for if the speed of the river increases, as it is liable to do the next time there is rain, then this sediment may well begin moving again.

The conditions required to set in motion the rock fragments or mineral grains at rest on a river bed are shown in Figure 4.6. Movement is caused by the fluid force of the water flowing over the grains. This force is proportional to the speed of the water, and it acts to both push the grains forward (F_d), and to lift them upwards (F_l), in the same way that moving air lifts an aeroplane wing. The grains will begin moving when this force is able to overcome both the force of gravity holding them down (F_g), and friction between the grains and the bed (F_f). Exactly the same principle applies, of course, to any fluid moving over sediment whether it be waves in the sea or wind across a sandy desert surface.

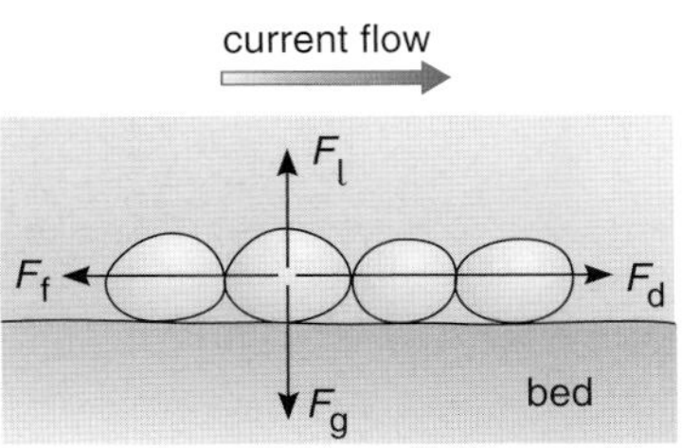

Figure 4.6 The forces acting on a sediment grain at rest on a bed beneath a water current. The force of gravity (F_g) acts to hold the grain down and the frictional force (F_f) impedes motion. However, the forwards push of the water (F_d) and the lift force (F_l) produce a net fluid force which both lifts the grains and propels them forwards.

This process is assisted when turbulent eddies are able to reach down among grains resting on the bed, which is easier when the bed is coarse-grained than when it is fine-grained.

The speeds at which grains of various sizes will begin moving in water, in other words at which erosion will begin, have been determined experimentally. The results are plotted on a graph of water current speed against grain size (Figure 4.7). There are several features of this graph which merit explanation before you can begin to interpret it.

First of all, the scales used for water speed and grain size are not linear, but are *logarithmic scales* (or log scales for short), which increase in powers of ten. This means that the intervals spanned by each division are not the same: water speed and grain size each increase by a factor of ten between one interval and the next. The reasons for using this sort of scale, rather than a conventional linear scale in which the interval between one division and the next is the same, e.g. 2, 4, 6, 8 etc., become obvious when you consider the range of grain sizes present in natural sediments; from minute clay particles to boulders (off the scale of Figure 4.7), and the range of speeds required to start them moving. If you were to try to plot the same data using linear scales, then you would run out of paper well before reaching even the sand-sized material.

Activity 4.1

If you are unfamiliar with log scales, you should complete Activity 4.1.

Secondly, the graph relates specifically to a water flow 1 m deep, so the relationships between current speed and grain size shown will not be exactly the same for flows deeper or shallower than this. This is because the deeper a flow is, the more likely it is to be turbulent and so more able to lift coarser grains than a shallower flow of the same speed.

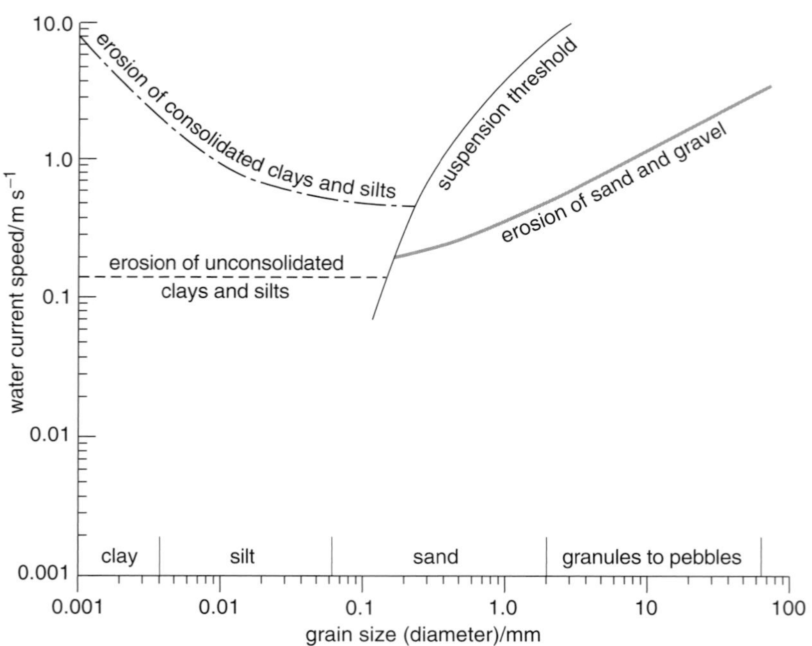

Figure 4.7 Graph to show the speed of a water current, flow 1 m deep, required to erode sediments of various grain sizes. The suspension threshold curve represents the transition from transport as bedload to transport in suspension.

The graph shows that for sand and other coarser-grained sediments, the coarser the grain size the greater the current speed required to initiate erosion. The curve representing the threshold speeds required to move (or erode) these sediments is a best-fit curve drawn through a range of experimental data. A second point to note from Figure 4.7 is that coarse-grained silts, sand, and sediment coarser-grained than sand, begin moving as bedload. The sediment is lifted into suspension only when the current speed reaches the suspension threshold shown on the graph.

Things are a little more complicated where muddy sediments composed of finer-grained silts and clays are concerned. The current speed required to erode these sediments depends upon the extent to which they have become consolidated. This is why two curves for erosion are shown: the upper one represents consolidated sediment and the lower one represents unconsolidated sediment.

Until now we have assumed that any sediment resting on a bed beneath water is totally free to move as individual grains. This is not necessarily true for muds. When muds are first deposited, they contain around 70–90% water by volume, and its rapid expulsion begins as soon as the next layer of sediment is deposited and exerts pressure on the material beneath. The volume of the sediment/water mix decreases as the degree of consolidation increases. Fine-grained sediments, especially clays which are flaky, also show **cohesiveness**, which means that they have a tendency to stick together. The finer grained they are, and the more they are consolidated, the more they stick together and so the faster the flow has to be in order to erode them. When the sediment is eventually lifted, it is not as individual grains, but as small clumps of grains, or even as mud clasts, sufficiently large and heavy to be transported as bedload. This property of muds explains the peculiar up-turn of the upper curve for consolidated fine-grained sediment in Figure 4.7. The lower curve is a theoretical extrapolation for unconsolidated muds. In reality, the erosion of muds may begin at flow speeds somewhere between those delineated by the two curves, depending upon the degree of consolidation that has occurred.

Question 4.2

(a) What, approximately, is the current speed required to put a grain 2 mm in diameter into motion?

(b) Suppose that over a period of time the water speed increases. When it reaches about 10 m s^{-1}, how will transport take place?

Although aeolian transport is not significant in the British Isles today, except in some coastal areas where there are extensive, wind-swept, sandy beaches, it has been widespread in Britain at times during the geological past and is found in other parts of the world. The relationship between sediment grain size and the wind speed required to move it is directly analogous to the relationship between sediment grain size and the water current speed. There are some differences, however, because of the differences in physical properties between the two fluids, in particular their densities and viscosities.

Question 4.3 What are these differences, and how will they affect the relationship between sediment grain size and the wind speed required for sediment movement?

As with water transport, silt-sized particles are lifted directly into suspension by the wind. Fine- to medium-grained sand will begin moving by saltation, and then be lifted into suspension as the wind speed increases – although it would take a gale force wind of about 20 m s^{-1} (72 km hr^{-1}) to lift sand grains 1 mm in diameter into suspension. Coarse-grained sand begins moving by surface creep and gravel coarser than about 4 mm in diameter does not move at all; it is left behind as a **lag deposit**. As wind moves across an expanse of loose sediment, **winnowing** takes place: finer-grained material is removed leaving the coarse-grained material behind. Given this selectivity, and the different rates at which transport by surface creep, saltation and in suspension occur, you can see why the wind is usually more effective at sorting sediments than is water.

4.1.3 Abrasion during transport

You will need access to your hand lens, grain-size chart, microscope, rock specimens and thin sections from this Section through to the end of Section 4.2.2.

The contrast between sedately saltating quartz sand grains and small rock fragments, both cushioned by water, and their more aggressively bouncing aeolian counterparts has important consequences for the particles themselves. Every time these particles strike other particles or rock surfaces over which they saltate, attrition occurs. The angular corners and sharp edges of grains or rock fragments are worn down a little so that in time they become progressively more rounded. Their degree of roundness is a textural property which becomes a measure of the time they have spent in transport: the longer grains and rock fragments are bounced around, the more rounded they become. You should not confuse this concept with how spherical a grain appears to be. The sphericity of a grain is a measure of how close it is to the shape of a sphere (Block 2 Section 7.1 and Figure 7.4).

❑ Bearing in mind the difference in viscosity between water and wind, will this make any difference to how quickly the grains become rounded?

■ Yes, it will. The much lower viscosity of air means that grains are not cushioned as they are by water. The impacts with each other and the bed beneath them are much harder and so rounding occurs very much more rapidly in air.

There is another factor to consider: it takes a higher wind speed than water-current speed to set grains of a given size in motion, so the speed of impact is correspondingly greater which leads to more rapid rounding of the grains. If you have ever walked bare-legged across a wind-swept beach, you will know just how effective abrasion due to saltation can be, from the unpleasant stinging when these same sand grains attack your skin.

We also find that during bedload transport in water larger sedimentary particles are rounded more quickly than smaller ones, because their momentum is greater and so the cushioning effect of the water is less. Freshly weathered angular rock fragments can become sub-rounded pebbles within a few tens of kilometres of rolling and bouncing in a river whereas it would take many thousands of kilometres to effect the same transformation in a small quartz grain. However, should these same quartz grains reach the sea, then their pace of abrasion would speed up dramatically. The constant to and fro motion by waves can bring about thousands of kilometres of particle movement in just one year. Even so, many end up no more than sub-rounded, because the relatively high viscosity of water still cushions them. By contrast, aeolian sands and larger beach pebbles are typically rounded to well-rounded (Plates 2.2 and 2.6).

As you may recall from Block 2 (Section 7.1), microscopic examination of the surface textures of water-borne and wind-blown quartz grains reveals another distinction between the two. Water-transported grains normally have a glassy appearance, like the crystals of quartz in a granite, whereas aeolian grains have a matt finish, similar in appearance to the frosting used to decorate some glass bottles (Plate 2.6a, b). Traditional wisdom has it that this frosting is the result of surface pitting caused by the high speed collisions among the grains in air.

From this discussion, you should have realized that there are now two textural clues we can look for to give us some idea about whether the particles in a sedimentary rock were transported by either wind or water at some stage in their history: grain roundness, and the surface textures of quartz grains. This need not be the most recent stage, though. For example, the wind may build dunes at the back of flat, exposed beaches, using sand derived from the adjacent beach. Although the most recent phase of transport for the dune sands would be by wind, the *distance* in transport may be insufficient for much frosting or extra rounding to take place. Similarly, if an aeolian sandstone is eroded, and the quartz grains are subsequently transported by water, they will not lose their frosting or become less rounded. This means that sedimentary grain textures can be inherited from a previous cycle of transport. For this reason, it is important to collate as many lines of evidence as possible before deciding how a sediment was transported and deposited. In other words, you must remember the principle of consilience (Section 2).

Activity 4.2

Now work through Activity 4.2 which gives you an opportunity to see if you can interpret textural clues.

We have now been able to find three independent lines of evidence pointing towards aeolian transport for RS 10: degree of rounding, frosting of quartz grains and the presence of feldspar.

❑ Are there any clues to the climatic conditions in which RS 10 formed?

■ The red colour: this is a ferric oxide coating, typical of that formed in a highly oxidizing environment (Section 3.3).

This would be consistent with a desert environment where wind transport is the dominant form of transport. However, you should not assume that red beds automatically imply *wind* transport and deposition in a desert environment. The Triassic sedimentary rocks that dominate the landscape of Britain in parts of the west and north Midlands, for example (see Plate 4.3), are largely the deposits of desert lakes, and also rivers flowing through desert areas. It is possible that the iron oxides in these sediments, and in other red beds in the geological record, formed by the decomposition of mafic minerals in oxidizing sediment pore waters *after* the sediment was deposited and buried.

As you can see, conclusions about depositional environments based on hand-specimen evidence alone can only ever be tentative, because such evidence provides only a partial picture. We explore other evidence for depositional environments in the next Section.

4.2 Sediment deposition in fluids

As you saw in the previous Section, for a given water or wind flow speed, a specific range of grain sizes will be transported, by either rolling and sliding (water) or by surface creep (wind), by saltation and in suspension (water and wind). This state of affairs is far from constant because rivers and winds vary in speed, sometimes on a daily or hourly basis. Today's raging torrent can become tomorrow's babbling brook, and tomorrow's howling gale might be next week's balmy zephyr. So sediment that is moved by water or wind one day may not be able to stay in transport another day, and then deposition will occur.

❑ You may recall that for a flow 1 m deep, a water current speed of around 0.5 m s^{-1} is required to set a sand grain 2 mm in diameter in motion. If the current speed increases and then decreases, do you think the grain will be deposited as soon as the speed drops to 0.5 m s^{-1} again?

■ No, it will have to drop *below* this speed.

At the beginning of Section 4.1.2, we said that to move a particle the water speed would have to be high enough to overcome the frictional and gravitational forces anchoring it to its bed. This speed is greater than that required to keep it moving once transport is underway.

The relationship between grain sizes and the water current speeds at which deposition will occur is shown in Figure 4.8. This Figure also shows the relationship for erosion that was illustrated in Figure 4.7, and includes the curve showing the transition between transport in suspension and bedload transport. This curve has been extended to show its intersection with the curve for deposition. Below the curve for deposition, no transport will take place. Above the curves for erosion, sediment resting on the bed can begin moving. In the area in between, sediment that has already begun moving will carry on being transported until the current decreases to speeds at which deposition can occur. As you can see, the difference in current speeds required to erode and deposit grains is greatest for fine-grained sediments such as silts and clays.

- ❑ Why is it that finer-grained sediments (silts and clays) require higher current speeds to transport them compared to coarser-grained sediments?
- ■ These sediments are cohesive, even when unconsolidated.

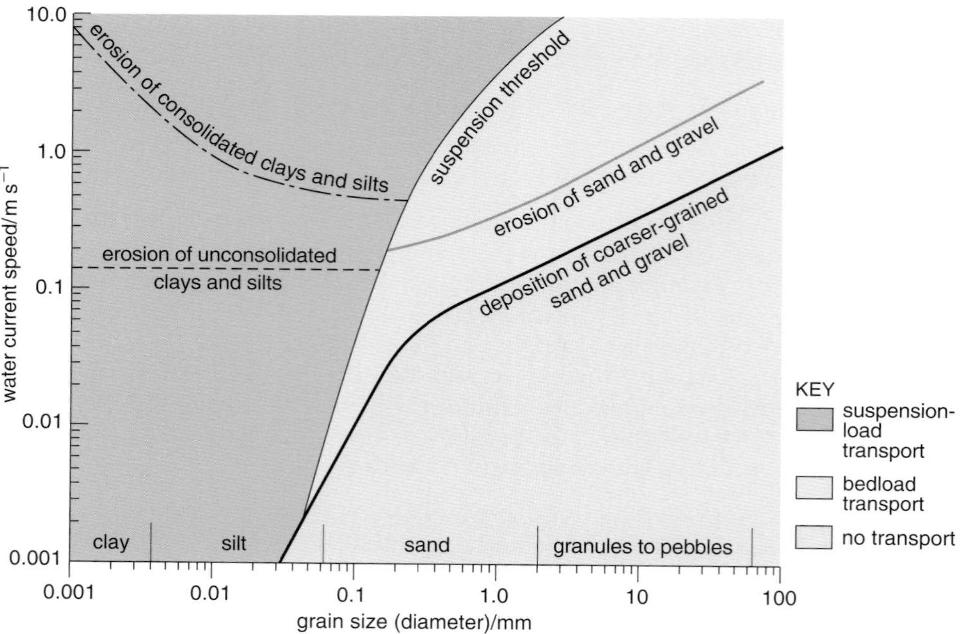

Figure 4.8 A composite graph to show the water current speeds of a flow 1 m deep required to erode sediments of various grain sizes and to deposit them. The dark blue area shows the conditions for which transport is as suspension load, the light blue area shows the conditions for which transport is as bedload, and the grey area shows the conditions for which no transport will occur.

Figure 4.8 also shows that all sediment finer-grained than coarse-grained silt is not only lifted directly into suspension as soon as transport begins, it is also deposited directly from suspension when transport ceases, and forms **suspension deposits**. When sediment coarser-grained than silt begins moving, it is transported initially as bedload, although it may be lifted into suspension as the current speed increases. As the speed decreases, the sediment must re-enter a stage of bedload transport again before finally coming to rest as a **bedload deposit**. The dark blue shaded area to the left of the suspension/saltation curve shows the grain size/speed conditions under which *all* transport occurs in suspension. The light blue shaded area to the right shows the conditions under which *all* transport occurs as bedload.

At this stage, it is important to note that the curve for deposition is based on experimental work using glass spheres, rather than naturally occurring quartz grains. Glass spheres have the same density as quartz, and sedimentary quartz grains are usually sufficiently equidimensional and well-rounded for shape difference to be insignificant. However, it is a different story for mica grains and clay minerals which are flaky and so offer a lot more resistance to water (or air) as they try to settle. Imagine dropping a ping-pong ball and a piece of paper of the same mass from the top of a tower on a totally windless day. The spherical ball would hit the ground first, because the flat sheet of paper would offer more resistance to the air as it fell. Sky divers take advantage of this difference by spreading out their arms and legs to become sheet-like, slowing their speed of descent before opening their parachutes. This means that for mica and clay to be deposited, current speeds have to be a lot less than for quartz grains of the same size.

From this discussion of deposition, you can see that while coarser-grained sediment is being deposited at any one time, finer-grained sediment and flaky grains carry on in transport until flow speeds decrease to the point where they can be deposited, too. Flaky minerals will carry on moving for some time after quartz grains of the same size have been deposited, and are often deposited along with fine-grained suspended sediment when flow speeds drop to a few mm s^{-1} (Figure 4.8). This process aids the sorting of sediments which we described in Section 4.1.2. The longer the time spent in transport, the more effective sorting of sediment is likely to be.

In some instances, both the coarsest- and finest-grained sediment available will be deposited in one bed with a gradational change between grain sizes. This is termed **graded bedding**. A bed may either contain the coarsest grains at the base and the finest grains at the top, in which case it is described as fining upwards; or, more rarely, it may show the finest grains at the base and the coarsest at the top, which is described as coarsening upwards. Examples of these features will be given in Sections 12–17 but you have already met coarsening upwards (reverse grading) in pyroclastic flows in Block 3, Section 6.3.2.

A second consideration is this: the grain size of the sediment that can be transported is related to the water flow speed. Hence we can use the overall grain size of a sedimentary rock as a qualitative measure of the **energy of the environment** in which the original sediment was deposited. Generally speaking, sediments of gravel size, or coarser, are considered to have been deposited in relatively high energy environments; medium- to coarse-grained sands represent medium energy environments, and fine-grained sands, silts and clays are described as having been deposited in low energy environments.

Although we have concentrated our discussion here on deposition from water, the same principles apply to aeolian deposition because air is a fluid. Dust that is fine-grained enough to be whipped up directly into suspension by the wind, is dropped when the wind stops blowing, to form suspension deposits. When sand saltating and creeping as bedload stops moving, it forms bedload deposits.

Activity 4.3

Now do Activity 4.3 which will test your ability to recognize the difference between suspension and bedload deposits, and the energy of the environment in which they are deposited.

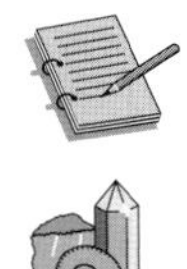

4.2.1 SEDIMENTARY FABRICS AND TEXTURAL MATURITY

The ability of moving fluids to sort sediments has an effect on the way in which the grains are arranged spatially within a sediment or sedimentary rock, i.e. its fabric (Section 1.1 and Block 2, Section 7.1). When the diversity of grain sizes is not very great, the grains come to rest on top of each other, rather like cherries in a jar or grains of sugar in a packet. Although there are spaces between the grains which are usually filled at least in part by finer-grained matrix material or post-depositional cement, if you were to remove this material the grains would still support each other. The framework of grains would not collapse. This sort of fabric is known as grain-supported (Figure 4.9a; see also Plates 2.2 and 2.3). When sorting is far less effective, so that there is a lot of matrix present in the

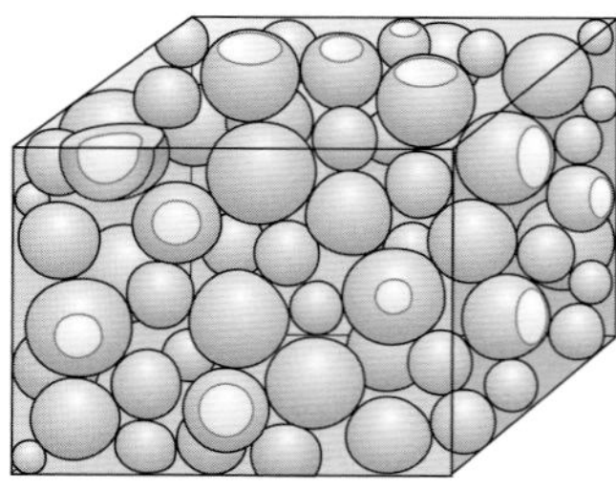
(a) grain-supported

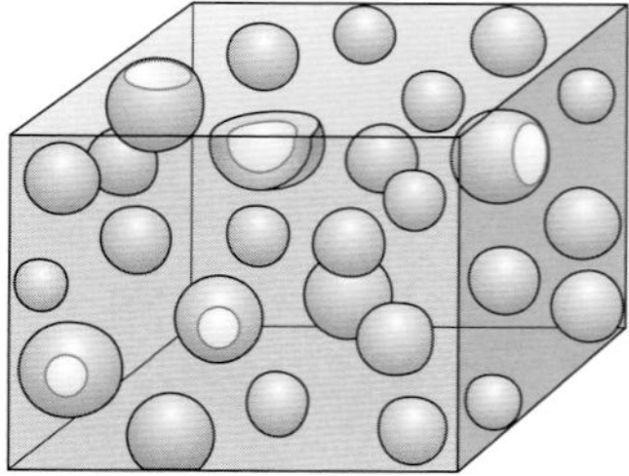
(b) matrix-supported

KEY
grains
matrix
projecting grain cut along face of cube

Figure 4.9 (a) Grain-supported fabric and (b) matrix-supported fabric in sedimentary rocks.

sediment, then the larger clasts may be supported by the matrix, rather than supporting each other. This results in a matrix-supported fabric (Figure 4.9b; see also Plate 2.4).

Another feature of fabric is the orientation of the grains. Elongated mineral grains and rock fragments that are deposited from strong unidirectional currents, often come to rest orientated with their long axes more or less parallel to the direction of current flow.

The two textural features of siliciclastic sedimentary rocks which you have already met, grain roundness and sorting, are used together with the proportion of clay matrix present in a rock (especially a sandstone) to define its **textural maturity**. Textural maturity increases as grain roundness and degree of sorting increase, and as the amount of clay matrix decreases.

As you know, the greater the time spent in transport by water or wind, the better sorted sediment becomes and the more rounded quartz grains and rock fragments are likely to be (Sections 4.1.2 and 4.1.3). Also the more likely it is that fine-grained silts and clays will be winnowed out, making them unavailable as matrix. As you will discover in Section 4.3, not all transport processes are as effective as flowing water and wind at sorting sediments and abrading grains, so the textural maturity of a rock can be a useful clue to help work out its history of transport and deposition.

Question 4.4 Explain which type of sandstone you might expect to be texturally less mature: one with a history of aeolian transport or one with a history of transport only by water.

Although water-borne sediments tend to be less texturally mature than aeolian ones, there is considerable variation. For example, the constant movement of quartz grains by waves means that beach sands often show a high degree of textural maturity, whereas **fluvial** sediments (i.e. those sediments transported in a river channel) vary more. For example, fluvial sediments deposited near to the source rock from which they were weathered are likely to be much less texturally mature than those deposited some distance away (Section 4.1.3). Do not forget, though, that you should not use textural maturity by itself to determine the depositional environment of a sedimentary rock. If the source rock is a sedimentary rock whose grains have already gone through one or more cycles of transport and deposition, a high degree of sorting and rounding could be inherited by the new sediment, regardless of how it is subsequently transported and deposited.

Activity 4.4

Now complete Activity 4.4 in which you will be able to practise describing the textural maturity of some of your Home Kit rocks.

Note, also, that textural maturity and compositional maturity do not necessarily go hand in hand. As you have just discovered (Activity 4.4), RS 10 is very mature texturally. However, it also contains some feldspar grains which are not totally decomposed (Activity 4.1), and so it is less mature compositionally.

Activity 4.5

You are now in a position to answer the remaining questions that we posed at the end of Section 2. Activity 4.5 will give you an opportunity to do this, and also help you to revise this first part of Section 4.

Activity 4.6

You should now be able to describe the grain sizes, textures, and fabric of the siliciclastic rocks in your Home Kit and to make qualitative estimates of their textural and compositional maturities, and the energy of their depositional environments. From all this, you should be able to make deductions about the conditions under which they were transported and deposited. Activity 4.6 will allow you to try to do this.

You have seen how the processes involved in sediment transport and deposition by fluids may affect the textures of sedimentary rocks. These processes also affect the way in which sediment is layered on a bed. Bedload deposition produces different patterns of layering from those produced by deposition from suspension, and the patterns produced by bedload deposition from unidirectional current flows are different from those involving wave action.

4.2.2 LAYERING IN SUSPENSION DEPOSITS

Look again at the curve for deposition in Figure 4.8 and the current speeds at which the coarsest silt grains will begin deposition from suspension. These are around 0.003 to 0.004 m s^{-1} (i.e. 3 to 4 mm s^{-1}). At these very low speeds, the flow is effectively laminar and so the sediment is able to settle out in neat flat layers. These layers may exhibit slight differences in grain size or mineral composition, reflecting slight differences in the energy of the depositional environment or differences in the type of material being deposited. These distinctive layers, less than a centimetre in thickness, are referred to as **laminae** (singular: **lamina**) and they give rise to a laminated structure within the sediments. Although laminae are common within suspension deposits, they are not restricted to these deposits and may occur in bedload deposits as well.

The geological literature is littered with terms describing the layering in sedimentary rocks, so it is important that you should understand what they mean. Box 4.1 summarizes the different terms.

Box 4.1 Sedimentary rock layers

The thinnest layers of sedimentary rocks, less than 1 cm in thickness, are termed laminae. Layers thicker than this are called beds. You were introduced to this term in Block 1. Beds are rock units with a sheet-like geometry, and their boundaries may be planar or irregular. Like 'laminae', the term 'bed' carries a thickness connotation, and so descriptions such as thinly-bedded or thickly-bedded have precise meanings. Thinly-bedded means beds 1–10 cm thick, medium-bedded describes beds 10–30 cm thick, and thickly-bedded refers to all beds more than 30 cm thick. Bed boundaries can be recognized by changes in mineralogical composition (e.g. a change from quartz-rich sandstone to clay-rich mudstone), by changes in the pattern of layering, or from evidence that periods of non-deposition or erosion have occurred. In some uniform lithological successions of beds, the only distinguishing features visible in the field may be the surfaces along which the rocks split naturally when hammered or quarried. Beds with similar features may be grouped together to form a **set**. Finally, stratum (plural: strata) is a general term used to describe any layer of rock (Block 1 Section 1.2). Its use does not imply any particular thickness, so it can be used to describe layers of rock ranging from a few millimetres to tens or even hundreds of metres in thickness, including sedimentary rock units shown on the Ten Mile Map.

4.2.3 LAYERING IN BEDLOAD DEPOSITS

Sediments deposited from the bedload often show rather more intriguing structures than do suspension deposits. This is because deposition does not always occur in flat layers; consider sand dunes in a desert or ripples on a beach, for example (Figure 4.10a). The collective name for features such as these is **bedforms**, of which there is a surprising variety, as you will discover shortly.

❑ What do you notice about the dip of the strata shown in Plates 4.1 to 4.3, and 4.9?

■ In each case the dip is quite steep, and in Plate 4.3 the dip varies in direction.

In Plates 4.1 to 4.3 and 4.9, dip has nothing to do with post-depositional folding and tilting of the strata, but results from the way in which the original sediments were deposited. The zones (or sets) of dipping strata in Plates 4.2 and 4.9, for example, are bounded top and bottom by surfaces that are either horizontal or much less steeply dipping. These surfaces reflect the regional angle of dip for these strata, and are bedding planes representing the general surface on which the sediment was deposited. The steeply dipping sets show **cross-stratification** because they dip locally at an angle to the surface on which the sediment was deposited. Cross-stratification represents the preservation of part of an original bedform. Before you can begin to interpret patterns of cross-stratification, you need to understand how bedforms develop.

4.2.4 BEDFORMS PRODUCED BY WATER CURRENTS

The easiest place to begin is with observations of experiments with sand grains in water flowing in one direction, as a current, using artificial channels known as flume tanks. To keep things simple we will fix the grain size of the sand at 0.5 mm diameter, the flow depth at 0.4 m, and see what happens as the flow is allowed to increase in speed.

To begin with, the sediment is flat and motionless, and the water above it quite still. As the water begins moving, gradually increasing in speed, the sand grains start to slide gently over the floor of the flume tank, and then to saltate. Eddies within the turbulent flow, or those caused by slight irregularities in the sediment surface, lead to the sculpting of the sediment surface into the shapes of ripples, like those shown in Figure 4.10a. These **current-formed ripples** are characteristically asymmetrical in cross-section, with the down-current slope steeper than the up-current one. In plan view, the early-formed ripple crests are fairly straight (Figure 4.10a).The ripples are about 3–5 cm in height with wavelengths (the distance between two adjacent ripple crests) of up to about 40 cm.

Question 4.5 With respect to the photograph in Figure 4.10a, which way was the current flowing that produced these ripples, and how can you tell?

These ripples are not static features: they migrate along the bed beneath them in the direction in which the current is flowing. How this takes place is shown in Figure 4.10b. Sediment is moved down-current both as bedload and suspension load. The coarser-grained bedload migrates up the shallow slope of the bedform and then avalanches down the steep slope where it is deposited temporarily. In Figure 4.10c you can see this coarser-grained material in the troughs between the large current-formed ripples, the crests of which are orientated from top to bottom of the picture. (The origin of the small secondary ripples running at right angles to the crests of the large ripples will be explained later.) Suspension load is carried over the ripple and some is deposited on the steep slope where eddies develop in the flow on the down-current side of the ripple. In this way, a series of laminae builds up parallel to the surface of the steep slope, inclined in the direction of current flow. The lines parallel to the steep slope in Figure 4.10b

(a)

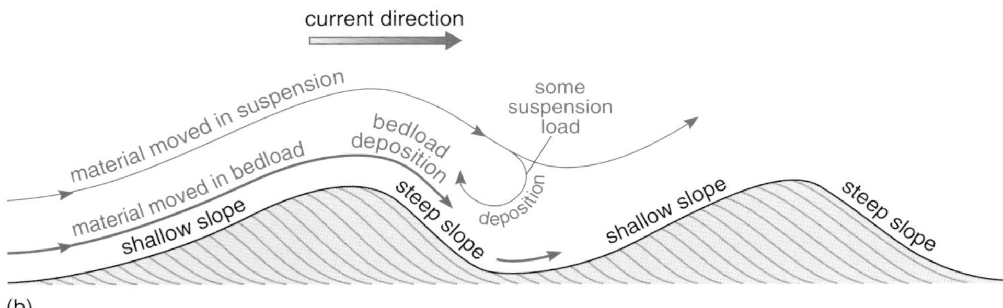

(b)

(c)

Figure 4.10 (a) Current ripples formed by water flowing over a beach. (b) The movement and deposition of sediment associated with a current-produced ripple, showing how erosion on the shallow slope and deposition on the steep slope lead to the migration of the ripple in the direction of current flow. (c) Large current-formed ripples (crests orientated from top to bottom of the picture) with coarse-grained sediment accumulating in the ripple troughs.

represent the build-up of successive laminae as the bedform migrates. It is cross-sections through the depositional slopes of bedforms that are preserved as cross-stratification.

If the same amount of sediment is eroded from the up-current side of the ripple as is deposited on the down-current side, the bedform will simply migrate across the bed in the direction of current flow. It would be preserved only if the flow stopped and the bedform were covered by fresh sediment. If sediment is in plentiful supply and deposition outweighs erosion, then sediment is deposited on both slopes of the ripple. However, more is deposited on the steep slope than on the shallow slope, causing the ripples to build upwards at an angle to the horizontal in the form of **climbing ripples** like those shown in Figure 4.11a and b.

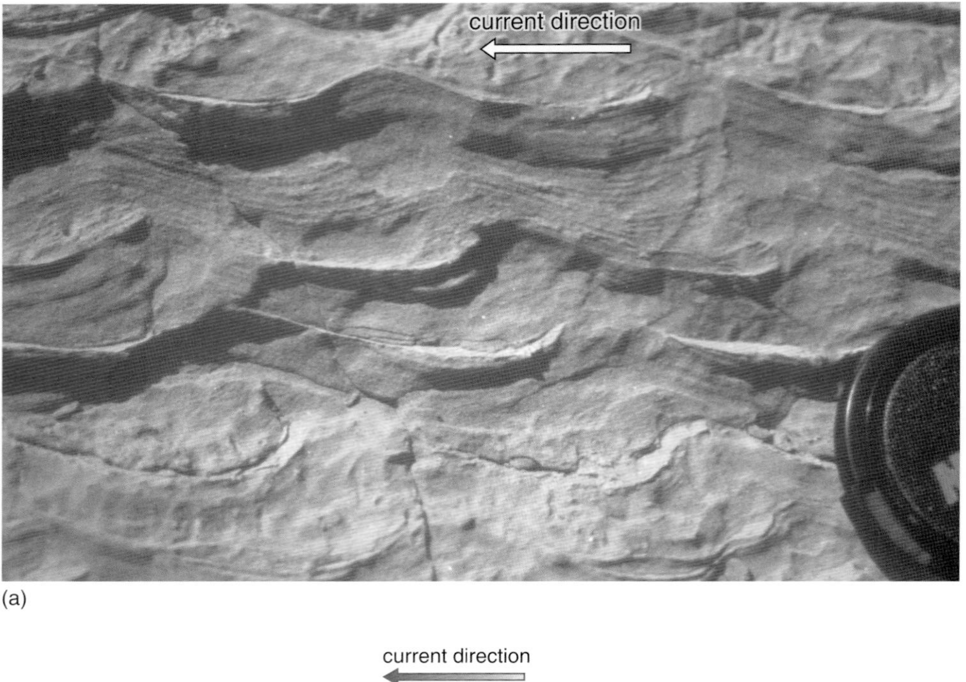

(a)

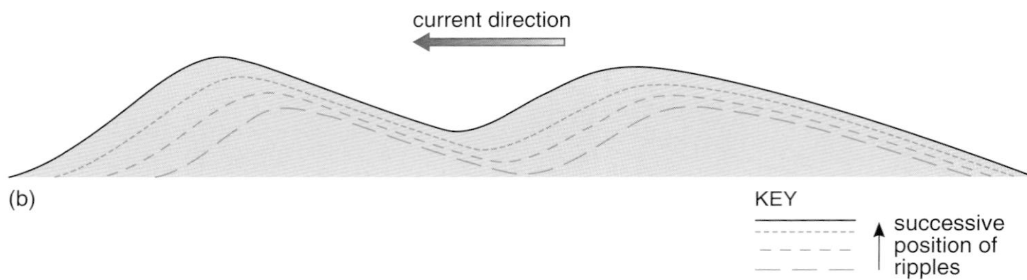

(b)

Figure 4.11 (a) Photograph of climbing ripples. (The scale is given by the rim of a camera lens cap visible in the bottom right-hand corner of the picture.) (b) Sketch to show the way in which climbing ripples build upwards and migrate laterally.

To return to our flume tank: as the water continues to gather speed the ripple crests become wavy (or sinuous) (Figure 4.10c) and eventually break up into curved-crested segments creating complex patterns like those shown in Figure 4.12. At still higher speeds, larger bedforms are produced, up to about 1 m high and with wavelengths of 0.5 m to 10 m (Plate 4.4). These are referred to as **dunes**. As these form beneath water we will call them *subaqueous* dunes,

Figure 4.12 Curved-crested ripples formed by water flowing over a beach.

to distinguish them from aeolian dunes in deserts. The cross-stratification in Plates 4.1–4.3 was formed by the migration of subaqueous dunes. Like ripples, these dunes can be straight-crested or curved.

If the flow speed increases still further, the subaqueous dunes are washed out and sediment sweeps over a flat surface once more. Any deposition at this stage occurs as planar deposition, building up flat layers of sediment which are preserved as **planar stratification**. If we had used a wider range of sand sizes in our flume tank, we would have noticed that the coarser grains were drawn up into thin ridges, about 1 mm high and some 10–15 cm long, to form distinct **current lineations** on the bedding planes, parallel to the direction of current flow. At speeds higher still, the flat bed changes into low relief, undulating mounds. These are known as **antidunes** because erosion (rather than deposition) occurs on the down-current side, and deposition occurs on the up-current side, causing the bedform (but not the sediment!) to migrate up-current. Antidune cross-stratification is therefore inclined in the opposite direction to the current flow, but is rarely preserved in the geological record because the bedforms are usually reworked as soon as the flow speed decreases. Consequently, it is reasonable to assume that any water-current-formed cross-stratification you see in the field represents deposition on ripples or subaqueous dunes.

The sequence of bedforms we have just described is illustrated in Figure 4.13, in relation to current speed. The planar deposition and antidunes produced at high flow speeds, are described as having formed in the **upper flow regime**, and those features formed at lower flow speeds as having formed in the **lower flow regime**. This two-fold division is useful because the types of bedforms produced, and their sequence of formation, vary according to the dominant grain size of the sediment, the current speed and the flow depth. So you cannot work out a current flow speed from the type of bedform alone.

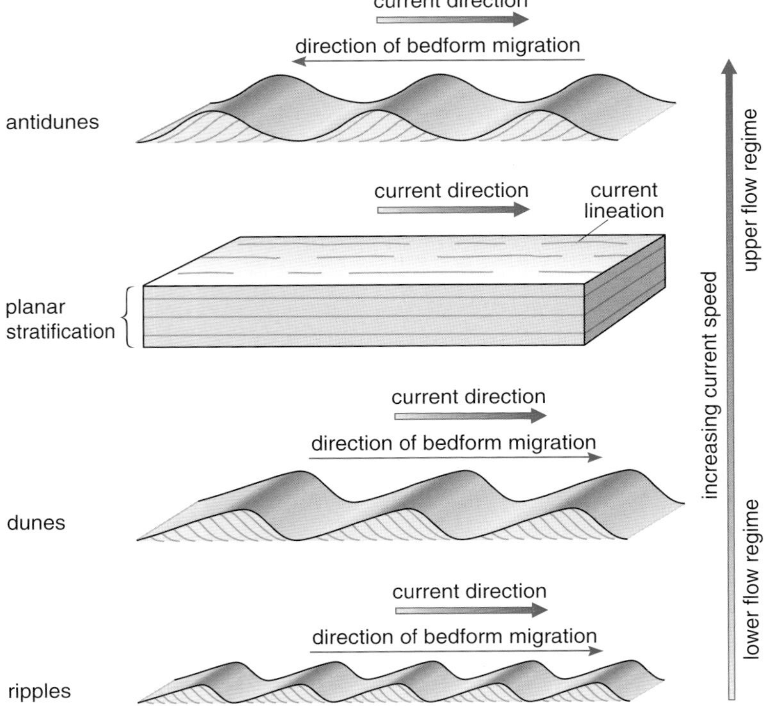

Figure 4.13 The sequence of bedforms developed in sands 0.5 mm in diameter in response to increasing current speed in a flow around 0.4 m deep. Cross-sections through the bedforms reveal the cross-stratification that is preserved, and its direction of inclination in relation to the direction of current flow.

The results of flume tank experiments using a wide range of grain sizes and current speeds, and flow depths of around 0.25–0.4 m, are summarized in Figure 4.14. This Figure uses log scales like Figures 4.7 and 4.8. The different shaded areas in

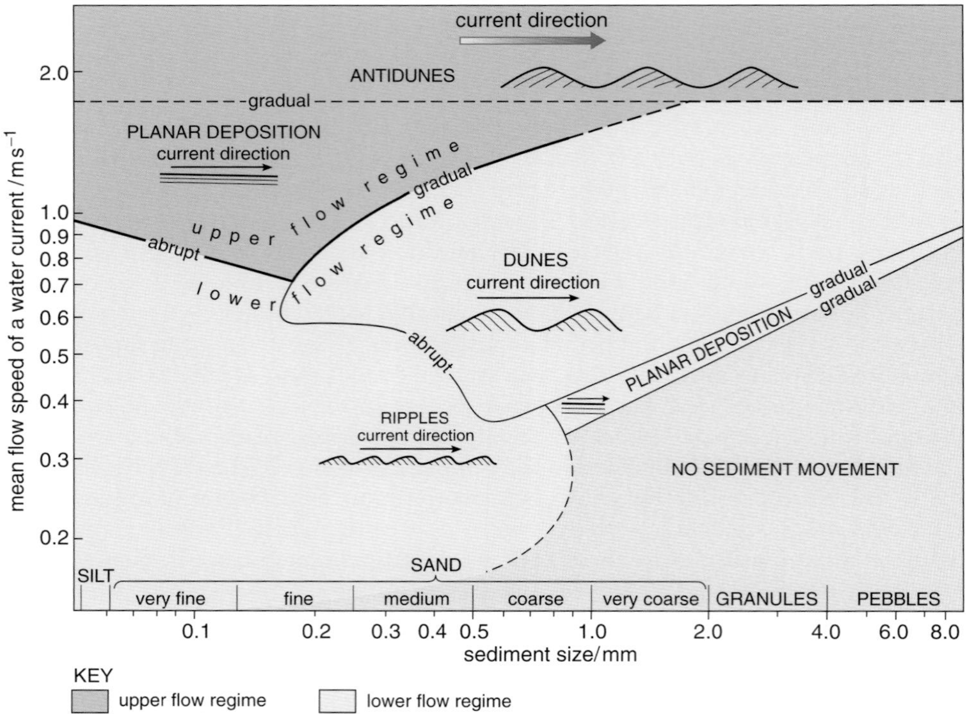

Figure 4.14 The relationship between current speed, grain size and subaqueous bedforms produced beneath a flow of around 0.25–0.4 m depth. This Figure is illustrated by a series of video clips in the processes section of the Sedimentary Environments Panorama.

the graph distinguish between upper flow regime conditions (dark blue), lower flow regime conditions (light blue), and conditions under which no sediment movement occurs at the flow depths used in the experiments (grey).

Question 4.6 For the flow depths shown in Figure 4.14:

(a) What is the coarsest grain size within which ripples will develop?

(b) What is the finest grain size in which subaqueous dunes will form?

The first thing to notice from Figure 4.14 is that when the sediment on the bed consists of very fine-grained sands or silts, there is no subaqueous dune formation in the lower flow regime. As current speed increases, ripples give way to planar deposition, producing planar stratification in the upper flow regime. When the bed is composed of coarse-grained sands, gravel or pebbles, the sediment is transported and deposited initially over a flat surface producing planar stratification within the lower flow regime, instead of ripples. As this planar stratification develops at slow flow speeds, it does not display current lineations like planar stratification formed in the upper flow regime. Upper flow regime planar stratification does not develop in sediment coarser-grained than about 2 mm. Instead subaqueous dunes give way to antidunes.

A second point to notice is that the change from lower flow regime to upper flow regime, or one bedform to another, is sometimes gradual, but at other times it can be rapid, depending on the grain size.

When flow depth is taken into consideration, the picture becomes even more complicated. It is found that although lower flow regime bedforms depend largely on grain size and flow speed, subaqueous dunes cannot form at very shallow flow depths because there is not enough room for them. Conversely, in very deep flows, such as those that exist within continental shelf seas, dunes may grow to heights of up to 18 m. These very large bedforms are more aptly termed **sand waves**. Furthermore, for a given flow speed and grain size, upper flow regime conditions are more likely to prevail in shallow flows than deep flows.

The observation that lower flow regime, current-formed cross-stratification dips in the direction of current flow provides another clue to help us unravel the transportational and depositional history of a sedimentary rock. We can use the dip of the cross-stratification to work out the direction in which the current that deposited the original sediment was flowing. From this **palaeocurrent** (i.e. 'ancient' current) direction, we can also get a general idea of the direction of sediment transport. If we put together a large number of palaeocurrent measurements, over a wide geographical area, they may give us an indication of the location of the sediment source and of how far the sediment travelled before being deposited. We can then put this information together with our knowledge of the textural and compositional maturity of the sediment to help clarify whether these features are likely to be totally original (source rock igneous or metamorphic, so only one cycle of erosion and deposition), or partly inherited (source rock sedimentary, so more than one cycle of erosion and deposition).

Interpreting sedimentary structures like cross-stratification in the field is rarely this simple. To begin with, the patterns of cross-stratification vary because of the variations in the shapes of bedform crests, as well as variations in current directions. We touched on this earlier, when we were talking about the sequence of bedforms that develops in response to an accelerating current.

❑ How do the shapes of these bedform crests vary?

■ Depending on the current speed, they vary from straight (Figure 4.10a) through to curved (Figure 4.12).

If bedform crests are curved or sinuous, then when the cross-stratification formed is viewed in an exposure at approximately right angles to the current direction, a series of intersecting troughs is seen evoking the description **trough cross-stratification** (Figure 4.15a). As you can see, the true dip of the stratification is visible only in an exposure viewed parallel to the current direction. By contrast, the cross-stratification produced by straight-crested forms appears like planar stratification when viewed at right angles to the current direction, hence the description **planar cross-stratification** (Figure 4.15b). Once again, the true dip is visible only in exposures cut parallel to the current direction. In both cases, if sections are viewed in between these two extremes, the angle of dip simply appears shallower than it really is. This explains why the upper cross-stratified set in Plate 4.3 dips at a different angle from the lower set. The current that produced the two sets changed slightly in direction, and the sets have been sectioned at different angles in relation to the direction of flow. So determining true palaeocurrent directions is fraught with similar problems to those you met when determining the true dip of strata in Block 1 (Section 4.3.1).

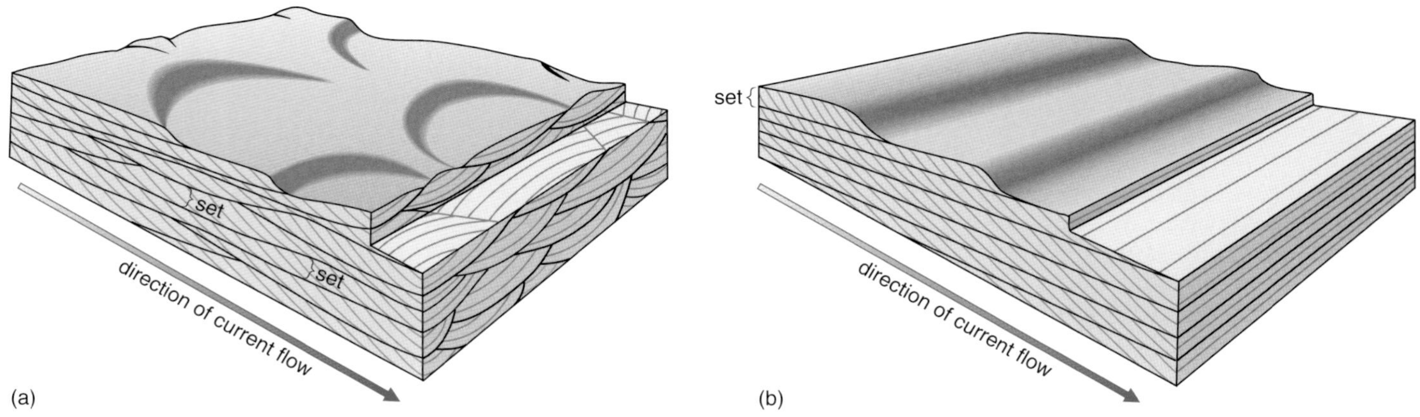

Figure 4.15 Block diagrams to show the patterns of (a) trough cross-stratification resulting from the preservation of curved-crested (or sinuous) bedforms, and (b) planar cross-stratification resulting from the preservation of straight-crested bedforms. The arrow shows the current direction.

Turn again to Plates 4.1 and 4.2. If you look carefully at these Plates, you will see that the cross-stratification is all inclined in the same direction. Now turn to Plate 4.5 which shows several cross-stratified layers, as you look from the top to the bottom of the photo. The cross-stratification in each set is all dipping in the same direction. Trace one of these sets across the photo from left to right as far as the geologist's arm, and note the direction of dip of the cross-stratification.

❑ What do you notice about the dip of the cross-stratification in the adjacent sets, above and below?

■ The cross-stratification in any one set is dipping in the opposite direction to that above and below it.

Unlike the situation shown in Plate 4.3, adjacent sets are separated by a horizontal plane.

❑ Bearing in mind that the direction of dip of cross-stratification in the lower flow regime reflects the direction of current flow, how might you explain this variation?

■ The direction of current flow must have changed.

The implication of this observation is that the current responsible for this cross-stratification must have systematically reversed direction.

❑ Based on your general knowledge, can you think of any environments in which current reversals are likely to occur on a daily basis?

■ Coastal and estuarine environments where, generally speaking, there are usually two tidal cycles a day, and so two incoming (flood) tides and two outgoing (ebb) tides a day.

Because of its distinctive pattern, this form of cross-stratification is referred to as **herring-bone cross-stratification**. You saw this form of cross-stratification in the video sequence *Coastal processes* (Activity 1.1), preserved in Carboniferous sediments. The precise bedform left by alternating currents will be determined by their relative strength and the rate of deposition of new sediment. If there is no deposition, a later current will partly or wholly erode the bedform left by its predecessor. If there is sufficient net deposition, then tidal currents of equal strength leave a symmetrical herringbone.

Even when no herring-bone cross-stratification is preserved, there are often other tell-tale signs that tidal influences have been at work during sediment deposition. To begin with, current speeds during the tidal cycles are not uniform. As you can see, reading from left to right across Figure 4.16, at low tide there is no water movement (slack water, current speed 0 m s^{-1}) but as the flood tide rises the current accelerates reaching a maximum somewhere between low tide and high tide. It then begins to decelerate until slack water is reached again at high tide. As the tide turns, this same process of acceleration and deceleration is repeated on the ebb tide, back to the slack water of low tide. The effects of

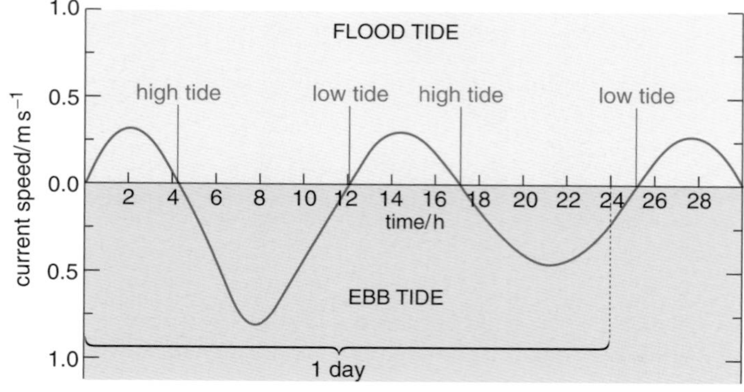

Figure 4.16 Variations in current speed with the daily tidal cycle in Chesapeake Bay on the eastern coast of the USA. The horizontal axis shows the number of hours that have elapsed since the tidal cycle began at 0 hours. Note that the tidal cycle in Chesapeake Bay is asymmetrical, with around 4 hours between low tide and high tide, and around 8 hours between high tide and the next low tide.

these regular, periodic variations in tidal flow are most likely to be seen in estuaries or on **tidal flats**, stretches of coastline where the **tidal range** (i.e. the difference between mean high and mean low tide) is high and wave action is low, so that tidal currents rather than waves are the main agents for sediment transport and deposition.

In these environments, the rising tide is likely to be carrying silt and clay (mud) in suspension but as the current accelerates, sand will begin to move as bedload. By the time the flow has reached its maximum speed, it is probable that ripples and subaqueous dunes will have developed.

Question 4.7

(a) As the current decelerates towards the slack water of high tide, what will happen to the clay and silt in suspension?

(b) What will happen to this same sediment as the current starts to accelerate on the outgoing ebb tide?

The coating of mud deposited at high tide may also protect the underlying bedforms from erosion so that they become preserved in the sedimentary record, complete with their **mud drapes**, as you can see in Plate 4.6. When you watched the video sequence *Coastal processes* (Activity 1.1), you saw recently deposited mud drapes on the tidal flats off Holy Island, and fossil examples in the Carboniferous sediments.

Tidal currents do not only vary on a daily basis. The *average* daily strength of a tidal current varies with the lunar tidal cycle. During the periods we know as the 'new Moon' and 'full Moon', high tides reach their highest levels, and low tides their lowest levels. These are the so-called *spring tides*: the tidal ranges are larger than usual and so tidal currents must reach their greatest speeds at these periods. Across the period mid-way between two spring tides, high tides attain their lowest levels, and low tides their highest levels: these are the *neap tides*.

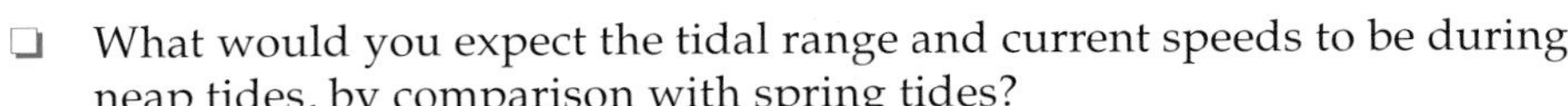

❑ What would you expect the tidal range and current speeds to be during neap tides, by comparison with spring tides?

■ Both will be lower than usual.

Cross-stratification representing deposition during the full lunar tidal cycle may show not only the mud drapes characteristic of the daily tidal cycle, but also a pattern of increasing and decreasing spacing between the sand/mud couplets (Figure 4.17). The more vigorous currents associated with spring tides lead to the deposition of thicker sand units and less well-developed mud drapes, whereas the more sluggish currents during neap tides result in thinner sand units and more obvious mud drapes. These features are known, in rather homely fashion, as **tidal bundles** (Plate 4.7a, b).

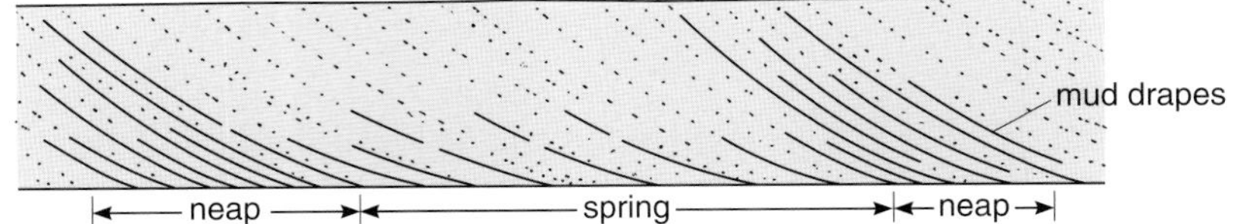

Figure 4.17 A schematic cross-section through tidal bundles formed by variations in current speed with the lunar tidal cycle. The dotted lines represent sand deposition.

Question 4.8 Two neap tide periods are represented in Plate 4.7a. How many are represented in Plate 4.7b?

4.2.5 WAVE-PRODUCED BEDFORMS

In order to appreciate how bedforms are produced by wave action you need to understand, first, how sediment is transported by waves. Waves are produced by friction as wind blows over the surface of a body of water. The movement of

water associated with waves is fundamentally different from that associated with a flowing current. A plastic bottle floating in the sea, offshore from the zone of breaking waves, bobs up and down as a wave passes. The wave moves on and eventually breaks on the shore, but the bobbing bottle does not shift its position significantly. This is because a given parcel of water within the wave moves round in a circular orbit as the wave form passes (Figure 4.18a). The water parcel reaches the highest part of its orbit as the wave crest passes, and the lowest part as the trough passes. In deep water, the orbital paths decrease in diameter with depth until movement ceases. This happens at a depth equivalent to half the wavelength, and this depth defines the **wave-base**. If the sea-bed lies below the wave-base, wave movement does not affect the sediment on the bed.

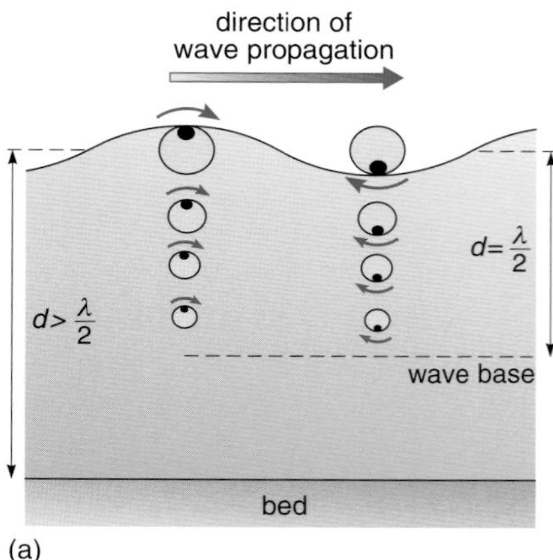

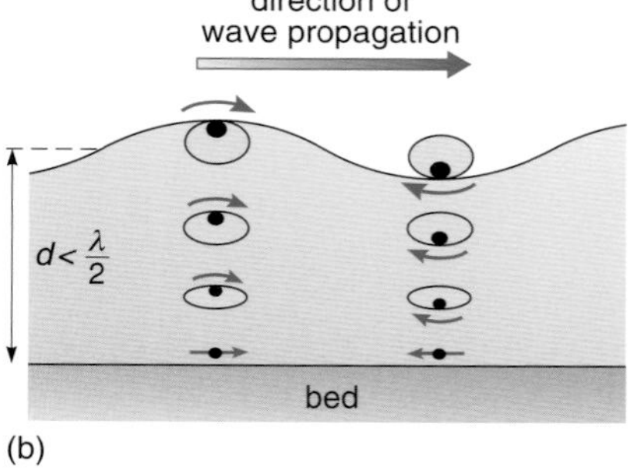

KEY
- position of water parcel
- direction of water flow

Figure 4.18 The orbital paths of parcels of water in waves associated with (a) deep water (depth (d) = half the wavelength (λ) or more), and (b) shallow water ($d < \lambda/2$).

However, in shallow water, where the water depth is less than half the wavelength, the moving parcels of water experience increasing frictional retardation with the sediment bed beneath them, which slows them down and progressively flattens their orbits. Adjacent to the bed the water simply oscillates backwards and forwards in a horizontal plane, with no vertical component at all (Figure 4.18b). As the water parcel moves to and fro near the bed, it reaches a maximum speed in one direction before decelerating, reversing and then accelerating to reach a maximum speed in the opposite direction.

Because of the friction with the bed, water in a wave can move the sediment resting on the bed, provided the orbiting parcels are moving fast enough.

❑ Look at Figure 4.18b. Which way will sediment on the bed be transported as the wave crest, and then the wave trough, pass over it?

■ The arrows on the orbital paths show that it will move in the direction the wave is travelling as the crest passes over (left to right), but in the opposite direction as the trough passes.

This seems to imply that sediment is moved to and fro on the same part of the bed all the time, making no progress in either direction. This, however, is not the case; net sediment movement (transport) occurs in both directions, but different grain sizes are involved in each case. From Figure 4.18b you can see that when an orbiting parcel of water is in a wave trough it is closer to the bed than when it is at a wave crest, so frictional retardation by the bed must be less when it moves to the right than when it moves to the left. This means that the speed of the water in a horizontal direction is greater during movement in the direction of wave propagation than in the opposite direction. However, the higher speed is maintained for *less time* than the lower speed because in each case the water parcel is travelling the same distance. Coarser-grained sediment is moved in the direction of wave propagation as bedload, with finer-grained sediment as suspension load. The coarsest-grained sediment cannot be returned when the flow reverses because of the reduced water speed. Because the lower speed is maintained for a longer time, there is a net gain of finer-grained material in the reverse direction, as well as a net gain of coarser-grained material in the direction of wave propagation.

This winnowing of finer-grained material by wave action leads to very effective sorting of the sediment. When the direction of wave propagation is perpendicular or oblique to a shoreline, there will be a net movement of coarser-grained material onshore, and of finer-grained material offshore. This process has implications for sediment zonation on beaches and shelf seas, as you will see in Section 16.

Working out the exact speed under which sediment of a particular size begins transport as the result of wave action is much more complicated than for currents because it depends on the orbital speed of the water. This is calculated from the length of time taken for a water particle to complete an orbit, i.e. to move from the top of its orbit as a wave crest passes, to the bottom of its orbit as a wave trough passes, and back to the top of its orbit as the next wave crest passes. The orbital speed of a wave depends to a large extent on the height of the wave and the water depth. It increases as the wave height increases. It also increases as the water depth decreases because the height of a wave increases as it moves from deep water to shallow water.

These relationships are too complicated for us to pursue any further. However, the important points to recognize from all this are that big waves have the capacity to move coarser-grained sediment than smaller waves; also, this capacity increases as a wave increases in height moving from deep water to shallow water. The orbital speed of a wave is at its maximum when the wave height is at its maximum, just before the wave breaks.

In the same way that the conditions necessary for current bedform development have been modelled using flume tank experiments, those necessary for wave bedform development have been modelled through wave tank experiments. If waves are allowed to pass gently over sand-sized sediment, then to begin with there is no sediment movement. However, as the orbital speed of the waves is increased, very small, low-relief ripples begin to develop as the grains roll backwards and forwards on the bed.

As the ripples begin to grow, sand is transported over the ripple crests (see Figure 4.19). As the wave crest passes, the flow of water is disturbed by the ripple so that eddies develop on the down-slope of the ripple where sand accumulates (Figure 4.19a). As the flow begins to reverse, the eddies are destroyed and some of the sand is suspended again (Figure 4.19b). As the wave trough passes, eddies now develop on the down-flow slope of the adjacent ripple where deposition occurs (Figure 4.19c). As the flow begins to reverse once

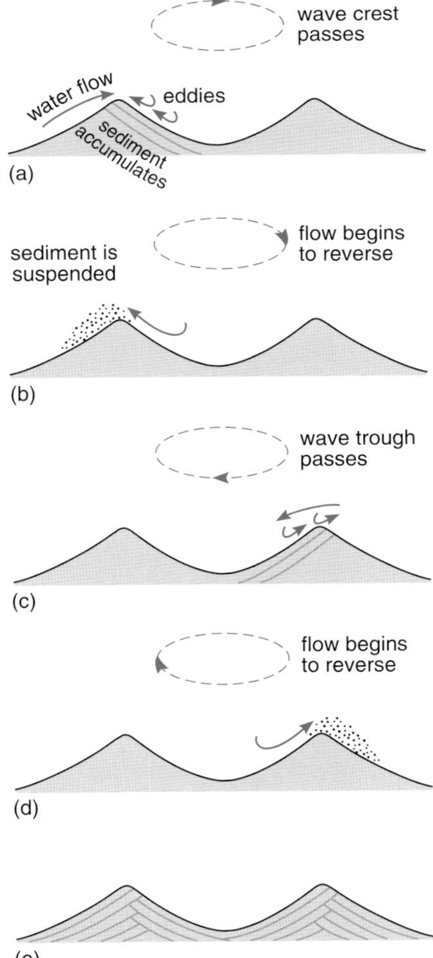

Figure 4.19 (a)–(d) Stages in the erosion and deposition of sediment associated with wave-formed ripples as a wave form passes. The dashed ellipses show the orbital paths of a parcel of water as a wave form passes over a rippled bed. The arrowheads represent the positions of a water parcel at various stages as the wave form passes over. (e) Cross-section through a wave-formed ripple to show a typical pattern of cross-stratification. (f) Wave-formed ripples preserved in sedimentary rocks from Dorset. Note that the laminae are better preserved on the right-hand slope of each ripple than the left-hand slope. This suggests that the orbital speed of the water was less in one direction than the other.

more (when the next wave crest approaches), some of the sand is suspended again (Figure 4.19d), and so the process is repeated for each ripple as wave forms pass. If the sediment contains a range of grain sizes, mineral types or shell fragments, then the opposing laminae built up on either side of the ripple crest will be easily picked out as cross-stratification (Figure 4.19e, f).

Wave-formed ripples can be distinguished from current-formed ripples in cross-section by their different patterns of cross-stratification (compare Figure 4.19e with Figure 4.10b). As bedforms, they can be distinguished by their shape.

❑ What is the characteristic shape of current-formed ripples?

■ They are usually asymmetrical in cross-section and their crests vary from straight, through sinuous to curved as current speeds increase.

By contrast, wave-formed ripples are usually symmetrical in cross-section, often with somewhat pointed crests, and concave-upwards slopes (see Figure 4.19e and the Carboniferous and modern day sedimentary deposits in the video sequence *Coastal processes* (Activity 1.1)). At low water speeds, they tend to be relatively straight in plan view, and to show some bifurcation (Figure 4.20). If sedimentation rates are very high, then wave-formed ripples may climb (video sequence *Coastal processes* (Activity 1.1)) in a manner analogous to the way in which current-formed ripples may climb (Section 4.2.4). We need to return to the wave tank to see what happens at higher water speeds.

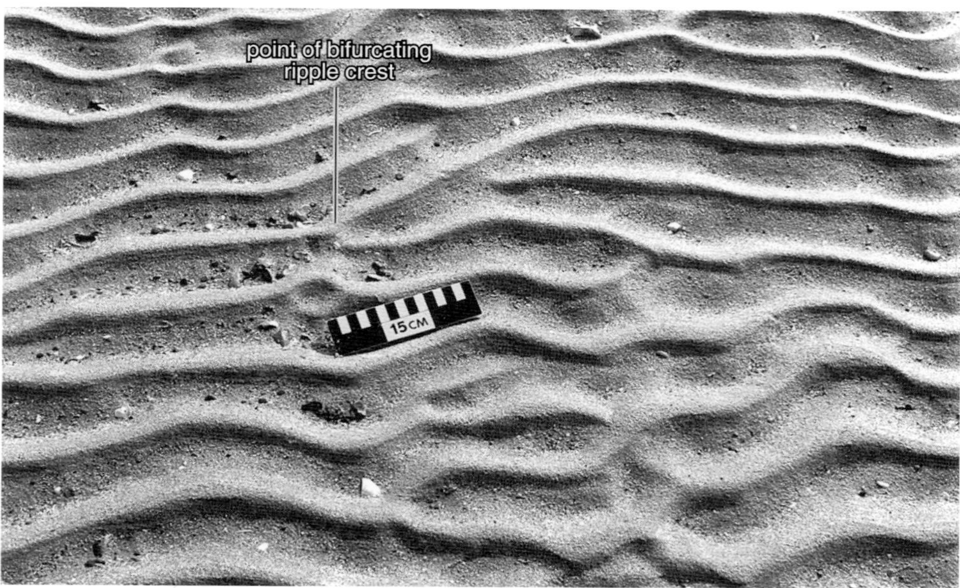

Figure 4.20 Wave-formed ripples on a beach. Note how some of the ripple crests bifurcate (i.e. split into two) when traced from left to right across the picture.

As the orbital speed of water passing over our rippled sand bed increases, the wavelength of the ripples increases and at the same time the ripples begin to break up into broad, low-relief hummocks. At higher speeds still, the ripples are washed out and planar deposition occurs, analogous to the planar deposition that occurs in the upper flow regime of current flow. The speeds at which this happens are grain-size dependent: around 0.4–0.5 m s^{-1} for finer-grained sand, and up to 1 m s^{-1} for coarser-grained sediment.

You are now in a position to understand the complex pattern of ripples shown in Figure 4.10c. The dominant bedforms are large current ripples with crests orientated from the top to the bottom of the picture. These were probably sculpted by the incoming tide. There are smaller, wave-formed ripples, with crests orientated perpendicular to the current ripple crests, playing piggy-back on the current ripples. The wave ripples probably formed at some stage around the slack water of the high tide period, when the only water movement would have been through wave action, with the crests of the wave ripples forming perpendicular to the direction of wave propagation. Their distinctive pattern has given rise to the description 'ladder-back' ripples.

There is another sedimentary structure that forms in response to wave action, but one which you will not see as a bedform in modern beach sediments. This is because it results from storm wave action, when orbital speeds exceed 1 m s^{-1}. It develops below mean low tide level, in water that is too deep for sediments on the sea-bed to be affected by normal, fairweather waves; i.e. they form below the **fairweather wave-base** (commonly abbreviated to FWWB), but above the **storm wave-base** (commonly abbreviated to SWB) (i.e. the maximum depth to which the sea-bed is influenced by storm waves). The bedforms consist of three-dimensional *hummocks*, up to 40 cm in height, spaced tens of centimetres to several metres apart and separated by intervening depressions known as *swales*. The resulting sedimentary structure is called **hummocky cross-stratification** (HCS for short), seen in cross-section in Figure 4.21a and Plate 4.8. As you can see, the structure consists of complex, low-angle, upward-curved laminae of sand which thicken slightly down into the swale. The laminae may be truncated at a low angle where another bedform has moved across the surface.

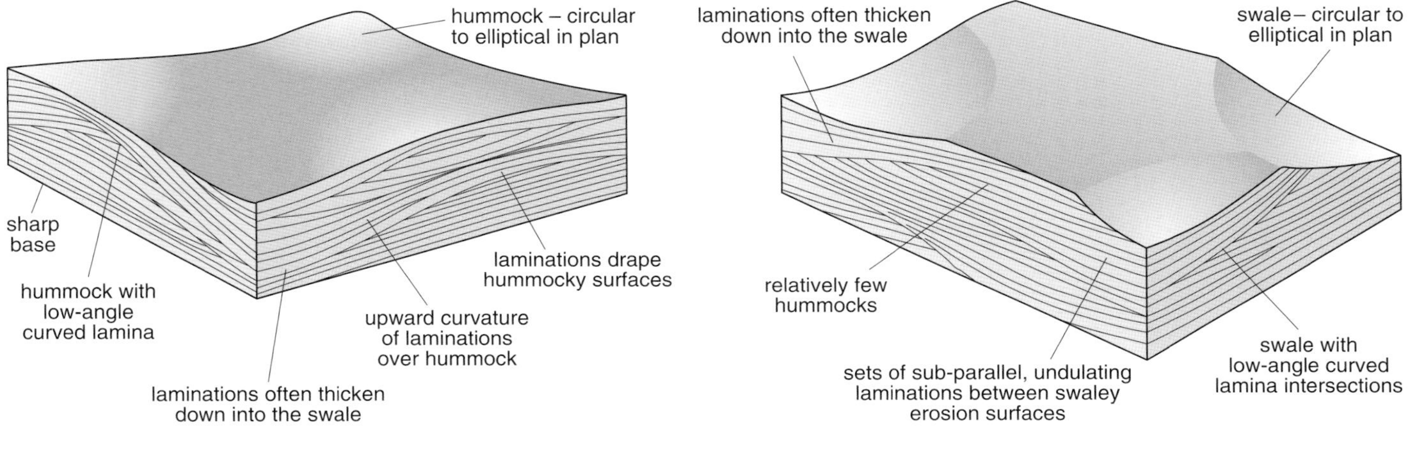

Figure 4.21 Three-dimensional block diagrams of (a) hummocky cross-stratification and (b) swaley cross-stratification, showing both the bedforms and the resulting sedimentary structures.

No one is completely sure how HCS forms because it has never been produced experimentally. However, its formation must be linked to the water motion produced by waves.

❑ Will storm waves give rise to essentially unidirectional or oscillatory water motion, and what is likely to be the amplitude and orbital speeds of such waves?

■ They will produce essentially oscillatory motion, and the amplitude of the waves will be high. This means the wave heights will be large, and so the orbital speeds will also be high.

❑ Does the water move at a constant speed as a wave form passes over the sea-bed?

■ No. You should recall that the water flow will reach a maximum speed in one direction before decelerating, reversing and then accelerating to reach a maximum speed in the opposite direction.

It seems that when there is a reasonably high sediment supply, sediment is picked up and suspended as the water flow reaches a maximum speed in one direction, but is rapidly deposited again as the flow decelerates prior to reversing to form hummocks which are then partially eroded as the next storm wave passes over. In slightly shallower water, just above the fairweather wave-base, the oscillatory movement of the water within waves leads to more erosion of the hummocks so that it is mainly just the cross-stratification in the swales that is preserved. In this case, the structure is known as **swaley cross-stratification** (SCS) (see Figure 4.21b). You saw HCS and SCS preserved in Carboniferous sediments when you watched the *Coastal processes* video sequence in Activity 1.1. Before moving on to the next Section, you may find it helpful to review this video sequence (Activity 1.1) which illustrates many of the processes described in this Section and Section 4.2.4.

4.2.6 AEOLIAN BEDFORMS

In Section 4.1 we were at pains to stress that air is also a fluid, so sediment movement by wind has much in common with water. However, we also pointed out that the differences in density and viscosity between the two fluids leads to some differences in sediment behaviour (Section 4.1.1).

> **Question 4.9** How do the modes of sediment transport by wind differ from those by water?

As you might expect, these differences in mode of transport lead to differences in the nature of the aeolian bedforms, and how they develop. There are other controlling factors as well. You may recall that the shape and size of a current-produced bedform depends on the flow speed and depth, and the grain size of the sediment on the bed (Section 4.2.4). The main controls on aeolian bedform morphology are wind speed, wind direction and duration, and sediment supply, as you will discover in Section 15.1. Morphology is not constrained by flow depth because air flows are vertically much more extensive than most water flows.

Despite various experiments with sands in wind tunnels, the process of aeolian ripple formation is not well understood. One suggestion is outlined in Figure 4.22. Sand grains begin to saltate as soon as the wind speed at a particular point on the bed is sufficient to start them moving. As aeolian sands are well sorted, the grains will all be fairly similar in size and so they will tend to saltate around the same distance, descending to hit the surface again in more or less the same place (Figure 4.22a). On impact, they set in motion the grains that were too coarse-grained to saltate. So the coarser grains move by surface creep (Section 4.1.1), and the impact makes the finer grains saltate again. As this process is repeated, a series of regularly spaced impact sites develops on the bed

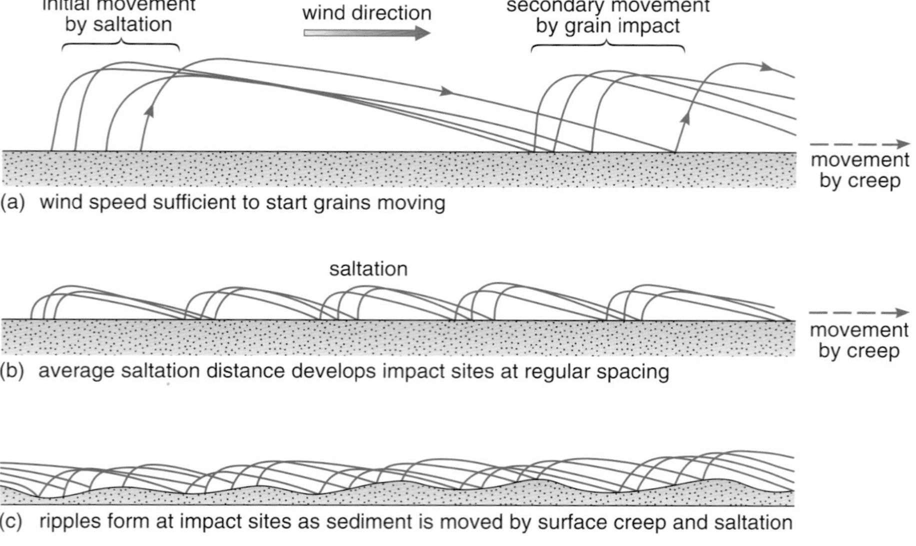

Figure 4.22 (a–c) are stages in the formation of aeolian impact ripples. Note that the scale used to demonstrate initial movement by saltation (a) is considerably larger than that used to demonstrate subsequent ripple formation (b and c).

(Figure 4.22b). Sand is pushed forward at these sites by surface creep and so **impact ripples** begin to form (Figure 4.22c). These are rarely preserved in the sedimentary record.

Finer-grained sand migrates rapidly over the ripples by saltation and some is deposited and trapped in the ripple troughs, but coarser-grained sand moves more slowly as it is pushed up the up-wind side of the ripples by surface creep. This process results in coarser-grained sand becoming concentrated at the ripple crests. This contrasts with current-formed ripples where coarser-grained material is located in the ripple troughs (Figure 4.10c). Impact ripples are slightly asymmetrical, like current ripples, but they are flatter and their crests are long and relatively straight with occasional bifurcations (Figure 4.23a).

Unlike subaqueous dunes, aeolian dunes do not grow from ripples in response to increasing wind speed. As wind speeds increase, ripples are 'washed' out and eventually replaced by a flat bed again. Aeolian dunes take much longer to develop than ripples and, provided there is enough sand available, may grow to spectacularly greater heights than the subaqueous varieties, because they are not constrained by flow depth (Figure 4.23b). Furthermore, actively migrating sand dunes may have migrating impact ripples superimposed on them (Figure 4.23a), and smaller dunes may grow on the backs of larger ones (Figure 4.23b).

(a)

(b)

Figure 4.23 (a) Impact ripples formed in a desert environment. Note the large aeolian dune to the right of the picture. The gullies visible on the steep slope of the dune formed as sand avalanched down the slope by grain flow. (b) Sand mountains with superimposed, migrating dunes (arrow points to an example) and impact ripples.

If aeolian dunes do not grow from ripples, then, given sufficient sand and a good strong wind, how do they develop? It seems that there has to be some obstacle in the path of the sand to begin with, such as an isolated clump of vegetation or a rock. As the airstream flows over the obstacle, a 'wind shadow' is created on the down-wind side, containing localized eddies where sand is deposited to build up drifts (Figure 4.24). When sufficient sand has accumulated the drift becomes independent of the original obstacle and starts migrating as a dune.

Saltating sand grains bounce their way up the shallow slope of a dune to the crest where they periodically avalanche down the steep slope by a process known as grain flow. In Figure 4.23a, you can see what appear to be shallow gullies on the steep slope of the dune; these are the sites of small avalanches. Thus, the dune gradually migrates in a way directly analogous to a subaqueous dune. We will explain the process of grain flow more fully in Section 4.3.1.

Question 4.10 How can you tell:

(a) which way the sand dune in Figure 4.23a and the arrowed dune in Figure 4.23b are migrating?

(b) what the dominant wind direction is in Figure 4.24?

Figure 4.24 Wind-blown sand drifts developing round clumps of seaweed on a wind-swept beach in Dorset, England. Wind direction is from left to right.

As with subaqueous dunes, preservation of the depositional slope may give rise to cross-stratification (Plate 4.9). If you compare the aeolian cross-stratification in Plate 4.9 with that produced by subaqueous dunes (Plates 4.1–4.3 and 4.5), you will see that there are some notable differences. To begin with, the aeolian cross-stratified sets are much thicker, commonly in the order of 5–10 m, whereas subaqueous cross-stratification is normally less than 2 m thick. Secondly, although the beds at the base of each set make only a low angle with the horizontal, they curve upwards more steeply than subaqueous cross-stratification. This is because the steep slope of an aeolian dune may be stable up to an angle of rest of about 35° (see Figure 4.23), whereas the angle of rest of subaqueous dune slopes is usually much less.

It is tempting to think that these differences might provide a universal means of distinguishing between aeolian and subaqueous cross-stratification. However, the steep slopes of aeolian dunes may be less than 35°, and the maximum angle will, of course, only be seen if the cross-stratification is sectioned parallel to the wind direction (analogous to the situation shown in Figure 4.15). Also, large-scale subaqueous dune-forms (i.e. sand waves) up to 18 m in height are known to exist on some continental shelf areas where large flow depths occur (Section 4.2.4). So it would be foolish to rely on the evidence of cross-stratification alone.

Question 4.11 What other evidence might you find that would help to confirm that a succession of cross-stratified sandstones was aeolian in origin, rather than marine?

Geologists would also examine the evidence from sediments occurring above and below the cross-stratified succession before coming to any conclusions about the origin of the cross-stratified sandstones (remember the principle of consilience, Section 2).

4.3 TRANSPORT AND DEPOSITION BY SEDIMENT GRAVITY FLOWS

Our discussion of transport and deposition so far has concentrated on fluids which have been relatively free of suspended sediment. Transport by fluid flow depends largely on the properties of the fluid. As the amount of suspended sediment increases, the viscosity and density of the sediment–fluid mix increase, and the nature of the flow begins to change. Eventually, gravity becomes the prime mover of the sediment, rather than the properties of the host fluid, and it is the properties of the sediment that maintain transport in some way. Because of their dependence on gravity, these sorts of flows are known as **sediment gravity flows**. There is a complete spectrum of flows relating to the proportions of sediment and fluid in the flow (Figure 4.25).

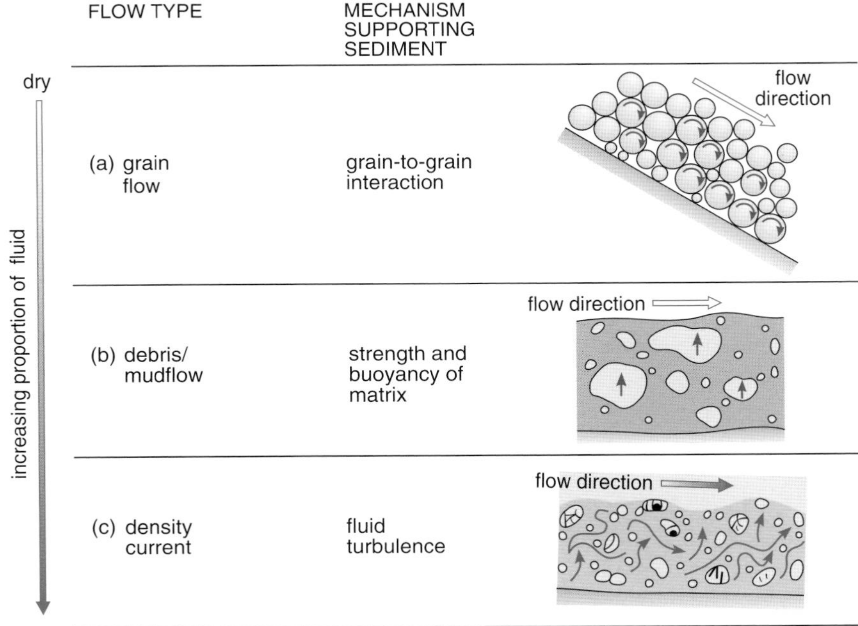

Figure 4.25 Mechanisms of sediment transport for different types of sediment gravity flows.

4.3.1 GRAIN FLOWS

At one end of the spectrum there is no fluid involved, beyond air or water that may be trapped between the grains, acting as a lubricant. These are called **grain flows**. They occur when sediment on a slope either exceeds its stable angle of rest, or is rendered unstable, and so begins moving under the influence of gravity. Once in motion, the flow is sustained by grains colliding and ricocheting, and rolling over each other (Figure 4.25a).

❑ What is the example of grain flow you have just met?

■ The avalanching of sand grains down the steep, down-wind side of an aeolian dune (Figure 4.23a).

There is no grain-size limit for grain flow which may encompass anything from sand grains to boulders.

4.3.2 DEBRIS FLOWS AND MUD FLOWS

When there is not enough water to produce turbulent flow, but there is sufficient water mixed with the sediment to form a viscous mass, it will flow plastically. Larger sediment particles are supported and sustained in transport by the high viscosity and density of the finer-grained matrix–water mix (Figure 4.25b). If the concentration of large grains is sufficient for them to come into contact with each other, a **debris flow** develops and the resulting deposit will be partially or totally grain-supported. You met very large-scale examples of debris flows in Block 3 (Section 6.4.4 and Plate 6.12) in the form of debris avalanche deposits resulting from partial collapse of a volcano.

Question 4.12 From what you have understood about how sediment textures may be modified by processes during transport, what degree of textural maturity would you expect debris flow deposits to display?

As debris flows very rarely extend beyond a few tens of kilometres from their source, the opportunities for sediment abrasion and sorting are further decreased. If the flow contains a high proportion of fine-grained material, such that the resulting deposits are matrix-supported, then it is generally referred to as a **mud flow**. You were introduced to mud flows initiated on the flanks of volcanoes when lahars were discussed (Block 3 Section 6.3.2). Both mud flows and debris flows have enormous transport and erosive capacities, travelling at speeds of 100 km hr^{-1} or more.

4.3.3 DENSITY CURRENTS AND TURBIDITY CURRENTS

Density currents are at the opposite end of the spectrum to grain flows. They contain sufficient fluid for grains to be sustained in transport by turbulent fluid flow once movement has been initiated (Figure 4.25c). However, they are driven in the first instance by the difference in density between the sediment–fluid mix (higher density) and the surrounding fluid (lower density). The denser fluid is able to flow down-slope due to gravity as a density current beneath the less dense fluid.

❑ You have already met one form of density current when we were discussing volcanic activity in Section 6.3.2 of Block 3. Can you recall what it is?

■ A pyroclastic flow (Block 3 Figure 6.14) in which the sediment is ash, pumice and other rock debris (lithic fragments), and the surrounding fluid is air.

More important for sedimentary geologists are the density currents that originate at the sea-floor. These are the main means whereby sediment deposited on the edge of the continental shelf (Block 3 Section 2.2.1), on the continental slope, and on offshore slopes, is transported down to the deeper ocean floor. Because they consist of a mass of turbid sediment and water, these density currents are given the specific name of **turbidity currents**.

The turbidity current is initiated when the sediment on the slope becomes unstable. As grains begin to move relative to each other, the water in the pore spaces between them seeps across the grain contacts. This lubricates the grains, reducing the friction between them. The sediment pile is now no longer fully grain-supported because of the thin films of water separating the grains, and it becomes fluidized. The whole sediment mass becomes effectively a high density, turbid fluid within which the individual grains are suspended. Because of its high density it can begin to flow down-slope beneath the surrounding lower density water, reaching speeds of 50 km hr^{-1} or more.

A turbidity current is not uniform along its length. As you can see from the experimentally generated current in Figure 4.26a, it can be divided into 'head', 'body' and 'tail' sections. It is the head and body that do all the work. In the early stages of flow, they travel at different speeds. Although the speeds of both are related directly to density contrasts between the turbidity current and surrounding water, the speed of the head depends also on its height, whereas the speed of the body is determined in part by the angle of the slope. The body travels faster than the head, so water and sediment are constantly being forced into the head area from the body. The pressure created in the head forces the additional fluid up and around, returning it back to the body (Figure 4.26b). The body, therefore, is a mass of turbulent eddies which maintain the sediment in suspension.

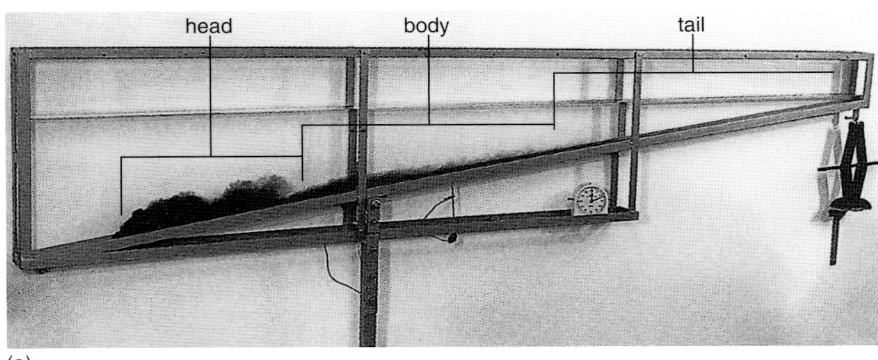

(a)

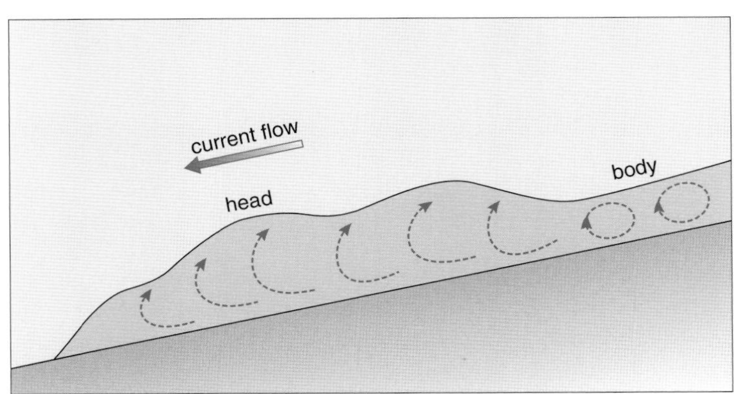

(b)

Figure 4.26 (a) The structure of a turbidity current generated when dyed, higher density salt water is allowed to flow down a slope in an experimental tank beneath less dense tap water (cf. Block 3 Figure 6.14). (b) The movement of water and sediment within the head and immediately adjacent body of a turbidity current is shown by dashed lines with arrowheads.

You can imagine how awesome a turbidity current must be, with a head maybe hundreds of metres high, thundering down the continental slope at speeds in excess of 50 km hr^{-1}. As you might expect, the head is a region of devastating erosion as the water, bearing sand and gravel derived from the continental shelf, is swept up and around. It rips up the underlying mud and shapes it into balls which are incorporated within the first sediments to be deposited as the current wanes.

Other common erosional features are flute marks and various tool marks. **Flute marks** are formed when spiral eddies develop in the flow as it passes over the irregular sea-bed. The eddies scour elongated hollows in the underlying muds which become shallower and wider down-current (Figure 4.27a). As sediment is deposited from the current, it fills the hollows so that the flute marks are preserved as reversed-relief flute casts on the basal bedding planes of the deposited sediments (Figure 4.27b).

Tool marks are also common. These are indentations of the underlying sediment

Figure 4.27 (a) Formation of flute marks due to scouring of muds by eddies at the base of a turbidity current. (b) Flute casts on the underside of a bedding plane in Silurian sedimentary rocks from Wales. (c) Various tool marks preserved as casts in sedimentary rocks from Poland. The line of circular marks at the top of the photograph was caused by a saltating fish vertebra.

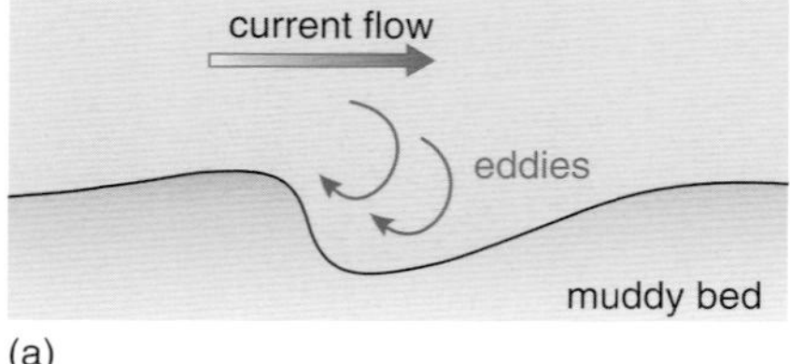

(a)

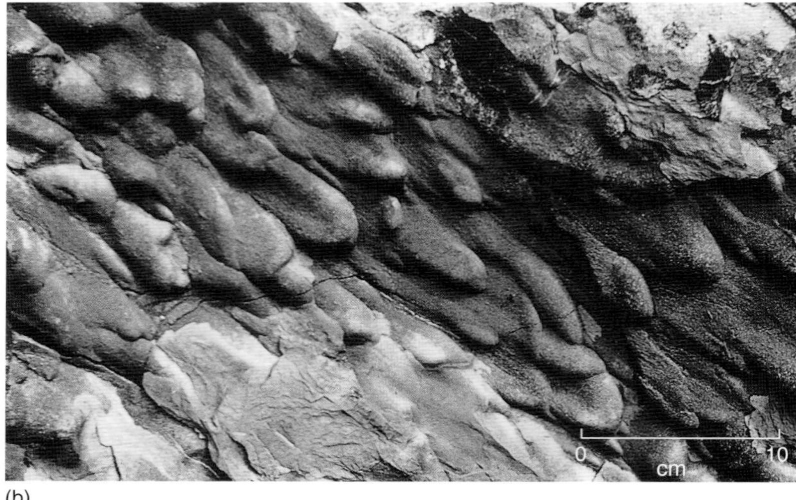

(b)

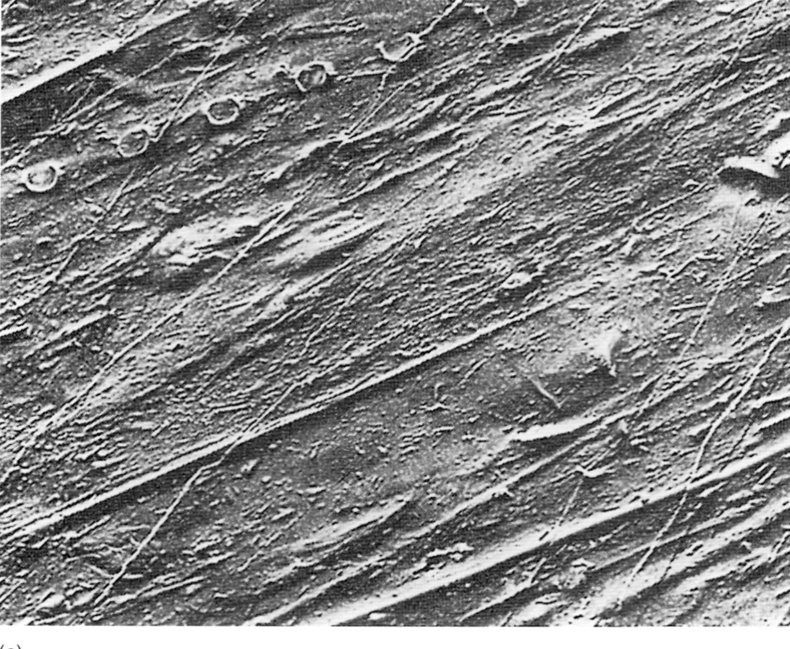
(c)

produced by objects or 'tools' being carried along in the bedload by the current. They may range from prod and bounce marks to linear grooves, caused by anything from small pebbles and shells to less likely objects such as driftwood and fish bones. Once again, these tool marks may be preserved as casts (Figure 4.27c).

Question 4.13 Explain which way the turbidity current was flowing that produced the flute marks preserved in Figure 4.27b.

Deceleration of a turbidity current begins as soon as the gradient of the slope begins to decrease (remember that the speed of the body, which drives the head, depends in part on the gradient of the slope). As a turbidity current decelerates, deposition begins from the body section, immediately behind the erosive head. We will look at what happens during deposition when we consider turbidity currents in their deep sea environmental context in Section 17.2.

4.4 Glacial transport

A third form of transport and deposition is by ice flow. Moving ice can be considered as an extremely viscous fluid in which the high viscosity maintains particles up to boulders several metres across in transport, either on the surface of the ice flow, within the flow, or embedded in its base. The sediment scraped off the base of the ice as it moves over the land surface (Plate 2.4), or deposited when the ice melts, is typically texturally very immature and lacks evidence of any stratification.

As landscapes dominated by ice, or glacial environments, are not investigated in this Block, we are not going to discuss glacial transport and deposition further. However, you should be aware that evidence of these processes is found in the British geological record. Most recently, and only about 13 000 years ago, Scotland was covered by ice. Some 14 000 years before that, an ice sheet covered most of Scotland, Ireland and Wales, extending south, beyond the English Midlands. You may recall that there are glacial sediments among the drift deposits shown on your Moreton-in-Marsh Sheet (Block 1 Section 3.2.2). The British tourist industry has good reason to be grateful for these glaciations because glacial erosion was responsible for carving out much of the spectacular mountain scenery in the Scottish Highlands, the Lake District and Snowdonia.

Activity 4.7

Now work through Activity 4.7 which will give you a chance to see in action some of the processes we have described in Section 4.

4.5 Summary of section 4

- Siliciclastic sediments are transported by fluid flow and sediment gravity flow.
- Transport by fluid flow depends upon the physical properties of the fluid, chiefly viscosity and density. Flow may be either laminar or turbulent, and the degree of turbulence increases as the viscosity of the flow decreases and the speed increases. Very low viscosity/density air flows (wind) are almost always turbulent and water flows become turbulent at very low flow speeds.
- Transport in fluids occurs as bedload by saltation, by rolling and sliding (in water only), and by surface creep (in wind only), and in suspension as suspension load.
- Sediment particles begin moving when the force of the moving fluid is sufficient to overcome both the force of gravity holding them down, and the frictional forces between the grains and the bed. Consequently, it takes higher speeds to erode grains of a given grain size than to deposit them.
- The grain size of sediment that can be transported by fluids is related to the speed of the flow and so can be used to describe the energy of the depositional environment.
- Significantly higher flow speeds are required to erode consolidated silts and clays than unconsolidated silts and clays, and to move sediment by wind rather than by water. The size of sediment that can be moved by waves increases with the orbital speed of the waves, which increases as wave height increases.
- Transport in fluids results in sorting of sediments and grain rounding. Aeolian (wind) transport is generally more effective than water at sorting and rounding, and prolonged wave action is generally more effective than fluvial transport. Aeolian transport also leads to frosting of quartz grain surfaces.

- The degree of sorting and rounding and the proportion of clay matrix present are used to define the textural maturity of sediments. Aeolian sediments and sediments subjected to intensive wave action are generally more texturally mature than fluvial sediments. However, textural maturity and quartz grain frosting may be inherited characteristics if the source of the sediment is a pre-existing sedimentary rock. The amount of matrix present in a sediment determines whether it is grain-supported or matrix-supported.
- Deposition from the bedload may result in the development of bedforms. The types and morphologies of bedforms produced by unidirectional water-current flow depend on grain size, flow speed and flow depth, and are classed as either upper or lower flow regime bedforms. Bedforms produced by wave action depend on the orbital speed of waves and water depth. The morphologies of aeolian bedforms are controlled by wind speed, direction and duration, and sediment supply. They are independent of flow depth. Bedforms produced by water currents, wave action and wind may be distinguished by their morphologies.
- Preservation of the depositional surfaces of bedforms results in cross-stratification. Water-current cross-stratification formed in the lower flow regime, and aeolian cross-stratification, are inclined in the direction of flow. Patterns of cross-stratification depend on the morphology of the bedforms and the consistency of flow direction. Those produced in tidal environments are influenced also by the varying strengths of tidal currents during the daily and lunar tidal cycles.
- Transport by sediment gravity flow is initiated by gravity but maintained by the properties of the sediment–fluid mix. There is a spectrum of flows relating to the proportion of fluid present.
- Grain flows are dry; debris and mud flows contain sufficient water to flow plastically, but not turbulently. Large particles are maintained in transport by the high viscosity and density of the fine-grained matrix. Debris flow deposits are typically texturally immature. Density currents are driven by contrasts in density between the sediment–fluid mix and the surrounding fluid, but the sediment is maintained in transport by turbulence. Subaqueous density currents are called turbidity currents and are highly erosive.

4.6 Objectives for Section 4

Now you have completed this Section, you should be able to:

4.1 Explain how the process of fluid flow depends on the properties of the fluid and describe how transport by fluids takes place, distinguishing between transport by water currents, waves and wind.

4.2 Explain how the speed of a moving fluid is related to the sediment grain size it is capable of eroding and depositing, and interpret and sketch graphs which illustrate these relationships.

4.3 Estimate qualitatively the energy of a depositional environment from the grain size of a sedimentary deposit or rock.

4.4 Describe the degree of sorting of a siliciclastic sedimentary rock, the roundness of the grains, the proportion of matrix and the fabric, from hand specimens and thin sections, and estimate the degree of textural maturity of the rock.

4.5 Explain how the processes involved in transport and deposition determine the fabric and textural maturity of a siliciclastic sedimentary rock, and infer these processes from the textural features and fabric of a hand specimen or thin section.

4.6 From photographs, recognize, describe and distinguish between the bedforms produced by water currents, waves and wind, and the resulting structures preserved in sedimentary rocks.

4.7 Explain, using sketch diagrams, how bedforms are produced and how factors such as sediment grain size, fluid speed, and flow depth and direction determine the types of bedforms that develop, and the type of cross-stratification that is preserved.

4.8 Use water-current-produced bedforms to determine the flow regime in which a succession of sedimentary rocks was deposited.

4.9 Explain how the process of sediment gravity flow depends on gravity and the properties of the sediment–fluid mix, and how the proportion of fluid in the flow determines its character.

4.10 Describe, using sketch diagrams, the structure of a turbidity current and some of the erosional features these currents produce.

Activity 4.5 gave you an opportunity to test your understanding of Objectives 4.3 and 4.4. Now try the following questions to test your understanding of some of the remaining Objectives.

Question 4.14 With reference to Figure 4.8:

(a) Given a flow depth of 1 m, what is the water-current speed required to begin eroding a sand grain 0.4 mm in diameter.

(b) When the grain begins moving, how will it be transported: as bedload or in suspension?

(c) How will the same grain be transported if current speed increases to 3 m s^{-1}?

(d) To what speed must the current decrease again before the grain can be deposited?

(e) How will the grain be deposited: from the bedload or from suspension?

Question 4.15 With reference to Figure 4.14, a water current flows at a speed of 0.4 m s^{-1} over a bed of gravel composed of grains 3 mm and more in diameter. The depth of the flow is around 0.3 m. What is the succession of bedforms you would expect to develop if the current speed increased to 2 m s^{-1}? How rapid would the transitions from one type of bedform to the next be?

Question 4.16 Look at Figure 4.16 which shows the variations in tidal current speed over a daily cycle in Chesapeake Bay, USA.

(a) How long into the daily tidal cycle shown do the extremes of (i) high tide and (ii) low tide occur?

(b) At which times shown on Figure 4.16 would you expect muds to be deposited and why?

(c) Assuming that the sediment in Chesapeake Bay is dominantly fine- to medium-grained sand, comment on the possibility that herring-bone cross-stratification might develop in the sediments deposited during a tidal cycle.

Question 4.17 Look carefully at the sandstone exposure in Plate 4.2.

(a) Assuming the exposure is a section parallel to the current flow that deposited the sandstone, and that the cross-stratification was produced by bedform formation in the lower flow regime, which way was the current flowing?

(b) Explain whether the water current varied in speed or was fairly constant.

5 SEDIMENTS FROM SOLUTION

In this Section, we begin by looking at the fate of the ions that end up in solution after chemical weathering. Rainwater that has not evaporated either seeps into the ground where chemical weathering continues and more minerals dissolve, or washes away down steep hillsides. Either way, much of this water ends up in streams and rivers, or percolates down through soil and rock layers, ultimately to enter lakes or the sea. Once in the seas and oceans, some of the dissolved ions are incorporated into existing sea-floor sediments and rocks. This is one of the reasons why the composition of seawater in the open oceans stays constant, instead of becoming ever more salty. However, we are more concerned with how dissolved materials come out of solution to form new sediments in their own right.

- ❑ There are two ways in which this might happen. The first is the result of biological processes. What is the other one?
- ■ Direct chemical precipitation.

5.1 BIOLOGICAL PROCESSES AND SEDIMENT FORMATION

Sediments which owe their origins to biological processes are known as **biogenic sediments** and the most important of these, geologically speaking, are some types of limestones. Limestones are composed predominantly of calcium carbonate ($CaCO_3$: Block 2 Section 7.2) and the process of their formation begins when Ca^{2+} is released into solution by chemical weathering of silicate minerals such as plagioclase feldspars and some amphiboles, and (more importantly) when pre-existing limestones are dissolved:

$$\underset{\text{as limestone}}{CaCO_3(s)} + \underset{\text{water}}{H_2O(l)} + \underset{\text{carbon dioxide}}{CO_2(g)} \rightleftharpoons \underset{\text{calcium ions}}{Ca^{2+}(aq)} + \underset{\text{bicarbonate ions}}{2HCO_3^-(aq)} \qquad (5.1)$$

As you can see from Equation 5.1, the weathering of limestone also releases bicarbonate (HCO_3^-) ions into solution.

Question 5.1 What is the other important source of bicarbonate ions in rain and surface waters?

Some of the bicarbonate ions partially dissociate to produce carbonate (CO_3^{2-}) ions:

$$\underset{\text{bicarbonate ions}}{HCO_3^-(aq)} \rightleftharpoons \underset{\text{carbonate ions}}{CO_3^{2-}(aq)} + \underset{\text{hydrogen ions}}{H^+(aq)} \qquad (5.2)$$

Under the appropriate conditions, the Ca^{2+} ions in seawater can combine with CO_3^{2-} to produce $CaCO_3$ (calcium carbonate):

$$\underset{\text{calcium ions}}{Ca^{2+}(aq)} + \underset{\text{carbonate ions}}{CO_3^{2-}(aq)} \rightleftharpoons \underset{\text{calcium carbonate}}{CaCO_3(s)} \qquad (5.3)$$

- ❑ What is the biological process that is illustrated by Equation 5.3?
- ■ Marine organisms extract Ca^{2+} and CO_3^{2-} from seawater, using these ions to construct shells and other skeletal parts made of calcium carbonate.
- ❑ What is the name given to grains that are the remains of shells and other skeletal material?
- ■ Bioclasts (see Block 2 Section 7.2).

After the demise of these organisms, which include everything from microscopic calcareous phytoplankton (e.g. unicellular algae) and zooplankton (microscopic animals) to larger animals such as clams, mussels, sea-urchins and corals, their discarded hard parts accumulate on the sea-floor where they may make up the carbonate sediments that will form the next generation of limestones. You will be introduced to some of these larger animals in Section 9. As long as seawater is sufficiently close to saturation with $CaCO_3$ to prevent calcareous bioclasts from dissolving, and there is insufficient siliciclastic sediment being supplied by erosion from adjacent land areas to dilute the bioclast content, then relatively pure calcareous sediments will accumulate.

There is another, rather less salubrious way in which biogenic carbonate mud can be preserved in marine sediments; this is as **faecal pellets**, the excrement of certain marine animals including shrimps and snails. These may be deposited *in situ* or, if sufficiently hardened, they may be reworked by waves or currents. Faecal pellets form rounded or ellipsoidal 'grains', typically 0.1–0.5 mm in diameter and are usually devoid of any form of internal structure (Figure 5.1a, b). The pellets may amalgamate with only a small amount of compaction, and then are indistinguishable from carbonate mud. As structureless ovoid grains can be produced by other means as well, for example by the alteration of sand-sized skeletal fragments, the term **peloid** is used to cover all grains of this description, regardless of origin.

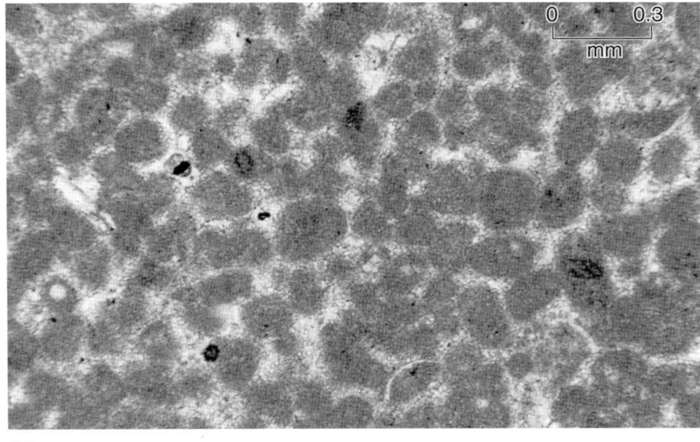

Figure 5.1 (a) Peloids with some skeletal debris, extracted from modern marine sediments. (b) A thin section of peloids in a Jurassic limestone from Dorset in southern England (viewed in plane-polarized light).

Directly analogous to the calcareous plankton are the siliceous phytoplankton and zooplankton, the remains of which accumulate to produce layers on the sea-bed. The post-depositional alteration of these remains produces bedded cherts, forming discrete bands ranging from a few millimetres to several metres in thickness. Silica in the form of quartz is only sparingly soluble in the pore waters between sediment grains, but biogenic silica is around an order of magnitude more soluble. With progressive burial, the siliceous skeletons are dissolved and then eventually re-precipitated as less soluble forms of silica which constitute chert.

Chert can also occur in nodular form (Block 2 Section 7.3). Chert nodules are produced when silica that was deposited in one place, as skeletal material, quartz grains, or silica-rich volcanic debris, dissolves and migrates through the sediments to be deposited elsewhere. Nodules vary in size from a few millimetres to several centimetres across and they may coalesce to form irregular layers, for example along bedding planes in limestones which form easy routes for migration of the silica-rich solutions.

Not all biogenic sediments form by the precipitation of ions in solution. For example, coal is formed from the accumulation of plant materials, usually in swampy conditions where vegetation is lush and the waters are anoxic. The lack of oxygen enables the plant material (which you can think of as carbohydrate, CH_2O) to be reduced to carbon, forming coal, rather than being oxidized to CO_2 which would escape in solution, or into the atmosphere.

5.2 CHEMICAL PRECIPITATION AND SEDIMENT FORMATION

Sediments can also be produced by direct precipitation from solution. One reason for this is that seawater or lake water may begin to evaporate. As it does so, it becomes supersaturated with respect to the various dissolved ions in turn, in other words the water contains more of these dissolved ions than is required for saturation. To begin with, $CaCO_3$ begins to precipitate, and when the volume of water has been reduced to a critical level (20% in the case of seawater), a sequence of salts crystallizes out, beginning with calcium sulfate as gypsum and its anhydrous form, anhydrite, and ending with sodium chloride as halite, and various other chlorides and sulfates of potassium and magnesium. Because of their origins, these rocks are known as evaporites (Block 2 Section 7.3).

Question 5.2 In Block 2, you observed that a characteristic feature of sedimentary rocks is their *fragmental texture.*

(a) (i) Can you recall what this means?

(ii) Would you expect evaporites to show fragmental texture?

(b) If you found evaporite minerals within a succession of sedimentary rocks, what could you infer about the climatic conditions under which the sediments were deposited?

Because $CaCO_3$ can be precipitated spontaneously from warm seawater, carbonate sediments can also form in this way. The mud-sized particles that are precipitated, typically less than 10 μm in size, are not composed of calcite, but are tiny crystals of aragonite, the other polymorph of calcium carbonate (Block 2 Section 4.7.1).

❑ Can you recall what this means?

■ It means that the aragonite and calcite have the same chemical composition, but a different crystal structure (Block 2 Section 2.3.3).

Chemically precipitated aragonite is only a small component of the aragonitic carbonate mud found on the sea-floor. The majority is biological in origin. The tissues of certain calcareous algae contain, or are coated with, microscopic crystals of aragonite which are released into sea-floor sediments when the algae die and their tissues decay.

Carbonate muds may also contain fine-grained calcite as well as aragonite. Marine organisms such as boring sponges, sea-urchins and fish like the parrot fish, attack calcareous bioclasts, other carbonate grains and even submarine exposures of limestone, generating fine-grained material. Similarly, the abrasion of bioclasts and grains as they are moved around on the sea-bed produces fine-grained sediment. After deposition and burial, the carbonate mud recrystallizes to produce very fine-grained calcite crystals, < 4 μm in size, known as **micrite**, a portmanteau abbreviation of two words: *micro*crystalline calc*ite*.

Question 5.3 What does the presence of micrite in a limestone suggest about the energy of the environment in which the limestone was deposited?

We will return to micrite in Section 6.

When warm seawater, supersaturated with $CaCO_3$, is agitated by waves or strong currents such that CO_2 is driven off, the spontaneous precipitation of $CaCO_3$ may occur directly onto small sediment particles which act as nuclei for the process. The reason for this precipitation can be understood by looking again at Equation 5.1. Calcium and bicarbonate ions are in equilibrium with undissolved calcium carbonate, carbon dioxide and water. If CO_2 is removed, the equilibrium is disturbed and the equation is driven to the left; i.e. Ca^{2+} ions combine with HCO_3^- ions leading to the precipitation of calcium carbonate.

This process is similar to what happens when 'hard water' derived from limestone areas is boiled. The water contains Ca^{2+} ions and HCO_3^- ions from the chemical weathering of the limestone. When the water is boiled, CO_2 is driven off, because it is less soluble in warm than in cold water, and so $CaCO_3$ is precipitated, forming the 'furring' inside kettles, for example. Precipitation of $CaCO_3$ is enhanced by the fact that it is less soluble in warm water than in cold water.

❑ Which sort of particles are likely to act as nuclei for calcium carbonate precipitation during deposition of sediments?

■ They are most likely to be small bioclasts, quartz grains or peloids.

As these nuclei are rolled to and fro over the sea-bed by water movement, layers of calcium carbonate build up round them through constant precipitation so that they grow in size. The end product of this process is a collection of creamy white, spherical to ovoid grains, typically around 0.5 mm in diameter. Because of their shape, these grains are called ooids, a word meaning 'egg-like' (Figure 5.2a, b) (Block 2 Section 7.2). Limestones composed predominantly of ooids are described as oolitic limestones or, less formally, as oolites (e.g. the Great Oolite and Inferior Oolite on your Moreton-in-Marsh Sheet, Block 1 Section 3.2.3).

(a)

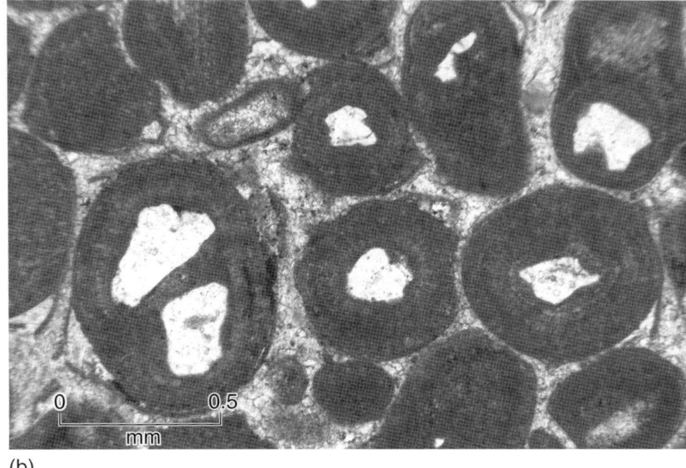

(b)

Figure 5.2 (a) Modern ooids from marine sediments. (b) Ancient ooids viewed in thin section (plane-polarized light).

Another theory suggests that some ooids are formed by biochemical processes. Bacterial activity within organic material in carbonate mud promotes $CaCO_3$ precipitation. The biochemical theory helps to explain how ooids can be found *in situ* within low energy micritic sediments.

At first glance, the ooids in Figure 5.2a seem pretty much the same as the peloids shown in Figure 5.1a apart from size, so how can we tell them apart? Attempt the following question to find the answer.

Question 5.4

(a) Describe the internal structure of the ooids in Figure 5.2b. How does this differ from that of the peloids in Figure 5.1b?

(b) Is there any evidence from Figure 5.2b that these ooids may have formed by the accretion of layers of $CaCO_3$ round a nucleus?

(c) Based on your understanding of how ooids form, which factor do you think is primarily responsible for their *shape*?

The answer to Question 5.4c shows that, unlike siliciclastic grains, the *shape* of ooids is unrelated to the length of time they have been in transport. The observation that they reach a maximum size of about 0.5 mm suggests that at this critical stage of growth they probably begin to suffer abrasion as fast as they accrete, which may help to explain why oolitic limestones may appear so well sorted. From this discussion, you can see that it would be inappropriate to apply the concept of textural maturity to all limestones. Many limestones accumulate *in situ* and contain particles ranging from mud to granules in size, or even coarser. Some even contain large fossil organisms, preserved in their life positions. It is best not to describe a limestone in terms of textural maturity unless there is good evidence that it consists predominantly of fragmented bioclasts that have been well sorted and rounded during transport.

Activity 5.1

You are now in a position to identify the grains in your Home Kit limestones, and to make deductions about whether they are biogenic or chemical. To do this, work through Activity 5.1.

Although we asked you in Activity 5.1 to concentrate on identifying only the grains in Activity 5.1, it is just as important to identify the intergranular material of a limestone if you want to work out the conditions under which it was formed. The issue of the role played by ions in the solutions found in pore spaces between the grains of buried sediments is addressed in Section 6.

5.3 BEDFORMS IN CARBONATE SEDIMENTS

In our discussion of bedforms in Sections 4.2.4 and 4.2.5, we alluded only to those formed in siliciclastic sediments. We would not want to leave you with the impression that bedforms are exclusive to these types of sediments. They can develop just as easily in carbonate sediments, especially those made up of sand-sized grains, such as ooids. The ripples shown in Figure 5.3 have developed in carbonate sediment. This means that some limestones may show cross-stratification which can be interpreted in the same way as cross-stratification in siliciclastic sediments.

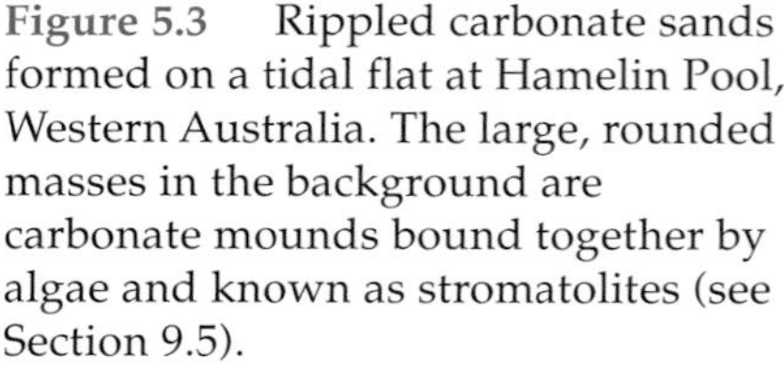

Figure 5.3 Rippled carbonate sands formed on a tidal flat at Hamelin Pool, Western Australia. The large, rounded masses in the background are carbonate mounds bound together by algae and known as stromatolites (see Section 9.5).

Question 5.5 Study the ripples shown in Figure 5.3 and explain whether you think they were formed by current or wave action.

5.4 Summary of Section 5

- Sediments may be formed by removal of ions from solution both by biological processes, such as the formation of skeletal material, and by chemical processes, involving direct precipitation of minerals.
- Accumulation of calcareous bioclasts and faecal pellets leads to the formation of biogenic limestones, while accumulation of the remains of siliceous phytoplankton and zooplankton leads to the formation of bedded cherts. The migration of silica-rich solutions through sediments leads to the formation of nodular cherts. Burial and preservation of abundant terrestrial plant material under anoxic conditions may result in the formation of coal.
- Extensive evaporation of lake water or seawater leads to crystallization of evaporite minerals. The spontaneous precipitation of $CaCO_3$ from seawater is related to supersaturation of the water with $CaCO_3$. The process is enhanced by removal of dissolved CO_2. It may also be promoted by the biochemical activity of bacteria in organic matter.
- Carbonate muds are mainly biological in origin, although some may be produced by the direct precipitation of $CaCO_3$. Precipitation of $CaCO_3$ around small bioclasts, peloids or mineral grains produces ooids. Post-depositional recrystallization of carbonate mud produces very fine-grained micrite. The presence of micrite in limestones is evidence for deposition in low energy conditions.
- Cross-stratification may be produced in carbonate sediments by the same processes that produce this sedimentary structure in siliciclastic sediments.

5.5 Objectives for Section 5

Now you have completed this Section, you should be able to:

5.1 Explain with the aid of chemical equations the origins of Ca^{2+} and CO_3^{2-} ions in solution, and the biogenic and chemical processes that lead to the production of carbonate sediments.

5.2 Outline the biological and chemical processes that lead to the formation of cherts and coal.

5.3 Outline the conditions under which evaporites form and explain how the textures of evaporites can be related to their mode of formation.

5.4 Describe the sources of carbonate mud in calcareous sediments and relate the presence of micrite in a limestone to the energy conditions under which the limestone was deposited.

5.5 Explain with the aid of chemical equations the factors favouring the precipitation of $CaCO_3$ from seawater, and the formation of ooids.

5.6 Describe limestones as either biogenic or chemical on the basis of their dominant grain composition.

Now try the following questions to test your understanding of Section 5.

Question 5.6 Extensive deposits of halite occur within the Triassic sands of Cheshire in NW England. What does this suggest about the geographic conditions (including climate) in Cheshire during the Triassic Period, and why?

Question 5.7 At the present day, we find that the abundance of oolitic and aragonitic sediments increases towards low latitude tropical and equatorial regions. On the basis of what you have read in Section 5, suggest reasons for this pattern.

Question 5.8 Oolitic limestones are locally abundant among Jurassic rocks in Britain and they sometimes exhibit large-scale cross-stratification. What can you infer about the energy of the environment and the climatic conditions in which they were deposited, and why?

6 From sediments to rocks

So far in Block 4, we have been considering only the processes that lead to the production of *sediments*. These are a far cry from the solid *sedimentary rocks* in your Home Kit. The transformation of unconsolidated sediments into sedimentary rocks, or **lithification**, is the last stage of the detective hunt we introduced in Section 2. Like the other stages we have already investigated, this one also leaves clues in the rocks which we can 'read', and so learn from them something about the processes that may have effected the transformation. These processes begin as soon as sediments are deposited and their burial commences. The term used to embrace all the changes that occur after deposition, but before sediments are buried sufficiently deep to be metamorphosed, is **diagenesis**.

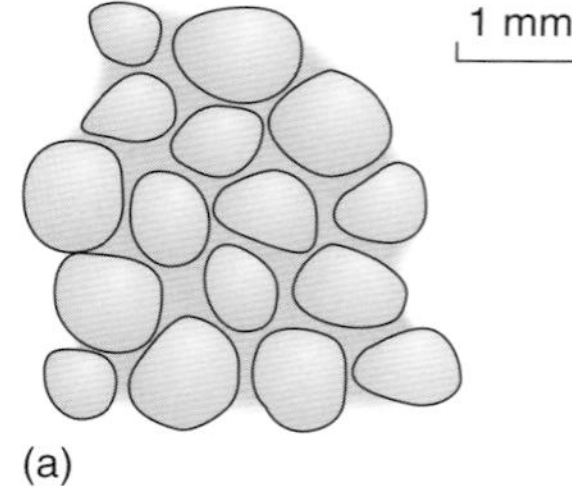

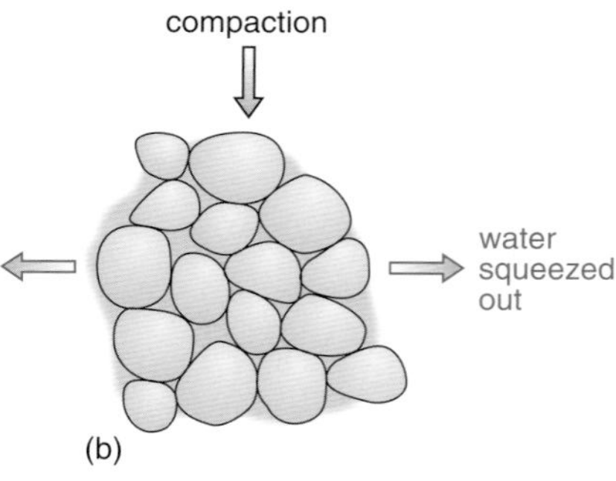

Figure 6.1 Cross-section through a sand deposit (a) before and (b) after compaction.

6.1 Compaction

Compaction is an important first step in diagenesis. Thirty per cent or more of the volume of a texturally mature sand deposit occurs as intergranular pore spaces. These are filled with water if the sediment was deposited subaqueously, or if it exists below the **water table** (i.e. the depth below which the ground is saturated with water). If the sediment is grain-supported immediately after deposition, a grain is in contact with one of its neighbours at one point only. These points of contact are not necessarily all visible in the 2D cross-section shown in Figure 6.1a (see also Plate 2.3). During compaction, the pressure of the overlying sediments packs the grains closer together and more efficiently, reducing the volume of pore space and squeezing out the pore water (Figure 6.1b). During compaction, any elongated or flaky grains such as micas, not already aligned parallel to the bedding plane (Figure 6.2a), become aligned. Under extreme pressures, mica flakes may even be bent as they are squashed, and mould themselves round the edges of quartz grains (Figure 6.2b).

When clay-rich sediments are first deposited, they may contain as much as 70–90% water by volume. On burial, both water loss and compaction are relatively rapid and much of the pore water is lost by the time sediments have been buried to about 1 km depth.

❑ Can you recall two factors that will help to consolidate muddy sediments during compaction?

■ Clays are flaky minerals and so will become aligned and flattened parallel to the bedding plane. Also, fine-grained, clay-rich sediments are naturally cohesive (Section 4.1.2).

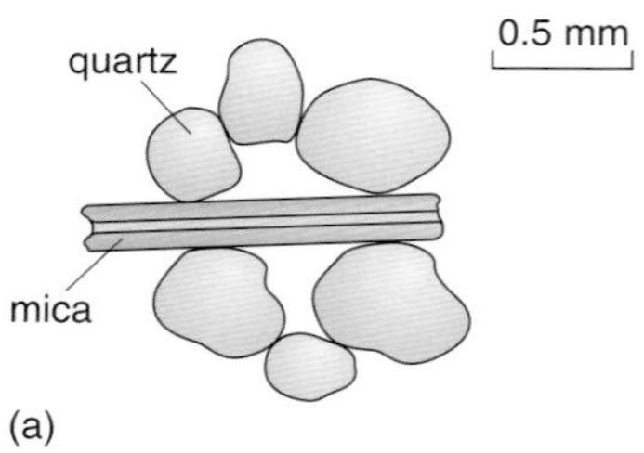

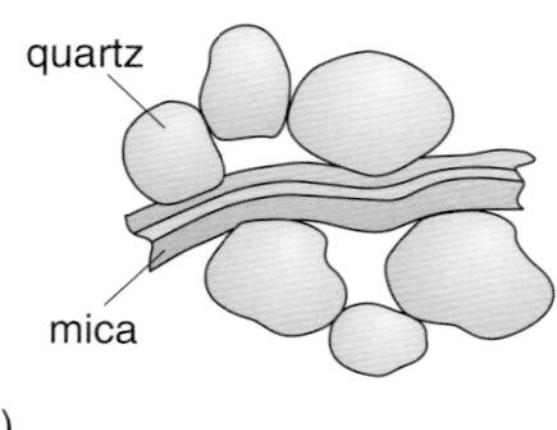

Figure 6.2 Deformation of a mica flake where it is bent around quartz grains: (a) before and (b) after compaction.

This alignment of clay minerals (Figure 6.3) may lead to a claystone or mudstone splitting easily into layers, like RS 15 in your Home Kit. Rocks which are fissile like this are known as shales (Block 2 Section 7.1).

As quartz sands are progressively buried, the pressure at the grain contacts increases until the quartz begins to dissolve. This **pressure dissolution** is more common in texturally mature sandstones than in those containing a significant amount of fine-grained matrix. The matrix helps to spread the load thereby reducing the pressure at the grain contacts. Pressure dissolution produces unusual quartz grain contacts (Figure 6.4). A grain may be simply indented where the adjacent grain has pushed into it, creating a concavo-convex contact (Figure 6.4a). If dissolution is more advanced then the contact may have a wavy, sutured appearance (Figure 6.4b) showing where one grain has become intergrown with the next on dissolution. Pressure dissolution and these sorts of grain contacts can also occur in limestones, such as those containing ooids or shell fragments.

6.2 Cementation in sandstones

Unless a sandstone contains a significant amount of muddy matrix, the precipitation of cement is required for the grains to stick together; compaction alone will not produce a lithified rock. This matrix can be original or, more rarely, can be composed of clays derived from the post-depositional breakdown of feldspars or other aluminosilicate minerals. To make matters more complex, clay minerals can also be precipitated directly from pore solutions, in which case they might be considered as cement, rather than matrix. The other very common cementing materials which are precipitated from solution within the sediments, are silica (as quartz), various carbonates (most commonly calcite) and more rarely iron oxides. Which of these occurs depends very much on the circumstances under which lithification occurs.

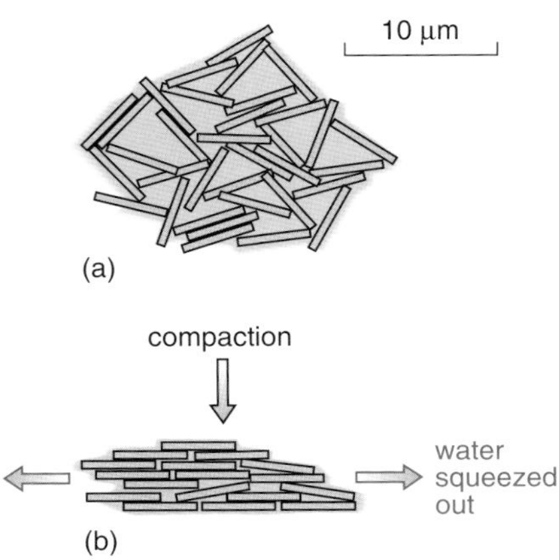

Figure 6.3 Cross-section to show the compaction of a claystone. (a) Clay flakes soon after deposition. (b) Clay flakes during compaction.

6.2.1 Silica cement

Silica cement is derived from more than one source.

Question 6.1 Recalling what you know about the processes that take place during compaction, suggest one possible source for this silica.

As pressure dissolution continues, silica is reprecipitated as quartz when the sediment pore waters reach supersaturation to form **overgrowths** on the existing quartz grains. The overgrowths are usually in optical continuity with the original grains; in other words, if a thin section of the sandstone is rotated between crossed polars, both the grain and the overgrowth go into extinction together (Figure 6.5b). Often, the rounded shape of the sedimentary grain is discernible in plane-polarized light because of thin coatings around the grain. For instance, in the case of RS 10, it is revealed by a thin, opaque 'dust line' of iron oxide (Figure 6.5a). If the overgrowth were to occupy all the original pore space in a sandstone, then without this dust line, the rock would appear fully crystalline and could easily be mistaken for a metamorphic quartzite.

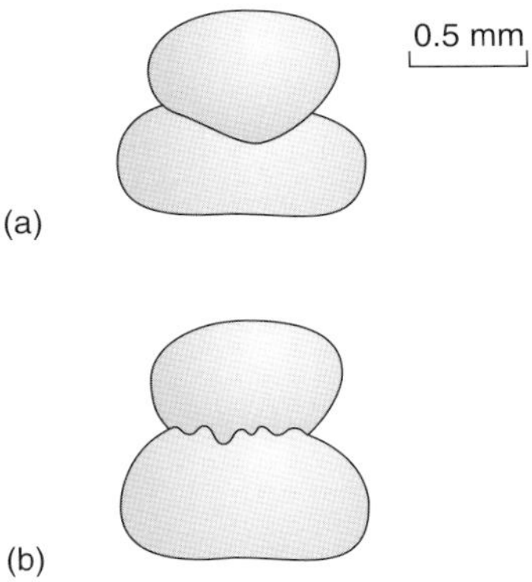

Figure 6.4 Sketch to show (a) a concavo-convex grain contact and (b) a sutured grain contact caused by pressure dissolution.

Question 6.2 What evidence would you look for in a thin section of a silica-cemented sandstone to show that pressure dissolution might have contributed some of the cement?

The greater the degree of compaction, the more likely pressure dissolution is to have occurred, and the smaller the volume of pore space left to be filled by cement.

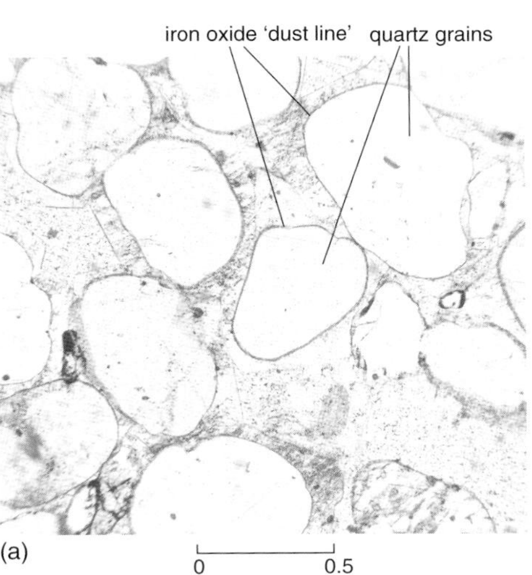

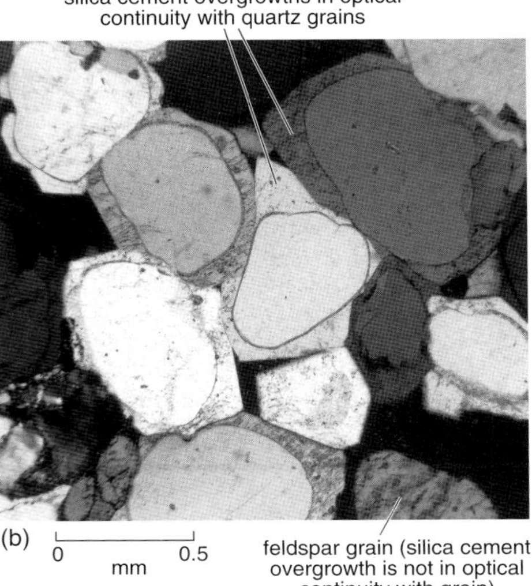

Figure 6.5 Thin section of a sandstone with silica cement viewed (a) in plane-polarized light. The grain can be distinguished from its overgrowth only by a dust line. (b) Between crossed polars. In the extinction position, grains and overgrowths appear as one angular crystal.

Other sources of silica cement include the chemical decomposition of mafic minerals and feldspars (Section 3.3), the dissolution of the remains of siliceous phytoplankton, zooplankton and other siliceous skeletal material in marine sediments (Section 5.1); and the silica in solutions that have migrated upwards from more deeply buried sediments where temperatures are higher, and the solubility of quartz is greater.

6.2.2 Carbonate cement

Calcite is the most common cementing material in sandstones, principally because it is more soluble than silica. It is derived from the dissolution of calcareous grains and micrite, and may be precipitated as large, single crystals, capable of enveloping several quartz grains at a time. Alternatively, it may occur as smaller crystals within pore spaces. Calcite cements are not restricted to marine sandstones, but are known to be forming in present-day river and desert sands, and also in soils. The greater solubility of calcite means that supersaturation of pore waters can occur in near-surface sediments where the acidity of ground waters also fluctuates regularly between acid and alkaline, leading to the rapid dissolution and reprecipitation of calcite. Other types of carbonate cements commonly occurring in sandstones are dolomite ($CaMg(CO_3)_2$) (Block 2 Section 7.2) and siderite ($FeCO_3$).

6.2.3 Iron oxide cement

Iron oxide rarely occurs as the sole cement in a well-consolidated sandstone. More usually it occurs in conjunction with silica or carbonates. It takes less than 1% of iron oxide to give a sedimentary rock such as RS 10 in your Home Kit a characteristic reddish-brown colour.

Question 6.3 Where does this iron come from in the first place, and under what sort of conditions would it be precipitated?

'Red beds' are common among sedimentary rocks deposited in continental environments, especially where highly oxidizing arid or semi-arid conditions prevail (Sections 3.3 and 4.1.3). In some cases, the iron may be carried in solution by upwards-migrating pore waters, to be precipitated as Fe_2O_3 in the oxidizing conditions within near-surface sediments. Red coloration in sediments is an excellent (but not infallible) indication of diagenesis, and probably deposition as well, in a continental rather than a marine environment.

6.3 Cementation in limestones

It should come as no surprise to you to learn that cements in limestones are mostly carbonates, because of the ease with which calcium carbonate is dissolved and reprecipitated. As with very fine-grained siliciclastic sediments, limestones rich in lime mud (i.e. micrite) can be lithified by compaction alone, although the micrite may undergo dissolution and reprecipitation during burial; therefore lithification is more pronounced in the early stages of burial in limestones. In limestones containing intergranular spaces, calcite is precipitated post-depositionally from pore waters to form larger crystals which act as a cement. This more coarsely crystalline, calcite cement is given the name **sparite** (an abbreviation of '*spar*ry calc*ite*'). Sparite is usually detectable in hand specimens of limestones by its translucent colour and highly reflective cleavage planes. In thin section (plane-polarized light), it is also translucent. Fine-grained micrite is usually white, cream or pale yellow in hand specimen, and dull in appearance. If it is impure because there are some clay minerals present, then it may appear grey. Wetting the hand specimen before examining it with a hand lens often helps to distinguish between sparite and micrite.

Question 6.4 If sparite crystallizes in intergranular spaces where micrite is absent, what might the presence of sparite in a limestone imply about the energy of the environment in which the clasts were deposited?

You need to exercise some caution when drawing this conclusion because crystals of calcite up to 100 μm in size may grow as the result of the dissolution and reprecipitation of micrite. This sparite could mislead you into thinking that the clasts were deposited in a high-energy environment, rather than a low-energy one. Examination of the specimen in TS usually allows the distinction between early and reprecipitated sparite to be made because ghost outlines of the original micrite crystals can be distinguished.

The behaviour of sparite cement in limestones is much the same as that of calcite cements in sandstones. Single crystals may enclose several grains, or several crystals may occupy a single pore space. Because the sparite has the same mineralogical composition as many of the limestone grains, it may also crystallize as overgrowths in optical continuity with the grains it encloses. This will only happen if the grains are single crystals of calcite, e.g. the skeletal plates of sea-urchins and sea-lilies (see Sections 9.3.2 and 9.3.6).

Dolomite may also form cement in limestones, and may even replace calcite grains soon after deposition. It has a more ordered atomic lattice than calcite, which appears to inhibit its direct precipitation from seawater. The reasons why dolomite crystallizes in place of calcite, or replaces existing calcite, are chemically complex and so we shall not explore them further in this Block.

6.4 Nodules

Nodules (or concretions) are intriguing diagenetic features of some sedimentary rocks (Block 2 Section 7.3). They are ovoid or irregular lumps of material that have a different composition from the sedimentary rocks in which they are found, and which formed at some stage after the host sediments were deposited. Nodules often grow around features such as fossils. Their composition reflects that of minor constituents within the host sediments which have been dissolved by percolating solutions, and then reprecipitated as aggregates at some point as the solutions migrated through the sediments.

❑ What are the examples of nodule formation that you met in Section 5.1 and Block 2, Section 7.3?

■ The formation of chert nodules, such as the flints in chalk, and carbonate nodules.

It is the dissolution of siliceous skeletal material in chalk that has led to the formation of the flints (Section 6.2.1). In other limestones, the silica may be derived from the dissolution of volcanic ash as well. Carbonate nodules you may meet include calcium carbonate and siderite nodules; these are commonly found in sandstones and mudstones.

Activity 6.1

You are now in a position to recognize in your Home Kit specimens some of the diagenetic features we have described. Activity 6.1 will give you an opportunity to try to do this.

6.5 NAMING AND CLASSIFYING SEDIMENTARY ROCKS

We have now travelled almost a full loop in the rock cycle. In Section 3 we began with rock exposures being broken down by weathering; in Sections 4 and 5 we saw how the breakdown products were transported and deposited to produce sediments; and in this Section we have ended with sediments being lithified to produce sedimentary rocks again. Now let's see how sedimentary rocks are named and classified.

❑ In Block 2 (Section 7) you were introduced to two ways in which siliciclastic sedimentary rocks are classified. What are these?

■ By grain size and by mineralogical composition.

6.5.1 CLASSIFYING SILICICLASTIC ROCKS BY GRAIN SIZE

Classifying *siliciclastic* rocks by grain size provides important information about the energy of the environment in which the sediments were deposited. The grain-size scale used for classification is reproduced in Table 6.1. As you may recall from Block 2, when describing sedimentary rocks using this scale it is important to quote the overall *range* of sizes present as this gives a measure of how well sorted the rock is. When classifying the rock, however, you should use the *volumetrically most abundant* grain sizes present. Note, also, that although names like 'clay' and 'sand' (by common association with quartz-rich sediments) may imply specific mineral compositions, when used as grain-size measures they can be applied to any type of sediment, even limestones. So it is not uncommon to read about gravel- or sand-sized bioclasts. You should also note that in Table 6.1 the sand class is further subdivided, into very coarse-, coarse-, medium- and fine-grained sand, to help refine the descriptions of sandstones. You are expected to be able to use the rock names shown in Table 6.1 (e.g. coarse-grained sedimentary rock, very coarse-grained sandstone) when you describe sedimentary rocks.

Table 6.1 Grain-size scale for siliciclastic sedimentary rocks. Note that it is the volumetrically most abundant grains in a sedimentary rock that determine its classification.

Grain size (most volumetrically abundant grains) [a]		Sediment name	Sand size	Sedimentary rock name	
>256 mm	(very coarse-grained sedimentary deposit)	boulders		conglomerate (rounded fragments) or breccia (angular fragments)	
64–256 mm		cobbles			
4–64 mm	(coarse-grained sedimentary deposit)	pebbles			
2–4 mm		granules			
0.25–2 mm (medium-grained sedimentary deposits)		sand	1–2 mm (very coarse)	sandstone	
			0.5–1 mm (coarse)		
62.5–250 μm (fine-grained sedimentary deposits)			0.25–0.5 mm (medium)		
			<0.25 mm (fine)		
4–62.5 μm (fine-grained sedimentary deposits)		silt		siltstone	mudstone[c]
<4 μm (very fine-grained sedimentary deposits)		clay[b]		claystone	

[a] μm = micrometre = 10^{-6} m. The grain sizes in this Table are arbitrary but not random: 256 mm is 2^8 mm, 64 mm is 2^6 mm, 62.5 μm is $(1/2^4)$ mm, and 4 μm is almost exactly $(1/2^8)$ mm.

[b] Clay can have two meanings: in terms of *grain size*, clay refers to grains less than 4 μm in size; in terms of *composition*, clay refers to certain types of sheet silicate mineral. However, most clay-sized particles in sedimentary rocks are, in fact, clay minerals.

[c] Shale, if fissile.

Activity 6.2

Now try Activity 6.2 to see if you can classify some of the siliciclastic sedimentary rocks in your Home Kit by their grain size.

Some rocks like RS 12 are difficult to classify according to their grain size; instead we can use their composition. This is discussed in Section 6.5.2.

6.5.2 CLASSIFYING SILICICLASTIC SEDIMENTARY ROCKS BY MINERALOGICAL COMPOSITION

In Block 2 (Figure 7.7) you saw how sandstones, mudstones and limestones can be informally named according to how much quartz, clay or calcite they contain. Names such as 'muddy sandstone', 'sandy limestone' or 'calcareous mudstone' give you a good idea not only of the grain size of a rock and which minerals it contains (as grains, cement or matrix), but also the relative proportions of the different minerals.

Question 6.5 What are the main minerals you would expect the following rocks to contain: a muddy sandstone, a sandy limestone, a calcareous mudstone? In each case, state which mineral you would expect to form the highest proportion.

Rocks like RS 12 contain not only sand to gravel-sized quartz grains and rock fragments, and maybe even feldspar grains, but a significant amount of mud matrix as well. This is why they are difficult to classify using grain size alone, or by using Figure 7.7 in Block 2. Rocks like this, with more than 15% by volume composed of matrix, are called greywackes (Block 2 Section 7.1), a name which implies usually both textural and compositional immaturity. To be classed as a sandstone, a quartz-rich rock should contain no more than 15% matrix.

But what about rocks like RS 18, which could be classified as a sandstone, because it contains sand-sized grains and little, if any, matrix, but which contain a high proportion of feldspar? These rocks are called arkoses (Block 2 Section 7.1), a name implying not only that the rock contains sand-sized grains, but that it also contains a lot of feldspar and so is compositionally fairly immature. Technically, a sandstone should contain 25% or more feldspar to qualify as an arkose. If it contains between about 5% and 25% feldspar, the name feldspathic sandstone is more appropriate. If there is less than 5% feldspar, then the rock is simply a sandstone, or quartz sandstone.

Activity 6.3

You should now be able to complete the Table left unfinished in Activity 4.6. Activity 6.3 helps you to do this.

6.5.3 CLASSIFYING LIMESTONES

While names such as 'muddy limestone' and 'sandy limestone' provide information on the purity of a limestone, they are not much help when it comes to describing the origin of the carbonate material, especially in relatively pure limestones; here a different form of classification is needed. One approach is to classify limestones according to the types of grains they contain, and their intergranular material.

Question 6.6 What are the three main types of grain and two forms of intergranular material that occur in limestones?

By using names such as 'bioclastic limestone' or 'oolitic limestone', we have already been classifying limestones by their grain types to some extent. We can now refine this classification by devising names which reflect both the dominant grain type *and* the intergranular material. This classification scheme is shown in Table 6.2. The prefix to each name describes the main grain type (*bio-* bioclasts, *oo-* ooids, *pel-* peloids), and the rest of the name describes the material between the grains, which can be either *micrite* or *sparite*. You are expected to know the names shown in Table 6.2 when you describe limestones.

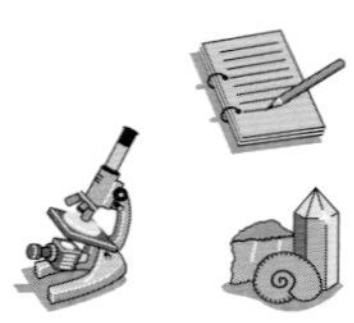

Activity 6.4

Now try Activity 6.4 which will give you practice in describing and classifying your Home Kit limestones.

Table 6.2 A classification of limestones based on composition of the grains and intergranular material.

Principal grain type	Limestone type*	
	cemented by sparite	micrite matrix
bioclasts (*bio-*)	**biosparite**	**biomicrite**
ooids (*oo-*)	**oosparite**	**oomicrite**
peloids (*pel-*)	**pelsparite**	**pelmicrite**

* Note that it is possible to find 'hybrid' limestones which contain significant amounts of more than one type of grain or intergranular material. The descriptive classifications of such limestones should reflect this mix. For example, a limestone containing bioclasts, together with a significant proportion of ooids and a sparite cement, would be classified as an oo-biosparite.

There is one other limestone type we should mention, because you will meet it when you come to look at carbonate environments in Section 16. It is called *biolithite* and it forms from the remains of living organisms, lithified *in situ* where the organisms lived. The most obvious examples are the limestones formed from the remains of corals, and other animals, that grew to form reefs. In Sections 7–9, you will learn not only about the organisms that built reefs, but also about the many others whose remains are found in limestones and other types of sedimentary rocks.

6.6 CONCLUDING COMMENTS

You should now have a clear idea of how and why sedimentary materials and the deposits they form bear some record of the processes that have shaped them *en route* to their site of deposition and, more importantly, what has affected them at the site of deposition itself, and even after burial. You should also be able to read these clues for yourself and to begin to use them to reconstruct the original depositional environments of sedimentary rocks. Sections 7–11 will examine what evidence we can find in fossils for processes and environments. In Sections 12–17, we will start to put all the evidence together so that the environment of deposition can be deduced.

6.7 SUMMARY OF SECTION 6

- Lithification begins after sediments have been deposited, and the diagenetic processes of compaction and cementation transform them into solid rocks. Sediments containing a significant amount of mud matrix may be lithified by compaction alone.
- The most common cementing materials are clay minerals precipitated from solution, silica, carbonates and iron oxides. Silica cements are derived from the silica released by pressure dissolution of quartz grains, the post-depositional decomposition of mafic minerals and feldspars, and the dissolution of siliceous organisms. Iron oxide cements are derived from the decomposition of mafic minerals. Calcite cements originate from the dissolution of calcareous clasts or micrite and occur as intergranular sparite.
- Pressure dissolution of grains leads to grain contacts which are concavo-convex or sutured, depending on the degree of compaction.
- Silica and calcite cements may occur as overgrowths in optical continuity with quartz or calcite grains, respectively.
- Siliciclastic sedimentary rocks may be classified by grain size and by mineralogical composition. Limestones may be classified according to their grain types and intergranular material.

6.8 OBJECTIVES FOR SECTION 6

Now you have completed this Section, you should be able to:

6.1 Describe with the aid of sketches some of the diagenetic processes that lead to the lithification of sediments, and recognize evidence of these processes in thin sections of sedimentary rocks.

6.2 Distinguish between matrix and cement in siliciclastic and carbonate sedimentary rocks, and recognize different types of quartz and calcite cements in thin sections.

6.3 Classify siliciclastic sedimentary rocks according to their grain size and/or mineralogical composition, using hand specimens or thin sections.

6.4 Classify limestones according to their grain types and intergranular material, using rock hand specimens or thin sections.

Now try the following questions to test your understanding of Section 6.

Question 6.7 Examine TS B (arkose, RS 18) and TS W (sandstone, RS 10) using your microscope, paying particular attention to the grain contacts and intergranular spaces. What can you tell about the degree of compaction the arkose suffered before cementation, relative to the sandstone?

Question 6.8 How would you classify the following sedimentary rocks, based on their grain sizes and/or mineralogical compositions?

(a) A rock containing around 80% clay minerals and 20% bioclasts.

(b) A rock containing shell fragments (50% of rock) bound together by an equal mixture of micrite and clay minerals.

(c) The rock shown in Plate 2.3.

(d) A rock containing 85% quartz with a dominant grain size of 250–500 μm, and 15% feldspar grains of the same size.

(e) A rock containing peloids in a micritic matrix.

7 Introduction to fossils

From its Latin root *fossilis* ('dug up'), the word **fossil** literally means any excavated object, but from around the early 18th century the term came to be mainly restricted to its present meaning – the remains or impression of an organism preserved in the geological record. The remains may contain original materials from the organism concerned, or they may have been entirely replaced by other minerals. Fossils can form in many ways, but their preservation has usually involved eventual burial in sediment – hence the association with being dug up. Less commonplace forms of entombment include woolly mammoths deep-frozen in Arctic ice and charred tree stumps engulfed in lava flows.

There are two widely recognized kinds of fossil: a **body fossil** comprises the (altered or unaltered) remains of an organism; and a **trace fossil** records the activities of an organism, such as burrows, borings, droppings or bite marks.

A common practical distinction used in the study of body fossils is based on size: a **macrofossil** is one large enough to be studied with the naked eye or a hand lens, while a **microfossil** is one best studied under a microscope, although microfossils are often visible with a hand lens, too.

A somewhat more cryptic category of fossils comprises residual chemicals derived from organisms, increasingly recognized through sophisticated analytical techniques. Such a vestige is called a **chemofossil**. Certain diagnostic breakdown products from cell membranes, for example, allow organic matter derived from higher plants to be distinguished from that of bacterial origin.

It is handy to have a 'dustbin' term for things that may be mistaken for fossils, but which in fact are not. Such an object is called a **pseudofossil** and examples may include anything from lumps of flint looking like bones (Figure 7.1) or bits of anatomy, to dewatering structures in sandstones that may resemble jellyfish.

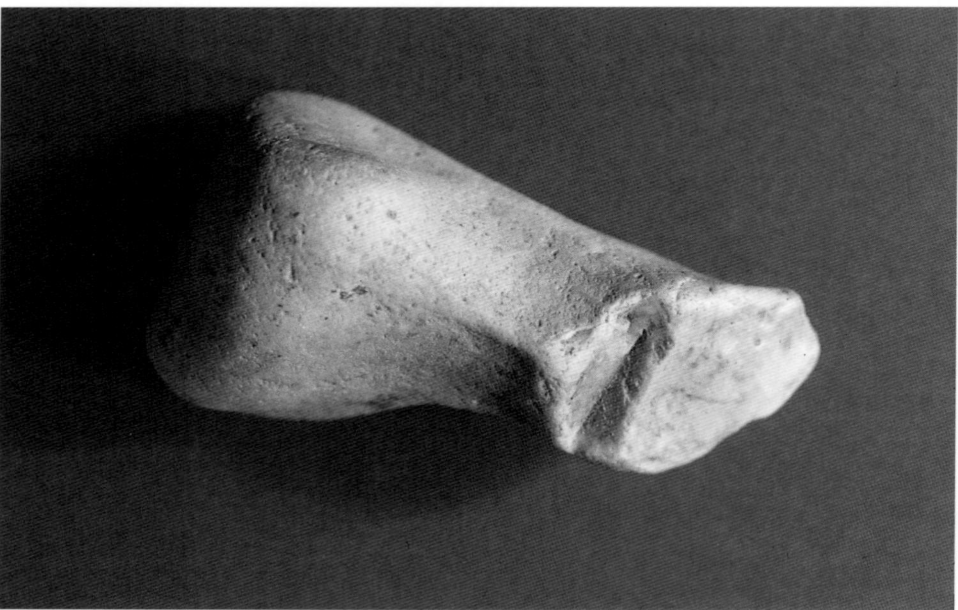

Figure 7.1 Despite its close superficial similarity to the end of a limb bone, this is merely a lump of flint and is hence a pseudofossil.

Fossils provide the only material record of life in the past. We can study them not only to discover what life was like, but, armed with knowledge of the biology and relationships of the organisms concerned, to infer much about the worlds they inhabited.

7.1 SCIENTIFIC VALUE OF FOSSILS

The scientific study of fossils is known as **palaeontology** (which means 'the study of ancient beings' in classical Greek). What we can infer from fossils may be outlined under three broad headings:

- *Biological and environmental reconstruction.* Here, the focus is on the organisms themselves, the aim being to reconstruct their **palaeobiology** – that is, their anatomy, life habits and manner of death and disintegration – and to use such insights, together with sedimentological observations, to help interpret the original depositional environments of the rocks in which the fossils are found.
- *Inference of patterns of evolution.* Comparative studies allow us to map out parts of the evolutionary tree of life, especially for extinct forms, for which fossils provide the only direct evidence. The patterns of evolution and extinction revealed may in turn yield further clues for detecting past environmental and climatic changes at all scales from the local to the global.
- *Stratigraphy and reconstruction of ancient geography.* The distributions of fossils in rock formations, combined with the insights on relationships mentioned above, provide crucial evidence both for the relative dating of strata (as noted in Block 1 Section 1.3.1 and Activity 1.1) and for past geographical configurations.

By providing such scientific insights, fossils may also serve practical economic ends. The oil, gas and coal industries, for example, make extensive use of fossils (particularly microfossils) in cores drilled from underground rock formations to help them date and correlate the rocks and thereby locate and exploit reserves.

In this Block, we concentrate on the first of the broad topics listed above, as space precludes exploration of the other topics within the scope of this Course.

As a brief trailer for how fossils can assist in geological interpretation, take a look at the fossil fish illustrated in Plate 7.1a. Notice how remarkably intact the coating of scales and skeleton is, pressed flat onto the bedding surface. This specimen comes from a finely laminated mudstone of Permian age that crops out a few miles south-east of Durham. The rock itself splits fairly easily along the thin depositional laminae (Plate 7.1b), sometimes revealing body fossils like that illustrated. Yet, apart from some faecal strands (Plate 7.1c), it contains no other trace fossils. It is dark grey when fresh, with some tiny crystals of pyrite scattered throughout.

From these few observations and some common knowledge concerning animal life habits today, we can already infer quite a lot about the original environment in which the sediments were deposited.

> **Question 7.1** What does the intact nature of the fossil fish in Plate 7.1a and the preservation of the lamination in the enclosing rock suggest about conditions for animal life at or near the sea-bed? What conclusion might you draw concerning animal life in the upper reaches of the water column?

The inferences from Question 7.1 in turn raise the question of what could have inhibited animal life on the sea-floor (and in the bottom waters).

❑ What factors can you think of that might have inhibited animal life in these bottom waters?

■ Several possibilities might come to mind: a lack of oxygen, extreme salinity, the presence of poisonous chemicals, extreme temperature or even some combination of these are all plausible.

How might the other observations help us here? An obvious candidate among the possibilities, given the presence of pyrite in the rock and the dark grey colour of the mudstone, is a lack of dissolved oxygen. It must certainly have been absent from the water permeating the muddy sediment, to have allowed the pyrite to form, so perhaps such anoxic conditions reached up into the bottom waters, too. Moreover, the lamination of the mudstone indicates minimal current activity, i.e. stagnant bottom water, so there would have been little circulation of any oxygenated water from shallower levels.

Already a model of the original environment is beginning to emerge, and it could be checked for its agreement with the implications of further observations – i.e. for consilience (Section 2). Evidence could be sought, for example, for some factor that might have inhibited water circulation at deeper levels (such as the ponding of relatively dense saline water on the bottom), so allowing anoxic conditions to develop there in the first place.

7.2 OBJECTIVES FOR SECTION 7

When you have completed this Section, you should be able to:

7.1 Define, and distinguish between, body fossils (macrofossils and microfossils), trace fossils, chemofossils and pseudofossils.

7.2 Outline the scientific value of fossils

Now try the following question to test your understanding of Section 7:

Question 7.2 Suppose that within a bed of mudstone you have found a layer of nodules which had been formed by cementation of sediment-filled horizontal burrows. On closer inspection, you find that in places the surfaces of the nodules have been encrusted by fully grown oyster shells and penetrated by small tubular holes (borings), though neither kind of fossil can be found in the intervening areas of mudstone.

(a) Categorize the different kinds of fossil described above.

(b) What can you infer from these fossils concerning the history of sedimentation? Was it continuous, or was it interrupted by erosional removal of sediment? Did the nodules form early on, while the clay was being deposited, or much later, following deposition of the entire formation?

8 INTERPRETING FOSSILS

So far, we have stressed the close link between process and product in the interpretation of sedimentary rocks. For example, by observing today how moving fluids variously pick up sedimentary grains, act on them during transport and set them down, we can infer the operation of similar processes in the past from corresponding compositions, textures and sedimentary structures seen in the rocks. You have been warned (in Section 4.2.1), however, that some features may be inherited from earlier cycles of transport and deposition, giving conflicting clues. So, for example, a sandstone of aeolian dune origin may have been eroded and the sand grains later redeposited with minimal change in shape or size in, say, a meandering river. Nevertheless, the inherited features can be unmasked as a faded 'memory' of earlier events because the environmental circumstances they imply are inconsistent with those inferred from other observations: there is a failure of consilience. Thus, our river deposit is likely to show other features contradicting an aeolian dune origin, such as poor sorting with the admixture of much larger particles (perhaps including plant fragments), and repeated sandstone to mudrock successions typical of deposition in meandering river channels (a characteristic pattern you will encounter later, in Section 13.2.1).

Organisms exhibit a more subtle mixture of responses to prevailing conditions and 'memories' of earlier ones, because they are the products of evolution. History is written into their genes. They combine a fundamental organization, inherited from ancestors, with specific modifications that suit them to their own particular circumstances (in the sense of promoting their prospects of survival and/or reproductive success relative to individuals lacking such traits). A modification of this kind evolves as a consequence of natural selection and is referred to as an **adaptation**. We can sometimes recognize the likely functional significance of adaptations in fossil organisms through uniformitarian comparison with living organisms (or even similarly constructed machines), and so find useful clues for interpreting past conditions. However, it is important to distinguish between evidence for comparable adaptation (evolutionary convergence) and that for common ancestry.

Consider the extinct woolly mammoth. Suppose there were no specimens with hair or soft tissues preserved. The skeletons alone would indicate merely some kind of elephant, identifiable as such from its massive ridged teeth, for example (Figure 8.1). These were inherited by the mammoth as the evolutionary

Figure 8.1 A fossil mammoth's tooth (about 15 cm long).

legacy of the ancestral elephant's adaptation to a rough vegetarian diet. Nothing in the mammoth's skeleton specifically reveals adaptation to cold conditions. Now, today's two surviving species of elephant happen to live in hot climates. With only the skeletons of mammoths available, we might be tempted to infer that Siberia, for example, must have been hot, too, when this extinct elephant lived there.

❑ What would be wrong with such a line of reasoning?

■ As noted above, we recognize the mammoth as an elephant merely because of features inherited from the common ancestor to all elephants. Though reflecting certain ancestral traits (such as diet in this case), they offer no reason why the mammoth should not have been adapted to cold conditions. Just because it was an elephant, it need not have shared the particular adaptations of living elephants.

Similarity of organization inherited from a common ancestor is termed **homology** – shared evolutionary 'memory', so to speak. As you see from the example above, it is an unreliable guide to the specific conditions experienced by past organisms. Homologous features may indeed become adapted to quite diverse functions, as illustrated by the wings of birds, the front legs of horses, the flippers of seals and our own arms, which are all homologous, being vertebrate forelimbs. At best, we may sometimes be able to make some generalizations from homologous features, where little modification is evident, such as inferring the retention of a rough vegetarian diet from the teeth of mammoths. On the other hand, homology *is* the essential key to recognizing evolutionary relationships, and thus classification.

Fortunately, there are fossil mammoths with preserved hair. Many unrelated mammals – such as the musk ox – have similarly evolved long shaggy coats (which we might wear) in very cold climates, so it is reasonable to infer that this kind of elephant might also have been adapted to such conditions. This idea could be tested by seeking other evidence from, say, association of the fossils with glacial sediments (i.e. by consilience). Similarity that arises as a consequence of independent adaptation to the same conditions by different organisms is termed **analogy**. Though superficially similar, reflecting their common function, analogous features may differ in basic structure, especially when they have evolved in organisms that are not closely related. Such is the case, for example, with the wings of birds and insects. Analogous features in more closely related organisms may be more similar in detail, however, if based on the same (i.e. homologous) original components. So, in the example of the different woolly mammals cited above, it is the exaggerated shagginess of their coats that is analogous among them; the hair itself (which they inherited from the ancestral mammal) is homologous. Analogy can offer precise clues as to how fossil organisms lived.

To interpret fossils correctly then, we need to know about the inherited common structural organization of each group of organisms – its **bodyplan** – in order to distinguish the signals of history from those of immediate function. That involves classification, the subject of the next Section.

8.1 OBJECTIVE FOR SECTION 8

Now you have completed this Section, you should be able to:

8.1 Explain the difference between homological and analogical similarities among organisms and recognize the value of each to the interpretation of fossils.

Now try the following question to test your understanding of Section 8:

Question 8.1 Ammonoids (as you will see in Section 9.3.4) are an extinct group of marine invertebrates that had a long tubular shell coiled up over the animal's body (Figure 8.2a). Though only distantly related, the living pearly nautilus has a similar shell (Figure 8.2b). Much of it is gas-filled, providing buoyancy and allowing the animal to swim around. The position of the coiled shell above the denser body also maintains the animal in a stable vertical orientation as it swims. Ammonoids are interpreted to have swum in a similar fashion. Although ammonoids and nautiluses apparently inherited a gas-filled shell from a distant common ancestor, it is likely that the coiling of the shell above the body evolved independently in the two groups. On the basis of this information, decide which of the following features should be considered homologous, and which analogous, between the two groups: (a) the ability to swim; (b) coiling of the shell; (c) buoyancy; (d) the maintenance of stability by the shell when the animal swims.

(a)

(b)

Figure 8.2 (a) A fossil ammonoid shell; (b) a nautilus shell. Both are *c*. 13 cm across.

9 Fossil classification and palaeobiology

This entire Section is closely tied to your study of the fossil replicas, labelled A–P in the Home Kit (from which you will also need the hand lens), and you will be making frequent reference to them, both while you follow the text and when you undertake the Activities. Therefore you should ensure that you can study this Section in a place where you can conveniently lay out and study the specimens. You may wish to check the results of your work, when you have finished the Section or as you work through it, against the annotated versions of the replicas in the 'Digital Kit' on DVD 1. Points you can check in the latter are indicated in the margin by the DVD icon. We have printed this here to show that the links are optional: you need only refer to the DVD if the point is not clear to you. Finally, in this Block we refer to fossil specimens by the abbreviation FS.

9.1 Classifying body fossils

To start with, you will be classifying the Home Kit specimens using a simple key system based on a few readily observable features. This is to help you become familiar with the specimens themselves and to learn to recognize other members of the groups they represent when we go into greater depth concerning their biology, in the next Section. Before proceeding, however, you should be aware of the main advantage and disadvantage of key systems. They make classification simple because you have to consider only one feature at a time. So, by going through a sequence of choices, you can eventually classify a specimen into one particular pigeonhole. The process is quick and easy, but it has one major drawback. As a consequence of evolution, one or more of the diagnostic features for a group may be modified or even lost in a given example, so an attempt to classify it using the key too rigidly would come off the rails, so to speak. It is therefore necessary in biological key systems to allow for frequent exceptions. As you become more familiar with fossils, you should become used to recognizing complexes of features in a given specimen, to cope with such exceptions and arrive at a correct classification.

The fossil specimens (FS) A–H all represent the shelly hard parts of marine invertebrate animals. These have been chosen because they are the sorts of macrofossil you are most likely to encounter in the field – at Summer School, for example.

❑ Why do you suppose that the hard parts of shelly marine invertebrates are the most commonly encountered macrofossils?

■ Environments of deposition are commonplace in the sea but scattered and often short-lived on land. The remains of marine organisms thus stand a greater chance of burial, and hence preservation, than do those of terrestrial organisms. Moreover, organisms with hard skeletal parts, especially the most abundant shelly forms, obviously stand the greatest chance of leaving resistant remains that might eventually be buried.

Activity 9.1

The starting point for the fossil identification key is to consider the number of discrete skeletal elements in each specimen. In Activity 9.1, you will examine a number of the fossil replicas in the Home Kit to determine whether the original skeletons were composed of one, two or more elements.

9.1.1 SPECIMENS WITH ONE SKELETAL ELEMENT

From Activity 9.1, you may have noticed that FS A, D and E can be distinguished further on their external shape and internal structure. Each shell was originally grown as an expanding tube, but while FS E is approximately straight, FS A and D are both coiled. The interior of FS A was filled with sedimentary matrix, smoothly plugging the shell opening, while FS D is an **internal mould** (i.e. the lithified filling of a fossil shell replicating internal surface features), so the site of its shell opening (the widest end of the coiled tube) is also solid. In all three cases, the soft parts of the animal that built the shell were originally housed inside the tube and projected from its open end in life.

❑ Inspect the sites of the original shell openings and classify each of FS A, D and E according to whether it is:

(a) subdivided by numerous thin projections running inwards from the shell wall, or

(b) unrestricted (other than by the sedimentary infill).

■ You should have classified the specimens as follows: (a) FS E; (b) FS A and D. The opening to the body space of FS E is subdivided by thin, projecting radial walls that extend towards the centre from the rim of the cone. By contrast, the sediment-filled areas corresponding to the shell openings of the other two specimens are smooth, indicating a lack of any such radial walls.

The radial walls visible in FS E are known as **septa** (singular: **septum**). They are a characteristic feature of the skeleton of most **corals**. FS E is from a **solitary** coral, the skeleton having been built by one individual. Some other corals are **colonial**, with a skeleton consisting of either openly branching or closely joined **corallites** each of which is like the skeletal tube of a solitary coral but joined to many other such tubes.

Now turn to FS D.

❑ What features on the internal mould of FS D might indicate that the body space occupied only part of the shell interior?

■ The specimen shows a series of intricately folded fine grooves encircling the coiled internal mould and spaced out at intervals along it, up to nearly half a whorl back from the site of the original shell opening. Since the specimen is an internal mould, these grooves must indicate where thin shelly partitions within the shell met the inner surface of the originally enclosing shell wall. The body could thus only have occupied the final half whorl of the shell, where these partitions evidently stopped.

Hence most of the shell cavity in FS D was divided into a series of chambers by transverse partitions, also called **septa** (to avoid confusion, it is worth remembering that the septa of corals are *radial*, while those seen here are *transverse*). The complex pattern noted in the question above, where each septum joined the inside wall of the coiled shell, is termed a **suture**. Evidently, the outer margin of each septum was thrown into a series of tiny folds, rather like the edge of a leaf of curly kale. Possession of such intricately folded septa is diagnostic of the shells of an extinct group known as **ammonoids**. Notice also that FS D is coiled in a single plane; this is common, though not universal, in ammonoids.

We cannot tell if FS A contained internal septa like those in ammonoids because only the outer surface of the shell is visible. In fact, it did not, and it is not an ammonoid. However, its asymmetrical coiling (coiled not in one plane, but as in a spiral staircase), is typical of the shells of most, but by no means all, **gastropods**, a huge group that includes not only snails, whelks and winkles, but also slugs, in which the shell is reduced or lost.

9.1.2 Specimens with two skeletal elements

In FS B and F, the shell consists of two parts (known as **valves**) which fit closely together. In life, the valves were hinged along one margin and the shell could be opened and shut: when shut, the shell entirely enclosed the animal's soft parts.

Both FS B and F show **bilateral symmetry**, i.e. the whole shell can be divided by an imaginary plane into two halves which are mirror images of each other. However, the two specimens can readily be distinguished by the different orientations of the symmetry plane with respect to the shell. Figure 9.1 illustrates the two kinds of symmetry.

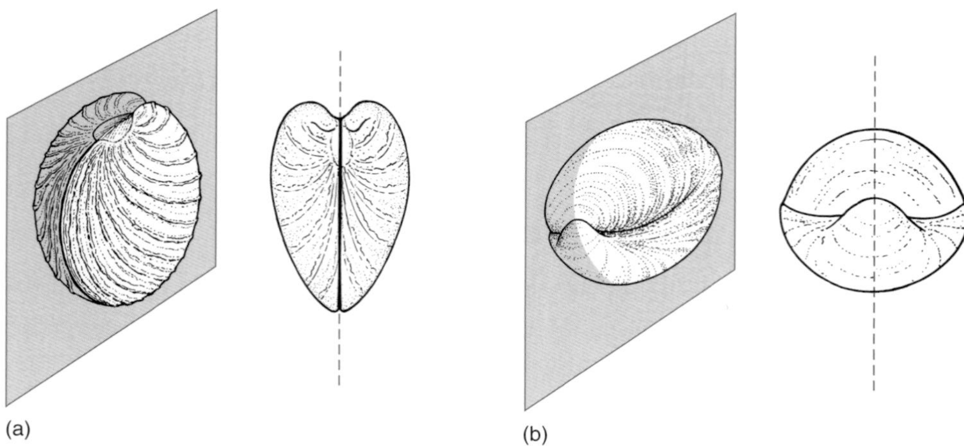

Figure 9.1 Contrast between the orientation of the plane of bilateral symmetry in (a) bivalves and (b) brachiopods.

❑ Classify FS B and F according to whether they show a plane of symmetry that (a) lies between the two valves so that each valve is a mirror image of the other (Figure 9.1a), or (b) cuts through both valves so that, although the two valves are differently shaped, each can be divided into two halves which are mirror images of each other (Figure 9.1b).

■ (a) FS B; (b) FS F.

The first kind of symmetry (Figure 9.1a) is typical of a group known as **bivalves** while the second kind (Figure 9.1b) characterizes **brachiopods**. There are, however, several exceptions. Oysters, for example, are bivalves that have distinctly asymmetrical valves. Nevertheless, those bivalves that lack the typical symmetry of the group never show perfect brachiopod symmetry (though scallops are bivalves that sometimes come close to that). A few brachiopods also deviate from the usual symmetry of their group.

9.1.3 Specimens with many skeletal elements

Symmetry again provides the next step in the key for FS C, G and H.

❑ Classify FS C, G and H according to whether they show (approximately) (a) bilateral symmetry of a row of transverse skeletal elements, or (b) radial symmetry around a central axis, with the shell composed of several sectors of similar elements.

■ (a) FS G; (b) FS C and H. The symmetries of FS G and FS C are obvious. That of FS H is less clear because only part of the specimen is visible. Nevertheless, if you look at it from the end where the stalk emerges, you can see that it shows a number of similar radial sectors of plates.

Now inspect FS G. Notice that there is a broad, mask-like **head-shield** at one end, bearing two prominent knobs which are eyes. Behind this is a central trunk region consisting of a number of transverse segments which were articulated in life. The **tailpiece** behind the trunk region is of similar construction, though here the segments were fused together.

❏ How can you tell that the tailpiece segments were fused together?

■ The smooth form of the rim running around the outside of the tailpiece segments shows that they were fused in one piece.

The central part of the body in FS G, from head to tail, is slightly raised and the two flanks on either side are mirror images of each other. This three-fold (trilobed) longitudinal division of the body gave the name of the group to which the specimen belongs – the **trilobites**. This specimen represents the shelly segments of horny 'skin' that covered the upper side of the animal. An external shelly skeleton of this kind is termed an **exoskeleton**. The skin of the lower surface and of the limbs lacked the inorganic salts responsible for the 'shelliness' of the upper skin and is therefore very rarely fossilized.

Now look at FS C and H.

❏ How many similar radial sectors can you detect in FS C and H?

■ In FS C, you should be able to count five sectors with large plates bearing two rows of large knobs, alternating with another five sectors with smaller plates bearing two rows of smaller knobs flanked by rows of small pits. If you look closely at FS H, particularly around where the stalk attaches, you will see that it could (and in fact it does) have five radially arranged sectors of plates, though one is hidden on the unexposed side of the specimen.

Hence both specimens effectively show approximate **five-fold symmetry** around a central axis. Both specimens belong to a major animal group known as the **echinoderms**, most members of which have approximate five-fold symmetry. Important exceptions exist, however, including many forms in which the shell has been modified during evolution so as to become bilaterally symmetrical, with a distinct front and back. In such forms the five-fold arrangement of plates remains, though the sectors are no longer radially symmetrical. This is a good example, incidentally, of the modification of homologous structures to suit new functions, as discussed in Section 8. We will discuss its adaptive significance later, in Section 9.3.2.

FS C and H are nevertheless clearly different. FS C is an **echinoid** (sea-urchin), one of a group of echinoderms that can move around. FS H, with its distinctive cluster of arms is a **crinoid** (sea-lily).

❏ What feature might suggest that FS H was not capable of moving around?

■ The stalk leading off from the base of the specimen suggests that the crinoid was attached to the sea-floor, in contrast to the free-living echinoid.

Now that you have sorted FS A–H into their respective groups, summarize your conclusions by working through Question 9.1.

Question 9.1 Working your way through the key in Figure 9.2, to remind yourself of the principles of classification used, allocate FS A–H to their appropriate groups, as indicated in the boxes shown at the foot of the key.

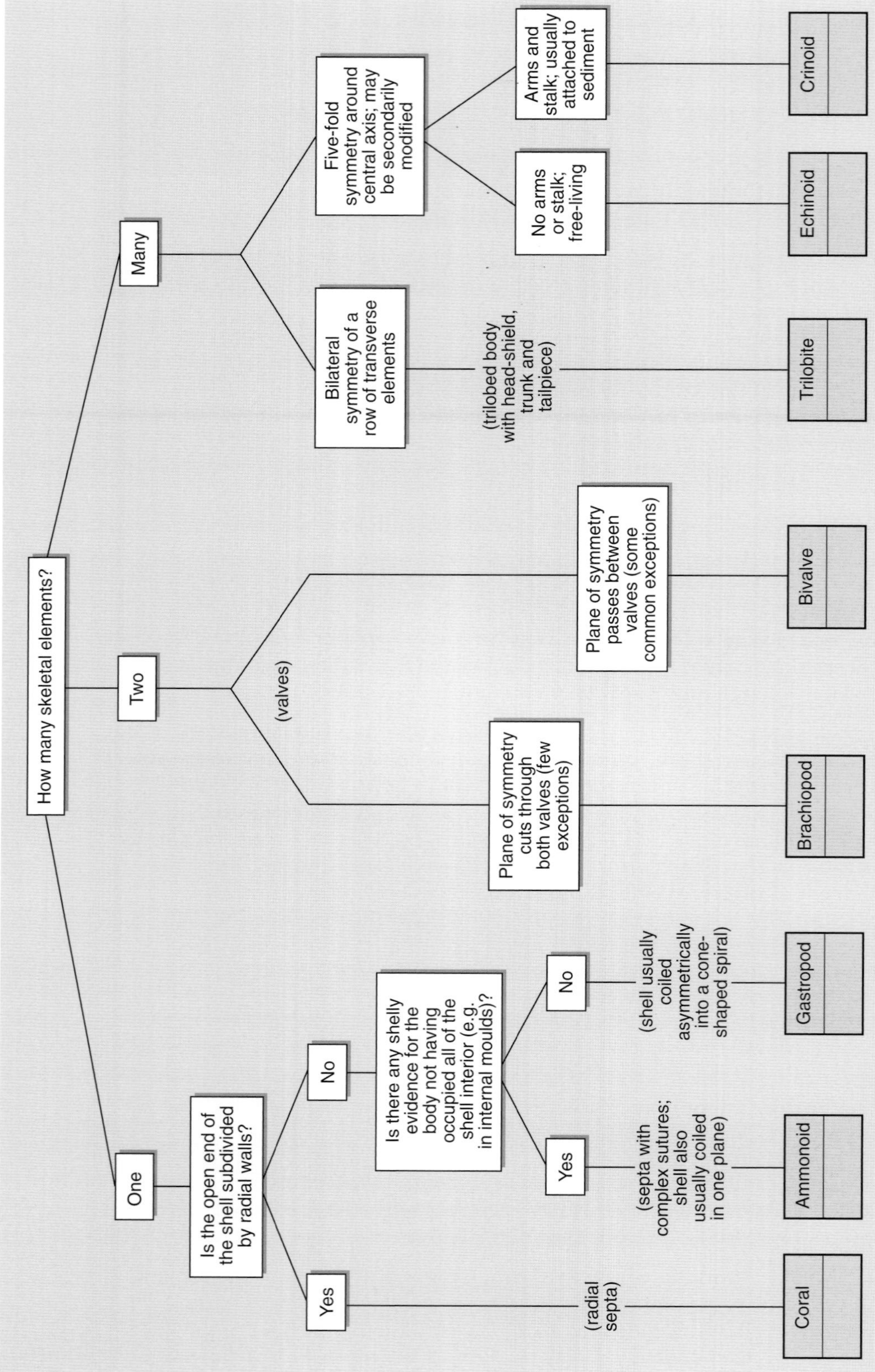

Figure 9.2 Key for classification of the marine invertebrate fossils discussed in Section 9. This Figure also appears on *Bookmark 4*.

9.2 MAKING SENSE OF BODY FOSSILS

Now that you are equipped with the basic knowledge of the appearance of some of the major groups of fossil invertebrates (Section 9.1), you should be able to recognize many of the invertebrate fossils that you see in museums or discover yourself. However, assigning a fossil to a particular group is only the first step in understanding it. In this Section, two aspects of body fossils will be studied that serve as introductory material for the rest of this Block. The two aspects are:

(i) the ecology of marine invertebrates and their fossil counterparts;

(ii) the shape and structure, or **morphology**, of fossils, especially the relationship between the surviving hard parts and the now vanished soft parts.

9.2.1 ECOLOGY OF MARINE ANIMALS AND THEIR FOSSILS

In studying FS A–M, three questions should be asked: Where did these organisms live in relation to the sea-bed? Did they move around (were they, for example, burrowers through the sediment, attached to rocks, or active swimmers)? How did they feed? Before tackling these questions, however, we need some idea of the range of possible answers derived from living organisms.

One way of describing the habitat of aquatic organisms is by reference to the sea-floor. Various possible living positions and habits are summarized in Figure 9.3. The **epifauna** are animals that live on the sea-floor, while animals that live entirely buried within the sediment are the **infauna**, although in many

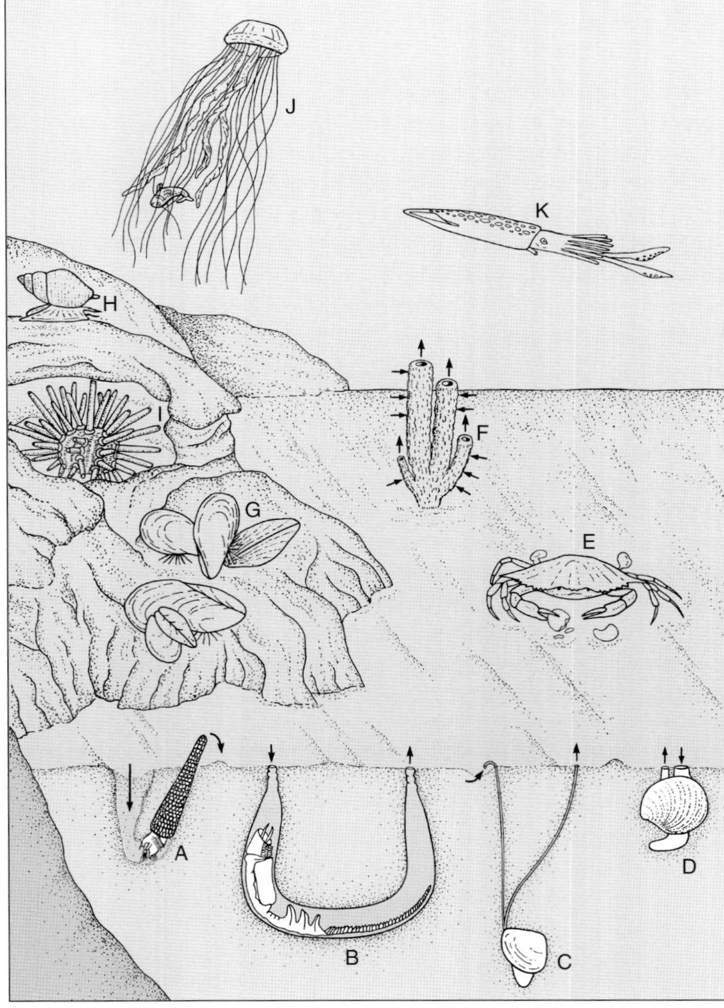

Figure 9.3 The life habits of adult marine invertebrates. The infauna includes vagrant worms (A) digging through the sediment, more sessile worms (B), and sessile to intermittently vagrant bivalves (C and D), all maintaining connections with the overlying seawater. The epifauna includes vagrant and sessile animals on both soft sediment and hard rocky surfaces. On soft sediment, a vagrant crab (E) and sessile sponges (F) are depicted, while on the rocks there are sessile bivalves (G), a vagrant gastropod (H) and an echinoid (I). In the overlying seawater, the pelagic animals include planktonic jellyfish (J) and nektonic squid (K). Most of these invertebrates have planktonic larval forms. Small arrows indicate feeding currents.

cases a connection is maintained with the overlying seawater. In either case, the animals may be **sessile**, habitually staying in one place, or they may be **vagrant**, frequently moving around. Sessile epifauna may just lie on, or encrust or otherwise attach to the surface, while vagrant forms move over it. Many infaunal animals are sessile, living in more or less permanent burrows, but other, vagrant, forms move through the sediment in search of food. Infaunal and epifaunal animals are classed together as bottom dwellers or **benthos**.

Not all kinds of sea-floor are equally accommodating. Solid rock impedes the establishment of an infauna, though you may know that some animals have evolved the ability to bore into solid rock. Epifaunal animals might be expected to be rare in water-rich soupy mud into which they would sink. Some members of the epifauna, however, are adapted to live on soft sediment with the aid of stilts, spines or broad thin shells or soft parts that spread the load. Both infauna and epifauna usually have little problem invading muddy sand, which can both be penetrated by burrowers and support surface dwellers.

In discussing the infauna and epifauna, we distinguished between sessile and vagrant animals in both cases, but this needs qualifying: infaunal animals that are sessile (dwelling in permanent burrows) are usually able to move in an emergency, and in some cases no strict division can be drawn between vagrant and sessile infauna.

Pelagic animals live above the sea-floor and its benthos (Figure 9.3). Pelagic animals that float or swim so weakly as to be at the mercy of currents and winds are collectively known as the **plankton**, whereas active swimmers comprise the **nekton**. Much of the plankton consists of minute animals, such as the floating larvae (juvenile forms) of benthic animals, as well as algae. The larger plankton consists mostly of various types of jellyfish, some with gas-filled floats. The nekton includes fish and other animals, especially squid.

Like most classifications involving living organisms, the one given above is not neat and decisive, as many animals do not invariably fall into one category or another. For example, the common prawn buries itself during the day (i.e. is infaunal), but at night emerges to join the epifauna while it feeds. Other epifaunal animals will bury themselves during low tide. Despite the difficulties in pigeonholing some species, this classification by habitats is widely used, especially as some groups have almost constant habitats. As you will see, adult brachiopods (FS F and L), for example, are practically always epifaunal and sessile.

A final point is an important distinction regarding distributions of benthic and pelagic animals. Benthic species often have a rather patchy distribution across the sea-floor. This is due to factors such as variation in sediment type and the amount of suspended sediment in the water, as well as clustered settlement of their larvae. Pelagic animals, by contrast, may be independent of bottom conditions and have wider geographical ranges. Furthermore, while the dead remains of the benthos may be transported by bottom currents, distances of transport are usually fairly limited. However, the carcasses of pelagic animals may float considerable distances, sometimes far beyond the range of the living animals.

In addition to divisions based on habitats, animals can be classified by their feeding methods (Table 9.1), though again there is much variation. Many **predators** will actively pursue vagrant prey. Others feed on sessile prey animals, which have to rely on such defences as protective shells or unpalatable taste as a deterrent. **Scavengers** will attack corpses and other decaying animal material, but some scavengers will turn predator if the opportunity arises. Similarly, at times some predators will scavenge.

Table 9.1 Principal feeding types in marine invertebrates.

Feeding type	Habits
predators	hunt and consume live animal prey
scavengers	consume dead and rotting animal material
grazers	scrape food, usually algae, off sediment or rock surfaces, or graze on large plants
suspension feeders	strain food particles from seawater
parasites	long-term attachment to a host, which serves as a food source: the host is usually much larger than the parasite
deposit feeders	eat sediment (swallowers) or select particulate grains (collectors)

Certain sediments are rich in residual organic fragments, or **detritus**, including faecal matter and associated bacteria that coat detritus and sediment grains alike. The detritus and particularly the bacteria are the food source for **deposit feeders**. Because the total surface area of the grains is much larger in fine-grained sediments, deposit feeders prefer muddy sediments to coarse-grained sands. Two sorts of deposit feeder are distinguished by the degree of selectivity shown during feeding. **Collectors** select a limited size or type of particle, whereas **swallowers** are less discriminate and ingest sediment wholesale, passing out processed sediment in large quantities, as in the abundant worm casts you may have seen on tidal flats.

Seawater contains microscopic suspended food particles that may be strained off, or trapped on an array of tentacles, by **suspension feeders**. In open water, microscopic plankton, especially unicellular algae, make up the bulk of their food material, though close to the sea-floor there may be sizeable quantities of detritus resuspended by currents. Commonly, the feeding organ of suspension feeders consists of a grid or fan of tentacles with abundant **cilia** (singular: **cilium**; minute, regularly beating, hair-like structures projecting from cell surfaces) that both generate a water current and help to trap food particles from it with the aid of secreted mucus. The mucus is then drawn into the mouth. Many suspension feeders have sorting mechanisms to bypass or reject unwanted particles, but in excessive turbidity the animal will stop feeding.

❑ Suspension feeders are often rare in fine-grained sediments. Why might this be so?

■ Fine-grained sediments (muds) would tend to clog the feeding organ.

Grazers are animals that rasp or scrape organic mats, especially those composed of algae, from sediment or rock surfaces. They may also feed on larger plants.

The final feeding type to be considered is parasitism, where one animal directly uses another organism as a more or less permanent food source. **Parasites** differ from other feeding types in that it is important the host remains alive. Death of the host may also kill the parasite, though in some cases the host finally dies only after the parasite has completed its own growth stage. Parasites often have parts of their anatomy simplified (e.g. the gut may be absent), but may have complicated life cycles. Although present on virtually all living organisms, surprisingly few examples have been documented from the fossil record.

As with the classification based on habitats, the feeding classification is not rigid because of many borderline cases. The overlap between predators and scavengers has already been noted. Another borderline case is that between suspension feeders and animals that trap very small live prey, as is the case with many coral polyps, for example.

Although these two classifications based on habitats and feeding methods are not foolproof, they are simple and particularly applicable to fossils. They can be amalgamated to give a matrix showing all possible combinations (Table 9.2). You will be filling in Table 9.2 as you work through the next Section, so you may find it useful to insert a marker to aid quick reference. Note that because the 'habitat' for a parasite is usually its host, parasites are not included in this Table.

Table 9.2 Combined feeding/habitat classification for marine animals.

Habitat		**Feeding type**					
					deposit		
		predator	scavenger	grazer	swallower	collector	suspension
pelagic	nekton						
	plankton						
epifaunal	vagrant						
	sessile						
infaunal	vagrant						
	sessile						

Some combinations in Table 9.2 are, of course, rather unlikely. Thus a wholly pelagic animal could not be a deposit feeder, although deposit feeders might occasionally swim from one feeding site to another. The same argument applies to grazers other than those forms that drift around with floating seaweed. An infaunal suspension feeder might also seem unlikely, but if you refer to Figure 9.3 again, you can see that many infaunal animals maintain connections with the overlying water and so can draw in suspended food.

9.3 INTERPRETING FOSSILS AS LIVING ORGANISMS

Several questions should occur to you when looking at FS A–M. The fossils, of course, formed from the hard skeletons, but what were the soft parts like? How did the animals move and upon what did they feed? In this Section, we shall concentrate on building up a picture of what the soft parts of the animals looked like, and then we shall link the functions of both the specific organs and the entire animals to their original ecology.

We have chosen two groups for detailed study – the bivalves and the echinoids. One advantage of looking at these groups, as with most of those represented in the Home Kit, is that they have living relatives. Much of the original anatomy of the fossil specimens can thus be inferred by homology. This advantage will be apparent when we try to reconstruct the life and habits of the extinct trilobites.

With the bivalves and echinoids, the extent to which the former soft parts leave marks on the available hard parts will be stressed. However, many soft organs have no direct effect on the hard parts and their reconstruction must be based entirely on living relatives, or, in exceptional cases, preservation of the soft parts.

9.3.1 THE BIVALVE *CIRCOMPHALUS* (FOSSIL SPECIMEN I)

It is important to satisfy yourself that you can see the features of FS I that are described here (again, you will be able to check your observations against the 'Digital Kit' on DVD 1). The hand lens will be an essential aid, and Activity 9.2 will also help your observations by getting you to sketch certain features on an incomplete photograph; for this you will need a pencil. The Activity also includes an optional experiment on bivalve mechanics, for which you will need two matching jam jar lids (through which you will need to punch some holes using a hammer and a fairly thick nail), string, adhesive tape and a rubber (which you may need to cut down in size).

FS I is from a Neogene species of the genus *Circomphalus*, of which some other species are still living. We start by introducing some standard terms that are used to orientate animals. These will help guide your exploration of the specimen, rather like the bearings on a map, as well as assisting comparison with other animals. *Note that these terms are for anatomical comparison alone and do not necessarily indicate how the animal is orientated in life.* Some animals, for example, live with their anatomical 'backs' facing downwards (though this was not, in fact, the case with FS I).

In Section 9.1.2, you saw how in bivalves the plane of bilateral symmetry, giving two mirror images, runs between the two valves. This means that one valve is placed on each side of the body and the valves are said to be **lateral**. How is the rest of the animal orientated? In bivalves, the back, or **dorsal**, surface of the animal is where the valves are hinged, so that the opposite anatomical underside is **ventral** (Figure 9.4).

In the majority of living animals, the front end, or **anterior**, is marked by the head and mouth, while the anus opens towards or at the rear end, or **posterior**. The soft body of bivalves, bearing the mouth and anus (there being no distinct head) rots away after death. This means that in fossil bivalves various shell features must be used to identify the anterior and posterior ends (Figure 9.4). These features are introduced below.

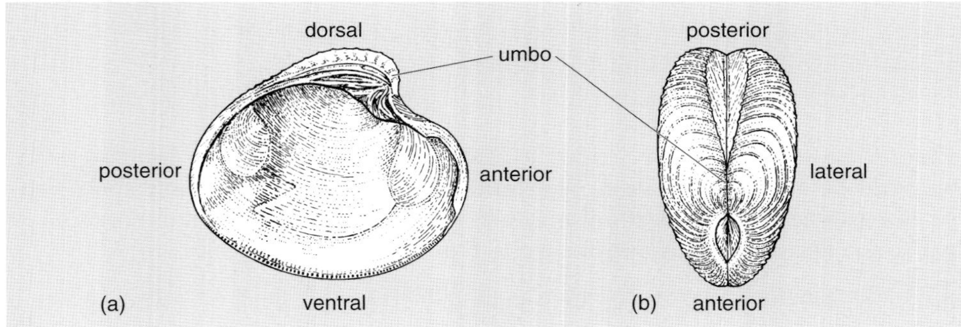

Figure 9.4 Orientation of a bivalve shell: (a) interior of left valve; (b) dorsal view of both valves.

FS I is just one of the valves of *Circomphalus*, that from the animal's left side (compare Figure 9.4a). The original shell was composed of aragonite (a polymorph of calcium carbonate, Block 2 Section 4.7.1), and it had only a thin outer layer of non-living organic material, though, as with the soft parts, this organic layer usually does not fossilize. The shell is thus an exoskeleton (Section 9.1.3). The interior of the shell is relatively smooth, with markings that, as you will see, are directly related to some of the soft parts.

The exterior of each valve is broadly convex, terminating dorsally in a blunt protrusion known as the **umbo** (plural: **umbones**). In most bivalve genera, including *Circomphalus*, the umbo is curved towards the anterior. The exterior has many sharp, broadly spaced ridges arranged concentrically around the umbo.

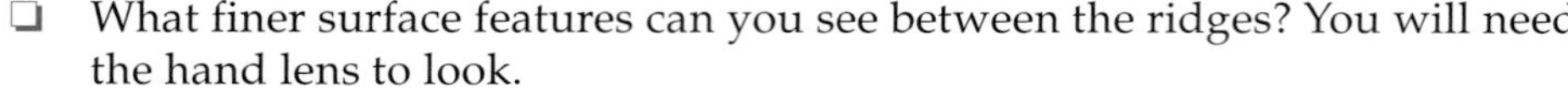

❑ What finer surface features can you see between the ridges? You will need the hand lens to look.

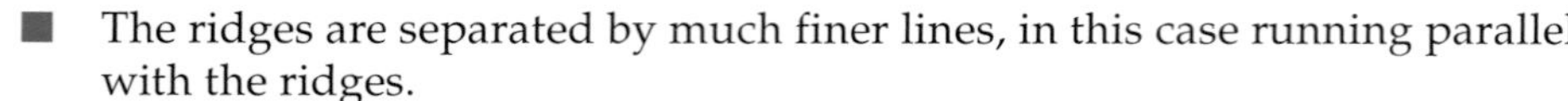

■ The ridges are separated by much finer lines, in this case running parallel with the ridges.

The shell was grown by progressive addition of material around its margins and over its inner surface. The fine lines you have just been studying mark the successive increments of this accretionary growth, and are called **growth lines**. Accretionary growth means that the entire history of an individual is recorded from the moment it starts producing the shell. In the case of *Circomphalus*, imagine subtracting successive growth increments from the margin, to leave progressively smaller growth stages, ending in the juvenile shell at the top of the umbo. In doing this imaginary exercise, you will see that the overall shape of the shell remained comparatively unchanged as it increased in size. However, in a few bivalves, as well as other groups, there may be a marked change in morphology between the juvenile and adult shells.

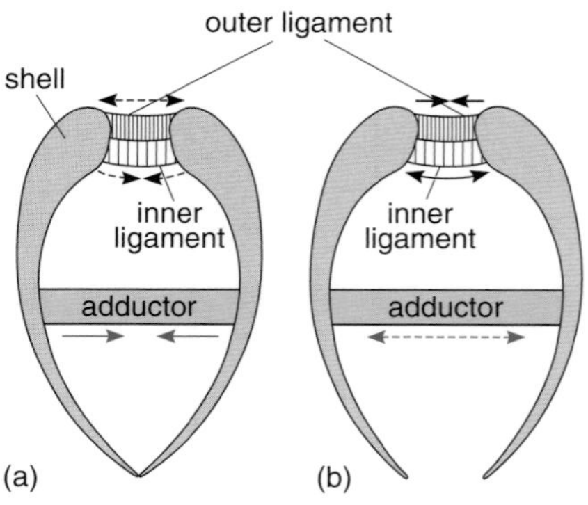

Figure 9.5 Counter-action of the ligament and adductor muscles in the closure (a), and opening (b), of the bivalve shell, seen in diagrammatic section across the valves. Solid arrows show the forces exerted, and dashed arrows show the forces imposed.

Look now at the inner surface of the shell. The dorsal area beneath the umbo consists of a platform that carries grooves and raised areas. This area marks the zone of hinging, or articulation, between the two valves of the original shell. In life the dorsal margins of the valves were connected by a strip of horny organic material known as the **ligament** (Figure 9.5). Although the ligament itself is not preserved here, its site of attachment to the valve can be detected as a shallow crescent-shaped groove – the **ligament groove** – curving back from the tip of the umbo, alongside the dorsal margin. When the valves were closed (Figure 9.5a), the elastic ligament was deformed like a C-spring, with its inner part compressed and outer part stretched. So, when the muscles (the **adductor muscles**) that closed the valves relaxed, the stressed ligament opened the valves (Figure 9.5b). The economy of this arrangement is that the bivalve avoided expending energy whilst keeping its valves open – a necessary posture for feeding.

Activity 9.2

Now do Activity 9.2 which provides further insights into the mechanics of shell opening and closing.

It should be apparent from Activity 9.2 that the **teeth** and **sockets** are integral parts of a bivalve's articulation. They guide opening and closing, and so prevent the two valves shearing past one another. The particular arrangement of teeth and sockets seen in FS I is one of several types found in bivalves.

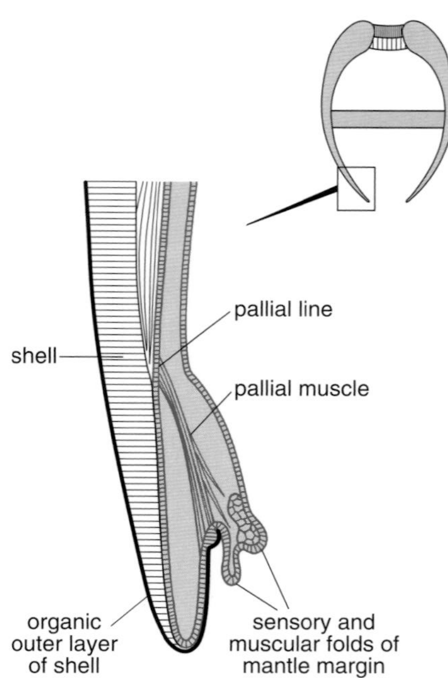

Figure 9.6 Section through the outer edge of one valve, and its mantle lobe, of a bivalve.

In life, the inner surface of a bivalve shell is lined with a sheet of tissue, known as the **mantle**, which secretes the shell. The mantle margins are complex and bear various sensory organs, as well as a muscular fold or flap in each valve (Figure 9.6), running around the inside of the valve margin. The folds act as 'curtains' and are used to control the size of the gap between the open valves. In many bivalves, the opposing folds are fused together. The points and extent of this fusion vary in different species, but their chief effect is to separate areas of water inflow and outflow.

The muscles that run into the mantle margins, especially the folds mentioned above, are attached to the inside of each valve, leaving an attachment scar known as the **pallial line**.

Question 9.2 Find the pallial line in FS I. Is it entirely parallel to the ventral margin of the shell, or is any deviation visible?

The posterior kink in the pallial line referred to in the answer to Question 9.2 is called the **pallial sinus**, and is an invaluable indicator of the original presence of

siphons, which are fleshy tube-like extensions of the cross-fused mantle (Figure 9.7a). The pallial sinus is invariably posterior and so is another useful guide for orientating the shell. Siphons are typical of infaunal bivalves and serve to connect the animal with the overlying water. If threatened by predators or inclement conditions, the bivalve withdraws its siphons and closes its shell. The space to accommodate the retracted siphons is marked by the pallial sinus. Thus, the absence of a pallial sinus indicates very short, or no siphons, with the bivalve living either on, or just within, the sediment.

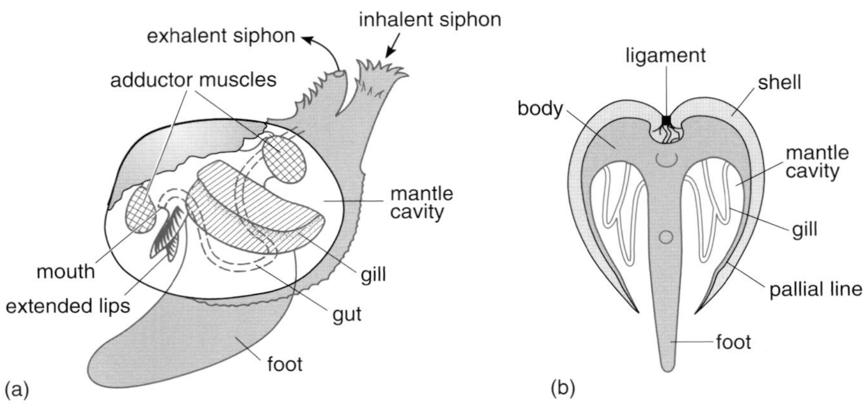

Figure 9.7 Diagrams of bivalve anatomy: (a) major soft parts in relation to the shell, in side view; (b) cross-section.

Typically, the siphons form an inhalent and exhalent pair (Figure 9.7a). The inhalent siphon serves either to draw in water with suspended food particles or, in some other bivalve species, to search across the sediment surface picking up detrital particles like a vacuum cleaner. The exhalent siphon removes waste material. The siphons also take in oxygenated, and expel deoxygenated, water.

While studying the pallial line and sinus, you may have noticed two roughly oval markings at the anterior and posterior ends of the shell interior. These are **adductor muscle scars**; they mark the sites of attachment of the adductor muscles that ran between the valves (Figure 9.5). As noted earlier, the adductor muscles shut the valves and so they act in opposition to the elastic ligament.

The remainder of the soft parts in FS I would have consisted of the body, with laterally suspended **gills**, extended lips and a **foot**, which is a muscular ventral extension of the body (Figure 9.7), attached to the inside of the shell by **protractor and retractor muscles** (Figure 9.8a). These muscles respectively extend, and retract, the foot. The bivalve body contains the gut and reproductive organs and is suspended from the dorsal part of the shell beneath the hinge area (Figure 9.7b).

❑ By reference to Figure 9.8, try to detect any markings corresponding to where the anterior muscles from the foot and body wall might have attached to the inside of the shell in FS I. (You will need to peer obliquely into the valve interior with the posterior rim towards you.)

■ By looking at the inside of the shell beneath the hinge area, you will see a small scar immediately behind the anterior adductor scar, where the anterior retractor muscle from the foot was inserted, followed by a row of tiny pits that show where small muscles from the body were attached.

Arising from both sides of the body would have been the large folded gills that hung like porous curtains in the space surrounding the body, or **mantle cavity** (Figure 9.7). At the anterior end of the gills four fleshy elongated lips would have surrounded the mouth, two of which are visible on Figure 9.7a. These and the gills together formed the feeding organ. We know from living bivalves that the beating of cilia on the gills produces a water current that draws water and food particles in through the inhalent siphon and across the gills, where food particles are trapped and then transported in a stream of mucus to the lips and thence to the mouth. Waste products from digestion are expelled with the water that has passed through the gills and leaves via the exhalent siphon. Movements of the overlying seawater ensure a constant renewal of food, and in most circumstances living *Circomphalus* can remain stationary.

❑ Can you suggest any events that could have disturbed the living position of *Circomphalus*?

■ During storms, either extensive deposition or erosion of sediment would have forced *Circomphalus* to move upwards or downwards.

When necessary, then, *Circomphalus* would have burrowed through the sediment. Burrowing in bivalves consists of a cycle of distinct stages involving the interaction of the shell and the foot. You noted the scar where the anterior retractor muscle was attached in FS I, a little earlier. The scars for the protractor and posterior retractor muscle cannot be so clearly distinguished.

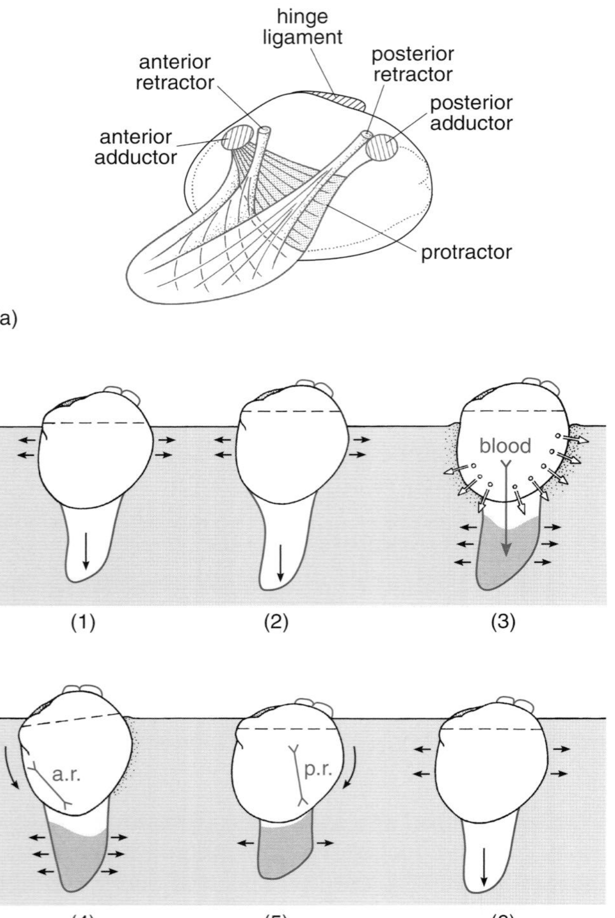

Figure 9.8 Burrowing in bivalves: (a) musculature associated with the foot; (b) burrowing cycle. The dashed band across the bivalve shows its burrowing progress. (1) and (2), adductor muscles relaxed and valves open, pressing against the sediment, as the foot probes downwards; shell slightly lifted during probing. (3), sudden contraction of shell, expelling water and loosening surrounding sediment; foot dilated with blood. (4) and (5), dilated foot anchors animal as anterior retractor (a.r.) and then posterior retractor (p.r.) contractions drag shell down in a seesaw manner. (6), bivalve ready to restart burrowing cycle.

In the first stage of the burrowing cycle, the foot extrudes between the valves, and probes into the sediment (Figure 9.8b 1, 2), as the protractor muscles contract around it. The valves, kept open by the elastic ligament, tend to wedge and anchor the shell. Next, the adductor muscles contract, thereby both driving blood into the foot, so dilating its ventral portion, and expelling water from within the mantle cavity into the surrounding sediment, so loosening it (Figure 9.8b 3). The siphons are closed at this stage so that the water is not expelled through them. Contraction of the foot retractor muscles then drags the shell downwards. The anterior retractor contracts (Figure 9.8b 4) before the posterior retractor (Figure 9.8b 5) so that the bivalve seesaws downwards. Once the shell has been drawn down, the adductor muscles relax, the valves open and a new burrowing cycle may follow (Figure 9.8b 6).

❑ Now study the overall shape and roughness of your specimen of *Circomphalus*, and try to decide whether it was a rapid burrower.

■ The relatively swollen shape of the valves combined with the prominent concentric ribs suggest that rapid movement would have been prevented by high friction, so this species was probably a sluggish burrower.

In some bivalves, the external ornamentation is so arranged as to help burrowing, but some of the most rapid burrowers have smooth compressed shells that can slice through the sediment.

Question 9.3 Thinking back to the discussion of life habits (Section 9.2.1), how would you now classify *Circomphalus* in Table 9.2 (on p. 80)?

Bivalves are a very large group, and although the majority of species are (or were) burrowing suspension feeders, like *Circomphalus*, other modes of life are also well represented. Attachment to the sea-floor, for example, is quite common, either by means of anchoring threads of organic material secreted by the foot (e.g. mussels), or by direct cementation of one valve (e.g. oysters). Scallops mostly attach in the first way, though some lie freely on the sediment surface and can actually swim for short distances by clapping their valves, if menaced by predators such as starfish. The characteristic bilateral symmetry of burrowing bivalves is frequently lost in such epifaunal types. Indeed, in the case of the swimming scallops, secondary adaptation of the shell for symmetrical streamlining of the clapping valves has superficially invested it with a brachiopod style of bilateral symmetry (cf. Figure 9.1b), though internal features, such as the ligament, unmask it as a bivalve. Burrowing has also been replaced in some instances by boring into hard materials, such as rock, stiff cohesive clay or wood. Nor are all bivalves suspension feeders: there are several kinds of deposit feeders (collectors), especially among primitive forms, and even a few predators that suck small worms and other such prey into the mantle cavity with highly modified gills. In addition, non-marine bivalves, such as freshwater mussels, may be familiar to you. Thus, a wide variety of habitats is encompassed, though careful consideration of the morphology of fossil forms may allow one to identify adaptations to specific environmental conditions.

9.3.2 The echinoids *Pseudodiadema* (fossil specimen C) and *Micraster* (fossil specimen J)

The lower surface of FS C is not well preserved and has been encrusted by tubes made by worms. Otherwise, it shows many well-preserved details, although you may need the hand lens to help you see them clearly. For Activity 9.3, which follows on immediately from your study of this specimen, you will also need FS J.

As an introduction to echinoids, the Jurassic genus *Pseudodiadema*, of which FS C is an example, is suitable in several ways. It shows fairly generalized features (i.e. it mostly lacks specialized modifications unique to itself). Also, relatives of this Jurassic sea-urchin are alive today, so that a reconstruction of the form of the soft parts can be made fairly accurately.

The key you completed in Question 9.1 included two important features characteristic of practically all echinoids. These are (i) the five-fold symmetry (modified to bilateral symmetry in some forms) and (ii) a skeleton composed of individual plates which are closely joined, giving it the superficial appearance of a single unit. A hollow skeleton such as this, composed of many interlocking pieces, is known as a **test**. As with *Circomphalus*, the skeleton is composed of calcium carbonate, but in this case the polymorph is calcite rather than aragonite. The mode of secretion, moreover, is rather different. Instead of the calcium carbonate being deposited *above* a layer of mantle tissue, plates are secreted *within* a thin layer of soft tissue. This means that the echinoid skeleton

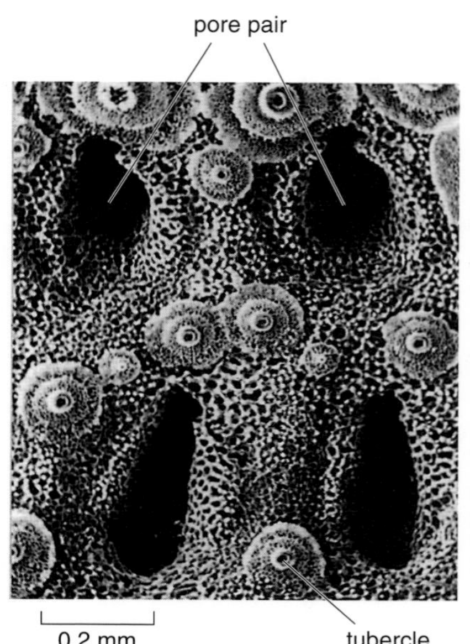

Figure 9.9 The sponge-like structure of the echinoid endoskeleton revealed by an electron microscope.

is strictly internal and so is an **endoskeleton**, not an exoskeleton. Moreover, each echinoid plate is not solid, but is riddled with fine pores so that under the microscope the skeleton resembles a sponge-like three-dimensional meshwork (Figure 9.9). In life, this meshwork is filled mostly with a fibrous protein known as **collagen**. It is the collagen fibres that bind the plates together in life; the calcite of the plates themselves is not fused. This type of meshwork skeleton is unique to echinoderms. Its texture is usually preserved in fossil material which means that even small fragments can be distinguished from the skeletal debris of other groups.

A curious characteristic of echinoderm plates, which is very convenient for identification in thin section, is that each plate is a single crystal. Thus, when seen between crossed polars, the whole plate goes into extinction at the same time, despite the pores running through it. You may recall noting this fact when you carried out Activity 6.1.4 in Section 6.4. *Echinoderm plates consist of single, optically continuous crystals, which are murky-looking in thin section because they are full of pores.*

The radial appearance of *Pseudodiadema*, whereby all sides look rather similar, means that the terms used to orientate the animal have to be somewhat different from those used for bilaterally symmetrical animals such as *Circomphalus*. The flatter surface, with the mouth, is called the **oral surface**, and opposite that lies the domed **aboral surface**. In life, the oral surface (Figure 9.10b) was applied to the sea-floor.

Figure 9.10 Echinoid anatomy: (a) aboral view; (b) oral view; (c) cross-section.

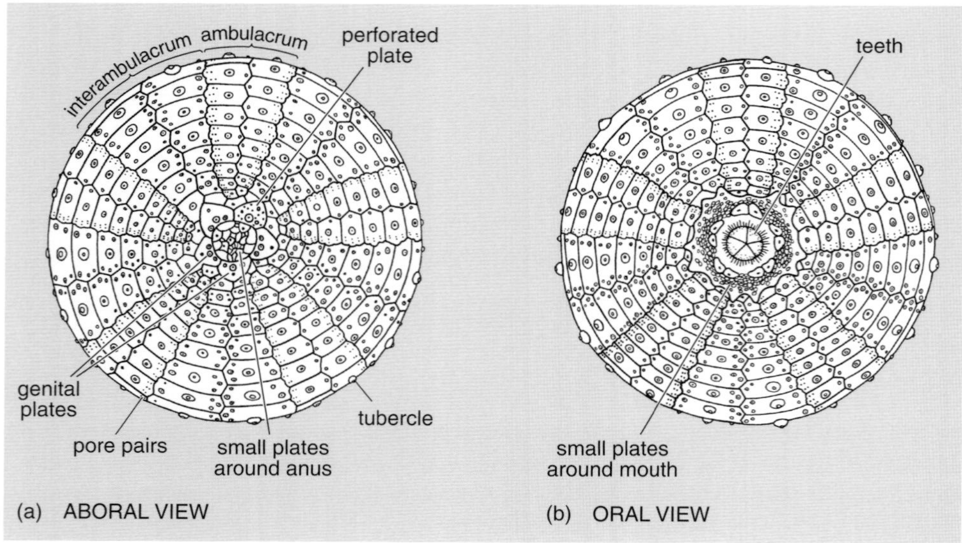

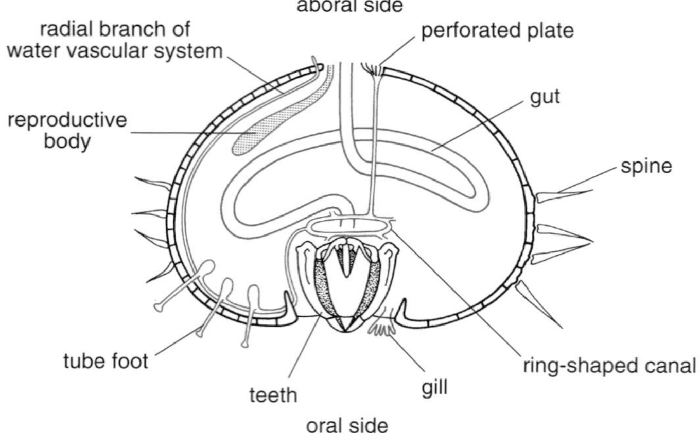

Question 9.4 Now look down on to the domed aboral surface of FS C (corresponding to Figure 9.10a). How many *types* of radial zone can you make out on the basis of the ornamentation, and how are they characterized?

The wide zones referred to in the answer to Question 9.4 are termed **interambulacra** (singular: **interambulacrum**), and the narrower and more subdued zones, **ambulacra** (singular: **ambulacrum**) (Figure 9.10a). The knobs (of all sizes) are known as **tubercles**. Both ambulacra and interambulacra show the five-fold symmetry, and we shall discuss their significance later.

In *Circomphalus*, the valve you studied was effectively intact. The specimen of *Pseudodiadema*, however, has suffered some loss of hard parts. The centre of the oral surface is occupied by an expanse of featureless material that represents infilling sediment. In life, this area was covered by a flexible membrane studded with plates, and with the mouth opening in the middle (Figure 9.10b). The mouth contained a complex structure that included a circle of five teeth operated by a series of muscles (Figure 9.10c). This feeding apparatus probably was used to rasp algae from rocky surfaces, and/or sessile prey such as coral polyps. The smaller hole on the aboral surface was also formerly occupied by a group of plates that surrounded the anus (Figure 9.10a). In both oral and aboral regions, the rotting of the membranes released the embedded plates and, with the loss of teeth, sediment could enter the skeleton.

The other major change in the skeleton of FS C has been the loss of all the **spines** that formerly were attached to each tubercle and so covered the animal (Figure 9.10c). In living echinoids, the spines act in defence, and in some species they are poisonous. They are also used in locomotion, to lever the animal across the sea-floor. The largest spines were on the most prominent tubercles (on the interambulacra). By looking closely at the skeleton, you will see numerous other smaller tubercles that mark the former position of small spines.

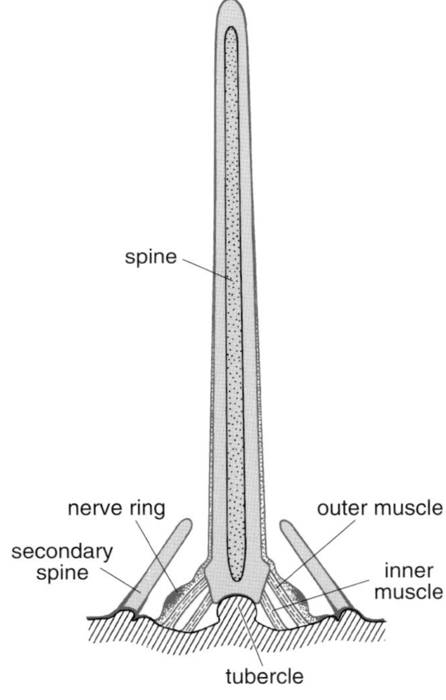

Figure 9.11 Longitudinal section of an echinoid spine and tubercle.

The spines are often found isolated in the sediment and it is only in exceptional circumstances that they remain attached. Inspect one of the large tubercles on FS C together with Figure 9.11. The joint between the spine and its base is a ball and socket, with the spherical tubercle acting as the ball. The central hole in the tubercle marks where a strand of tissue ran to the spine. The spine was moved by sets of muscles that ran from its base to the outer margin of the tubercle. Look closely at the outer margin of a tubercle and you should see that it forms a flattened ring. The spine muscles were attached onto this ring and consisted of an inner and outer layer (Figure 9.11). If the spine itself was touched, the inner muscle clamped the spine rigidly to its base. If, however, the animal was touched elsewhere, the outer muscles would swivel the spine towards the disturbance. After death, and rotting of the muscles, the spines would tend to fall off the test. Notice how the size of the tubercles, and hence the spines, decreased towards the oral region so that *Pseudodiadema* was not kept away from the sea-floor and its food.

Echinoids and their relatives are unique in possessing arrays of hydraulically operated tentacles connected to a system of fluid-filled canals inside the body known as the **water vascular system** (Figures 9.10c and 9.12). This consists of a ring-shaped canal around the gut, which gives rise to five radial branches, each one running up the inside of an ambulacrum. The ring canal is connected to the exterior of the animal by a tube leading to a single perforated skeletal plate (the **perforated plate**). The perforations allow the pressure between the water vascular system and the seawater to remain at equilibrium.

❑ Look down on the aboral surface of FS C again, and, using the hand lens, try to identify this perforated plate amongst the ring of plates immediately surrounding the central hole.

■ Surrounding the central hole are several small plates. You should be able to see that there are five larger plates with another five smaller plates wedged

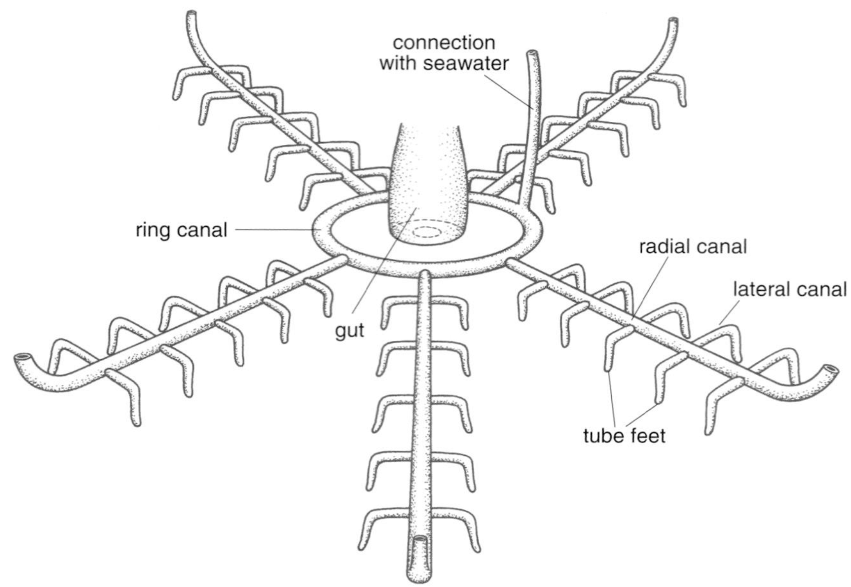

Figure 9.12 Water vascular system of a generalized echinoderm. The mouth is shown facing downwards, as in echinoids (in crinoids, it faces upwards).

between them. One of the larger plates is slightly bigger, more convex, and has a porous appearance. This is the perforated plate.

Each of the radial branches of the water vascular system bears numerous short soft tentacles called **tube feet** that project through the skeleton into the surrounding water (Figure 9.10c).

❑ See if you can detect the holes through which the tube feet emerged in the plates of the skeleton in FS C.

■ By looking at the edges of one of the ambulacra you can see a line of small paired pores running down each side.

Each *pair* of pores identified in the answer, called a **pore pair**, led to *one* tube foot outside the test, connecting it with the water vascular system inside. The tube feet could be inflated by hydrostatic pressure produced by the action of muscles around the bulb-shaped bases of the tube feet, and flexed by small muscles around them. You can see from the distribution of pore pairs that each ambulacrum must have had two rows of tube feet, one on either side of its entire length. Tube feet have a variety of functions in echinoids, including feeding and locomotion.

As you have already seen, the mouth and anus were opposite one another on the skeleton. The intervening gut was a coiled structure (Figure 9.10c). The interior of the skeleton of echinoids is divided by thin membranes into a series of fluid-filled cavities. The bulk of the interior is formed by one large cavity, but smaller cavities surround the anus and mouth region. Suspended within the large cavity are five reproductive bodies (one of which is shown in Figure 9.10c). Look again at the aboral surface and the ring of ten plates around the anus. You should see that the perforated plate and three of the four larger plates have a small pore; the other plate (almost opposite the perforated plate) seems to be abnormal in that it has two pores. Well, none of us is perfect. The reproductive bodies released eggs or sperm via these pores.

Question 9.5 From the above description, classify *Pseudodiadema* in terms of habitat and feeding as discussed in Section 9.2.1. Try to assign it to its appropriate box in Table 9.2 (p. 80).

Having read this far, you may be wondering whether the comparatively detailed reconstruction of the two fossils you have been studying can be extended into some principles of general use in palaeontological work.

Activity 9.3

In Activity 9.3, we aim to show how understanding one example from a fossil group allows you to deal with almost any other example from that group. You will need FS J, which is the echinoid *Micraster* from the Chalk, of late Cretaceous age, and so is younger than *Pseudodiadema*.

Because of the various modifications to the body plan that you have just been observing, *Micraster* is an example of an **irregular echinoid**, in contrast to the **regular echinoids** such as *Pseudodiadema*.

❑ Recall how and upon what *Pseudodiadema* fed. Do you think *Micraster* behaved in the same way?

■ Almost certainly not; it would be difficult to imagine large jaws like those in *Pseudodiadema* (Figure 9.10c) protruding from the mouth and being able to seize food.

In fact, *Micraster* was probably a deposit feeder. The lower lip was equipped with spines against which specialized feeding tube feet scraped in food.

The difference in positions of mouth and anus between *Pseudodiadema* and *Micraster* results from the evolutionary forward migration of the mouth, and rearward migration of the anus, from their originally central positions, during successive stages in the evolution of *Micraster* from earlier forms. Some of the stages in this change in position of the anus and mouth from species to species may be seen in other fossil echinoids. Moreover, in the development of modern irregular echinoids this change is repeated in the early stages of each individual's growth, as the anus moves from its central aboral position. In studying the aboral surface of *Micraster* where the five ambulacra meet, note that the small plates, including the perforated plate, have not accompanied the anus on its migration, but have remained central.

❑ Study the pore pairs in the ambulacra of *Micraster* and decide how they differ from those of *Pseudodiadema*.

■ The pore pairs of *Pseudodiadema* are circular and separated by a tiny ridge. In *Micraster*, although the inner pore (nearest the midline of the ambulacrum) is more or less circular, the outer pore is generally more slit-like, so the tube foot was probably flattened in cross-section. Also the tube feet of *Micraster* did not extend onto the lower surface, although the ambulacra can be traced as rows of plates extending beyond the pore zones.

In comparing these details of the ambulacra, you should also have noticed that the broad area with tubercles between the two rows of pore pairs in *Pseudodiadema* is absent in *Micraster*.

❑ In *Pseudodiadema*, the former position of the spines is marked by the tubercles. Remembering that the size of the tubercles is roughly proportional to the original spine size, what can you infer about the size of the spines in *Micraster* and any variation around the skeleton?

■ The tubercles of *Micraster* are far less prominent than those of *Pseudodiadema*, and indeed the spines were much thinner and shorter in *Micraster*, forming an almost hair-like coating over the animal. The largest tubercles of *Micraster* occur across the oral surface, and attached to them were relatively large flattened spines that were used in locomotion.

We could carry on this comparison between the two echinoids almost indefinitely, but now the principal points of similarity (the homologies) and the differences will be apparent. What, however, is the reason for the difference between regular echinoids such as *Pseudodiadema* and irregular echinoids like

Micraster? The fundamental reason is connected to the shift from an epifaunal mode of life in regular echinoids to an infaunal burrowing mode of life in the irregular echinoids. Consistent movement in one direction makes bilateral symmetry more efficient for burrowing than a radial symmetry. The positions of the mouth and anus in *Micraster* are correlated with feeding and waste disposal in a burrow. Recall also the evidence from tubercle size and pore-pair morphology for the size of the ventral spines and the type of tube feet. Both the spines and tube feet are modified according to a burrowing existence. As we noted earlier, in Section 9.1.3, this is a particularly striking example of the adaptation of a bodyplan to suit a new way of life.

9.3.3 GASTROPODS (FOSSIL SPECIMENS A AND K)

You are probably familiar with gastropods. Today, they are among the commonest marine animals (e.g. sea-snails, whelks, limpets, sea-slugs) and moreover are abundant in terrestrial environments (snails and slugs).

The gastropod body is partially covered over its dorsal surface by an over-arching sheet of tissue, the **mantle**, which lines and grows the shell (when present). The body has two main components (Figure 9.13): (i) the **visceral mass**, with the internal organs such as the gut, reproductive and excretory systems. This is permanently housed within the shell, and is generally connected via a narrow waist to (ii) a broad fleshy **foot** and conjoined **head**, which is equipped with a mouth, eyes and sensory tentacles. The animal creeps around by producing muscular waves that pass along the base of the foot.

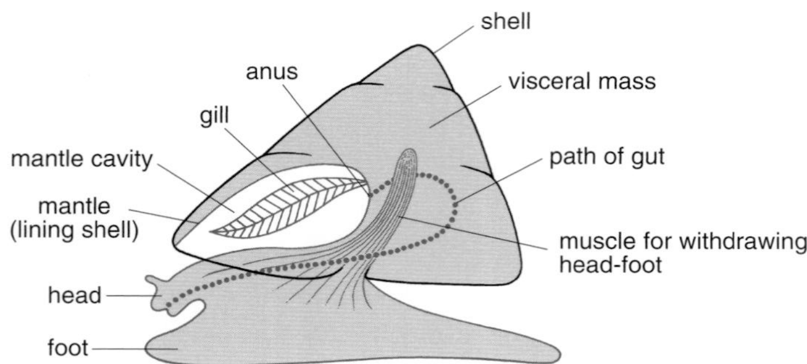

Figure 9.13 Principal features of gastropod anatomy, seen in diagrammatic side view.

The distinction between the visceral mass and the conjoined head and foot is exaggerated by a peculiar process called **torsion** that occurs during the larval development of each individual: the visceral mass (and with it the mantle/shell) is twisted around so as to become 'back to front' with respect to the head-foot beneath. Torsion introduces a marked asymmetry to the gastropod body, making it rather difficult to visualize the orientation of the shell with respect to the foot. In Figure 9.14, plan views of the original form of the head-foot in FS K and A are shown, with the position of the overlying shells indicated by dashed lines. By placing the specimens on the diagrams as indicated, you will gain an idea of how the shells would have been borne by the two animals that produced them.

The wide skirt-like extension of the mantle over the body produces a large space between the mantle itself and the visceral mass and upper part of the head-foot. This is the **mantle cavity** (Figure 9.13), and it connects with the outside around the free margin of the shell opening. The mantle cavity plays a vital role in the gastropod's life, for it houses the gill(s), and various sensory organs, as well as the anus. It also allows protective withdrawal of the head, followed by the foot, brought about by contraction of a large muscle (Figure 9.13), running from the head-foot to the central column inside the coiled shell.

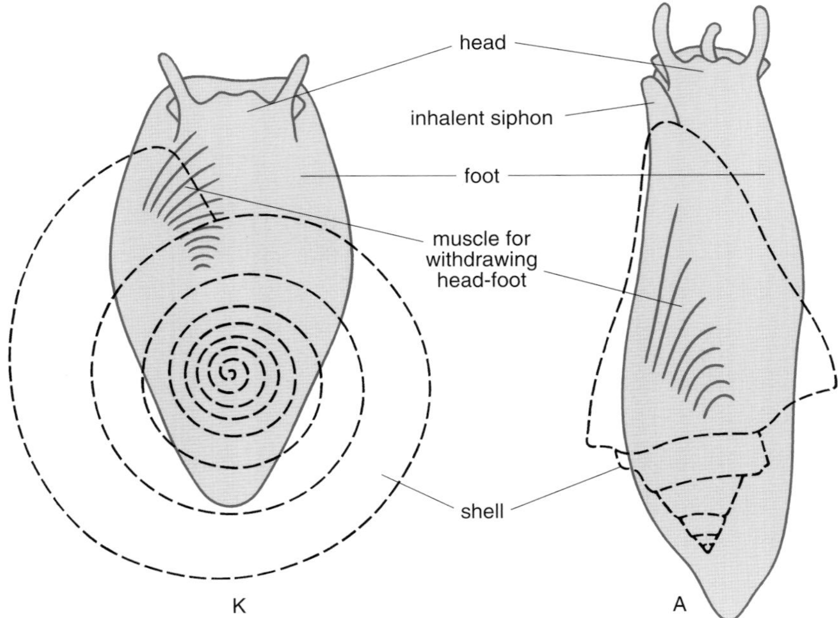

Figure 9.14 Orientation diagrams for FS K and A, seen in dorsal view.

Many of the differences between FS K and A are linked with the arrangement of the gills in the mantle cavity and with associated differences in their life habits. Some primitive marine gastropods have two gills, one on each side of the mantle cavity (Figure 9.15a). Water is drawn into the mantle cavity low down on both sides, flows over the gills (driven by the beating of cilia on their surfaces) and is exhaled upwards centrally, above the head. As the water leaves, it carries off faeces from the anus, which, as a result of torsion, is sited between the gills. In such gastropods, the exhalent site is usually marked by a pronounced indentation of the shell margin (Figure 9.15a). As the shell grows, former positions of this indentation are marked by a distinct track of indented growth lines around the middle part of the outer shell wall.

❑ Does either of FS K or A show such a track?

■ There is no track in FS A, but it is present in FS K.

The presence of the track on FS K means that the shell opening originally possessed an indentation, as in Figure 9.15a. Since the indentation itself is not visible on the margin of this specimen, the shell around the opening must have been broken off. The central position of the exhalent current indicated by the track in FS K in turn implies that inhalent currents entered from either side of the animal. Therefore this gastropod possessed two gills, as in Figure 9.15a.

Question 9.6 Would the indentation of the margin in FS K have brought the exhalent current closer to, or further away from, the head than if the indentation had been absent? You can best visualize this by placing the shell over the head-foot diagram in Figure 9.14 again.

The inhalent margin in FS K reduced the possibility of exhalent water, which would have been depleted of oxygen and which would have borne excretory products, returning to the inhalent sites, or to the head, in FS K.

The majority of marine gastropods have only one gill, situated on the left-hand side of the mantle cavity (Figure 9.15b). Water is drawn in anteriorly on the left-hand side of the head, to pass over the single gill and thence out posteriorly to the right of the animal. The anus is situated next to the exit, on the right of the animal. Thus, no central indentation nor corresponding track marks the outer part of the whorl. Furthermore, in some gastropods, inhalent water is channelled into the mantle cavity by a fleshy tube or **siphon** extending from the

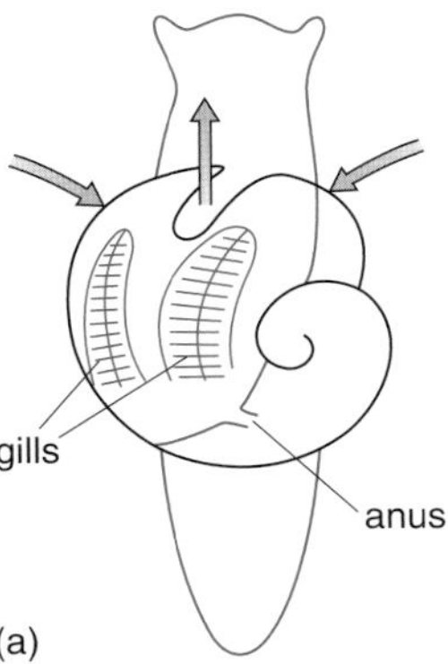

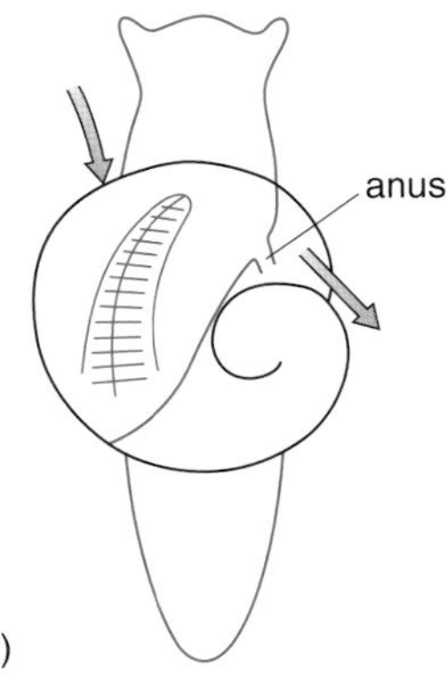

Figure 9.15 Arrangement of gills in gastropods, with water currents arrowed, shown in dorsal view: (a) two gills with exhalent indentation; (b) single gill.

mantle cavity. This is usually housed in a scroll of the shell margin that projects in the opposite direction from the coiled apex of the shell. Such a scroll is clearly visible in FS A. The presence of a single inhalent current associated with the scroll in FS A indicates that this animal would have had only one gill, as in Figure 9.15b.

Question 9.7 Summarize what you have been able to find out about the respiratory currents of FS K and A by placing the shells in their correct positions in Figure 9.14 again, and drawing onto the diagrams the likely inhalent and exhalent currents, by means of arrows.

You may have observed that the exhalent current of FS A would have been almost in line with the inhalent current on the opposite side; there would have been even less chance for exhalent water to have returned to the inhalent site than in FS K.

Now let us briefly consider locomotion and position on or in the sediment.

❑ Do you suppose either of FS A and K would have been capable of burrowing into soft sedimentary substrata? Think carefully about the organization of respiratory currents and the shell shapes (again you will find it helpful to position the specimens on Figure 9.14 for this).

■ It is most unlikely that FS K could have burrowed beneath the sediment surface. Submergence of its aperture would have immediately blocked the inhalent currents, and, besides, the broad shell would have presented an uncompromising impediment to burrowing. In contrast, FS A is a more likely candidate for burrowing. Its inhalent siphon would have allowed it access to overlying water even when the aperture was submerged. Note, furthermore, the streamlined form of the shell, with the spirally coiled apex pointing backwards: it is almost torpedo-shaped, offering little hindrance to burrowing (look particularly at it in profile, when placed over Figure 9.14b).

What can be said about the feeding habits of these two gastropods? Gastropods feed in a wide variety of ways, employing different modifications of the same basic feeding apparatus: the mouth contains a protrusive cartilaginous bar over which a tooth-studded ribbon of tissue (Figure 9.16a) called the **radula** is run to and fro. When the mouth and radula are applied to a food-bearing surface, food material is brushed or scraped into the mouth.

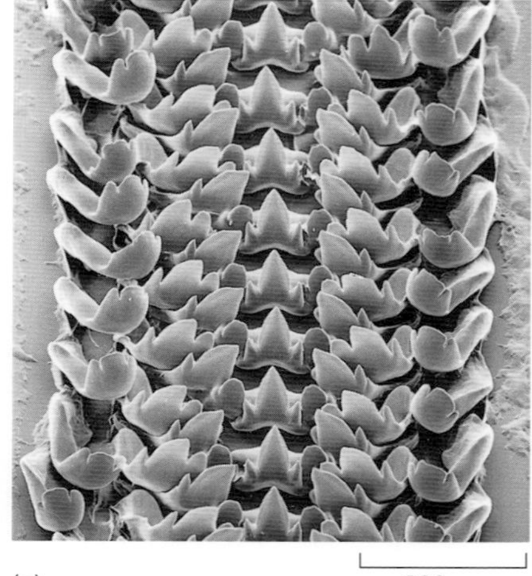

Figure 9.16 (a) Electron microscope view of a gastropod radula (from a grazing form); (b) boring made in a bivalve shell by the radula of a predatory gastropod (*c.* life size).

Many gastropods use the radula to graze on algal material growing on surfaces or on loose organic detritus and microscopic organisms on the sea-floor. Many others hunt vagrant prey, however, which they kill by smothering with the foot or even by driving into them specialized radula teeth furnished with poison. Certain predatory forms use the radula rather like sandpaper to drill holes through the shells of their prey (Figure 9.16b). A tubular trunk with a terminal mouth is then used to tear out the tissues of the shell's occupant. Most of the advanced gastropods that possess a well-developed siphon (as in FS A) are predators or scavengers, feeding in one or other of these specialized ways. As noted above, the siphon probably evolved in the first instance as an adaptation enabling gastropods to penetrate soft sediment. However, since the gill it supplies is typically equipped with a sensory gland that can 'sniff' the incoming water, the manoeuvrable siphon allows the gastropod to locate the source of any 'smell' with accuracy. Thus, in predatory or scavenging forms this apparatus seems to have become secondarily adapted as a means of locating and tracking their food.

Question 9.8 How would you now interpret the likely feeding habits of FS K and A? Drawing your conclusions together with those you reached earlier in this Section, enter them in the appropriate boxes in Table 9.2 (p. 80).

9.3.4 AMMONOIDS (FOSSIL SPECIMEN D)

The ammonoids are an extinct group that lived in Palaeozoic and Mesozoic seas. Nevertheless, there are other representatives of the broader group to which the ammonoids belong – the **cephalopods** – that are still alive today: these are the squids, cuttlefish and octopoi, in which the shell is reduced or lost, and the pearly nautilus, whose pretty shell decorates many a mantelpiece as well as showing obvious similarities with that of ammonoids, and which you will see on video when you attempt Activity 9.4.

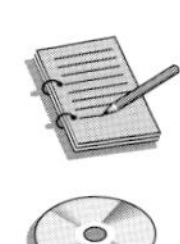

Activity 9.4

To give you a better insight into how the life habits of extinct ammonoids can be investigated, view the video sequence *Form and function of fossils* on DVD 3 as Activity 9.4, when it is convenient.

Ammonoid shells vary greatly in form. Some are relatively compressed from side to side, like FS D, and others are fat. Some are coiled so that later whorls overlap earlier whorls, while others are coiled with no overlap, rather like a coiled rope. The sides of the shells are commonly ornamented with radial ridges called **ribs**. Sharp, fairly evenly spaced ribs are visible on the earlier whorls of FS D. However, they get smaller on later whorls, so that they can be seen only as vestiges on the outer shoulders of the final whorl. Remember, however, from Section 9.1.1 that FS D is an internal mould, so that we cannot be certain from this specimen that the outside of the shell was rib-free. Notice also that the ribs do not continue around the outer edge of the coil of the shell. Instead, the outer edge of the coil in this specimen is marked by a sharply projecting ridge called a **keel**.

In contrast to the gastropod body, the ammonoid body occupied only a small fraction of the shell interior, the rest of the shell being divided into chambers by the septa (Section 9.1.1).

❑ How much of the shell was occupied by the body in FS D?

■ The body occupied just under half a whorl. The sutures that mark the outer edges of the septa stop just under half a whorl away from the aperture of the shell (Section 9.1.1). (You cannot see the whole of the body chamber on FS D because the specimen is incomplete near the aperture.)

In the absence of living examples, palaeontologists have reconstructed the probable anatomy and life habits of ammonoids on the basis of: (a) homology and analogy (Section 8), especially by reference to *Nautilus* (Figure 9.17), and to

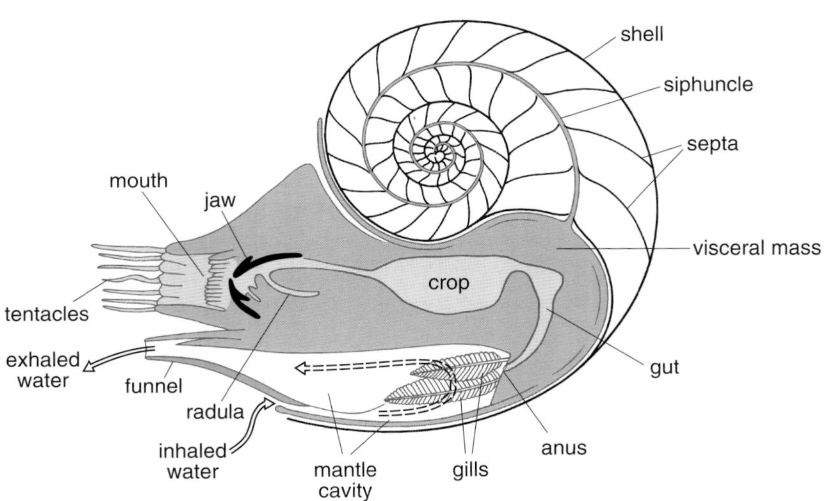

Figure 9.17 The anatomy of *Nautilus*, seen in diagrammatic section along the body.

artificial models whose mechanical properties can be measured; (b) rarely preserved soft parts or gut contents in ammonoid fossils; and (c) information on the distribution of fossil ammonoids and the other fossils and sediments with which they are associated.

In life, the chambers of the ammonoid shell were not entirely sealed off, for a membrane-bound tube called the **siphuncle** ran through them, connecting with the rear part of the body. As is the case in *Nautilus*, it is probable that the siphuncle of ammonoids was capable of withdrawing water from the chambers and introducing gas. The chambered shell thus seems to have served as a buoyancy tank (as it does in *Nautilus*), so ammonoids are believed to have been predominantly pelagic, like most other cephalopods.

We may further suppose that the ammonoids, like *Nautilus*, possessed a well-developed head with eyes and a mouth surrounded by a ring of tentacles. The mouth certainly was equipped with beak-like jaws as well as a narrow radula, comparable with, but simpler than, that in most gastropods; both jaws and radulae are known from ammonoid fossils. Behind the head would have lain the visceral mass, occupying the part of the body chamber on the inside of the coil. Most of the outer side of the body chamber was probably taken up by the mantle cavity, as in *Nautilus* (Figure 9.17). The gills, anus and various sensory and other organs would have been situated here, beneath the body. In living cephalopods, the mantle cavity serves as a sort of bellows into which water is drawn around a wide inhalent margin, and then pumped out via a tubular **funnel** situated immediately beneath the head. This provides for a form of jet propulsion at the same time as supplying water to the gills. It is thought likely that many, if not all, ammonoids were similarly capable of at least weak swimming by jet propulsion, though work on experimental models shows that even well-streamlined ammonoids (such as FS D) would have fallen far behind today's squid and high-speed fish in terms of swimming efficiency.

The range of shell forms in ammonoids, implying widely different swimming abilities, shows that their life habits were correspondingly diverse. Many species are interpreted as having been nektonic suspension feeders or predators on small prey, while others were possibly large-scale predators, scavengers or detritus collectors swimming near the sea-floor. These interpretations, are however, still largely speculative. By way of comparison, the living adult *Nautilus* lives in moderately deep water (up to a few hundred metres depth), swimming close to the bottom, and scavenging on the corpses of animals such as crustaceans and fish. It is also occasionally predatory.

One aspect of ammonoid palaeobiology – that of shell development – can be studied in detail. Since the shell was grown incrementally throughout the later larval and post-larval life of the animal, it reveals a 'potted history' of its development. As the shell opening grew forward through the addition of new material, so the body migrated forward to keep pace with it, and laid down the successive septa behind it. Forward growth of the shell opening and migration of the body evidently slowed down in mature individuals, for the final septa in these became more closely spaced.

❑ Using the criterion of septal spacing, does FS D represent a mature individual?

■ It seems reasonable to suppose that FS D comes from a mature individual because the last ten or so sutures (and therefore the septa they represent) are more closely spaced than earlier ones.

We are now in a position to look for differences between juvenile and adult features. To study the juvenile features of the shell, you need to look, of course, at the inner whorls, which were grown first.

❑ What is the most striking difference in ornamentation between the visible part of the juvenile shell and the adult part of the shell?

■ As you saw earlier, the juvenile shell (inner whorls) has fairly pronounced ribs, while the adult shell is much smoother with very subdued ribs. Note, however, that for a full comparison between the juvenile and adult portions, we would either have to find a juvenile specimen or strip away the outer whorls. You may have also noticed that the sutures on the inner whorls are less complex than those in the outer whorl, although this is not surprising as the septa are smaller.

These observations illustrate an important point when dealing with fossils of this kind: different growth stages can look very different and, since ammonoid species are typically identified on the basis of shell shape, ornamentation and sutural form, there is considerable scope for spurious separation of different growth stages into different species. To make matters worse, it is now known that at least some ammonoids were like some other animals in showing differences between the adult shells of the two sexes (**sexual dimorphism**, Figure 9.18).

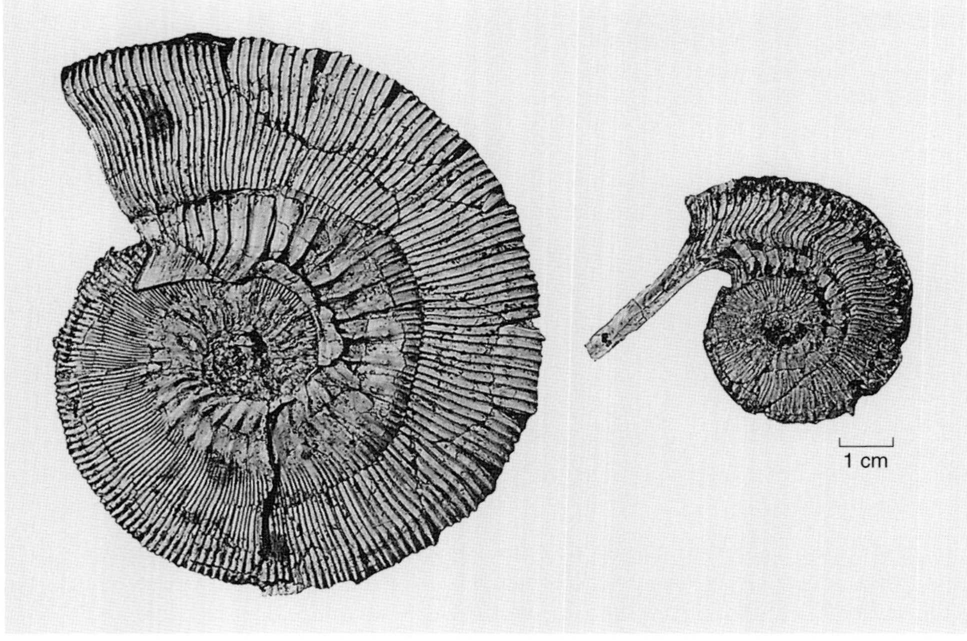

Figure 9.18 Sexual dimorphism in a species of the Jurassic ammonite genus *Kosmoceras* consisting of a large and a small form. Prior to their recognition as dimorphs, they had been identified as separate species. It is not certain whether the larger form represents the male or the female, although the latter seems more likely.

❑ How might a consideration of shell development help to pair up males and females from dimorphic species?

■ The juvenile shells (inner whorls) might be expected to be closely similar before sexual maturity.

Unlike the other specimens studied so far, we cannot confidently assert what the feeding habits of FS D itself were, though its streamlined shape suggests that it was a relatively good swimmer (in ammonoid terms) and thus nektonic. Because many ammonoids were nektonic, their shells tended to become widely distributed. This, coupled with relatively high rates of evolution, make them particularly suitable candidates for stratigraphical correlation. Ammonites are thus widely used for dating Jurassic and Cretaceous marine deposits.

9.3.5 BRACHIOPODS (FOSSIL SPECIMENS F AND L)

Brachiopods are a relatively small group today, but they are very abundant in the fossil record, particularly in the Palaeozoic. The brachiopod body is enclosed by the shell, which consists of two valves, one dorsal and the other ventral (Figure 9.19).

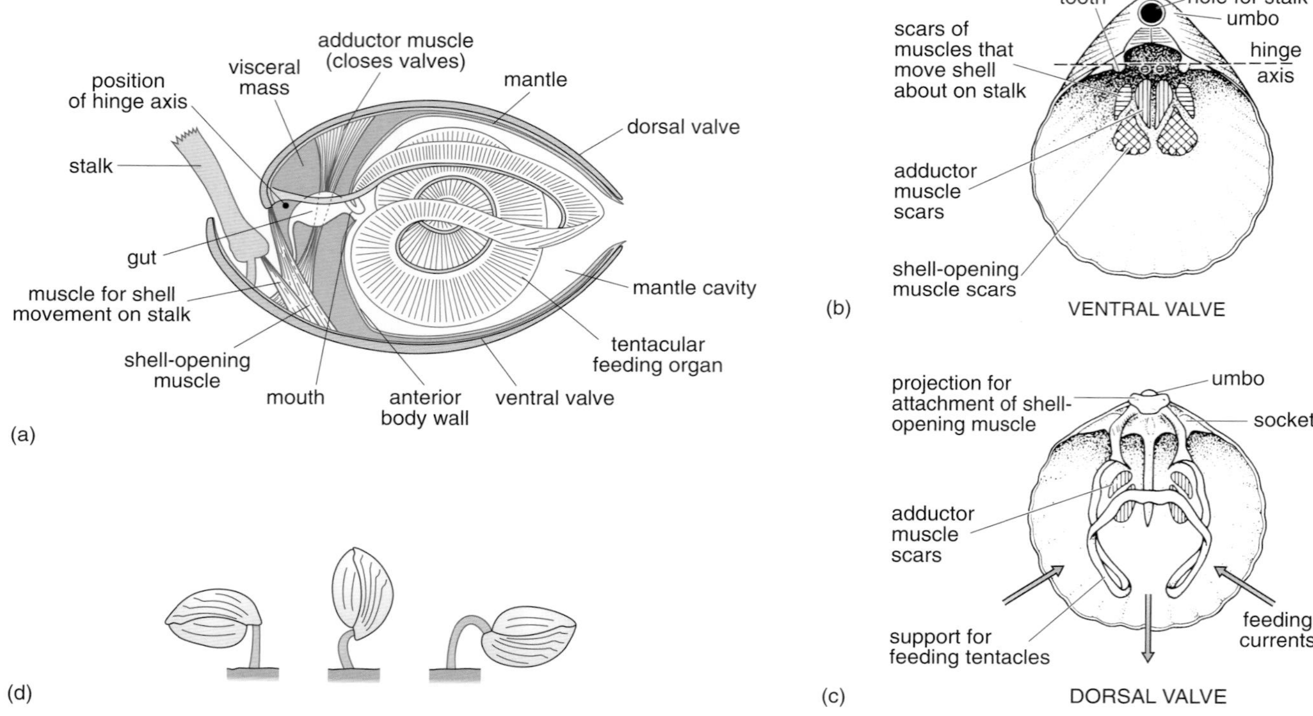

Figure 9.19 (a) Anatomy and shell of a living articulate brachiopod, in diagrammatic longitudinal section; (b, c) interiors of (b) ventral, and (c) dorsal, valves (with orientations of feeding currents indicated in c); (d) variety of possible life positions.

❑ How does this arrangement of valves differ from that of bivalves?

■ Bivalves have their valves on each side of the body, so that they are lateral (Section 9.1.2).

In brachiopods, the two valves articulate *posteriorly* and, in most examples, the hinge is formed by the projection of a pair of **teeth** from the ventral valve (Figure 9.19b) into a corresponding pair of **sockets** in the dorsal valve (Figure 9.19c). Such forms are referred to as being **articulate brachiopods**. Both FS F and FS L are of this sort but, since teeth and sockets are internal features, they are not directly visible in these specimens. Another small group of brachiopods lack hinge teeth and sockets, the valves being held in position entirely by muscles. These are the **inarticulate brachiopods**.

Anterior to the body is a large **mantle cavity**, lined by mantle tissue, and occupied by a convoluted bar of cartilaginous material that supports an extended row of tentacles (Figure 9.19a). In some articulate brachiopods, this bar has a shelly support (which may be folded back on itself) that is rooted in the dorsal valve (Figure 9.19c). For this reason, the dorsal valve is often referred to as the '*brachial valve*' – the word 'brachial' referring to this arm-like support. Beating of cilia on the tentacles creates a current from which food particles are trapped and passed to the mouth. A number of different feeding patterns have been described from living brachiopods and these are associated with much variation in the shape of the valves and feeding organ. In FS L, it is likely that the feeding currents entered the mantle cavity on each side of the shell and left from the middle of the anterior side (as in Figure 9.19c).

❑ Considering the positions of the inhalent and exhalent water currents, suggest an explanation for the overall form of the valve margins in FS L. (*Hint*: cast your mind back to your consideration of the respiratory currents in gastropods.)

■ The abrupt flexure of the central parts of the anterior valve margins keeps them markedly separated from those on each side. Since the central parts supposedly contained the exhalent site and those to the sides the inhalent areas, these large-scale flexures of the margins may be considered as an adaptation to promote separation of the inhalent and exhalent currents.

The shell is opened and closed entirely by the action of muscles (Figure 9.19a). **Adductor muscles** draw the valves together, while **diductor muscles** pull the valves open.

❑ How does this method of opening and closing the valves with two sets of muscles differ from that of bivalves, such as *Circomphalus* (FS I)?

■ Bivalves are similar in that they employ adductor muscles to close the valves, but they differ in using the stress stored in an elastic ligament, rather than muscular force, to open them (Section 9.3.1).

In life, most brachiopods are attached to the bottom by means of a horny stalk (Figure 9.19a, d), technically known as a *pedicle*, that projects out of the shell through a hole situated on the umbo of the ventral valve (Figure 9.19b). For this reason, the ventral valve is often referred to as the '*pedicle valve*'. Muscles which insert into the stalk allow the brachiopod to move its shell around (Figure 9.19d). Reduction or loss of the stalk occurred during evolution in some brachiopods, which lay on the bottom or nestled shallowly into it, and in some cases became tethered there by fine branches of the stalk projecting into the sediment.

FS F has a large and obvious hole for the stalk. This indicates that it was attached to a hard surface by means of a large stalk on which it could probably change its orientation. FS L, in contrast, has either a very small hole or none at all on its small umbo; it is not possible to tell which, as the umbonal tip of the ventral valve has been chipped off. Thus, if it had a stalk it must have been greatly reduced and could only have served to tether the shell to the sea-floor. It is thus likely that FS L nestled in loose sediment.

Question 9.9 Summarize what you have learned about the life habits of FS F and L by entering them in the appropriate boxes in Table 9.2 (p. 80).

9.3.6 Crinoids (fossil specimen H)

Like brachiopods, crinoids are less abundant and diverse now than in former times. In the Palaeozoic, their remains sometimes make up the bulk of whole beds of limestone, as in RS 24 (TS U) from the Home Kit. Crinoids, like other echinoderms, cannot tolerate significant departures from normal marine salinity, and have apparently always been marine.

Most crinoids found as fossils were evidently attached in life to the sea-floor by a stalk with a root-like **holdfast** (Figure 9.20). The majority of living crinoids have a greatly reduced stalk or none at all, and are capable of creeping around or even swimming. The few surviving stalked forms generally live in deeper water.

The crinoid body is housed in an enclosed cup at the top of the stalk (if present), and this cup contains most of the internal organs, including a twisted U-shaped gut, with the mouth in the middle of the cup's upper surface and the anus to one side of it (Figure 9.20). Surrounding the rest of the cup is a ring of branching arms showing a five-fold radial arrangement, although this is not particularly clear in FS H.

Figure 9.20 Basic anatomy of a crinoid.

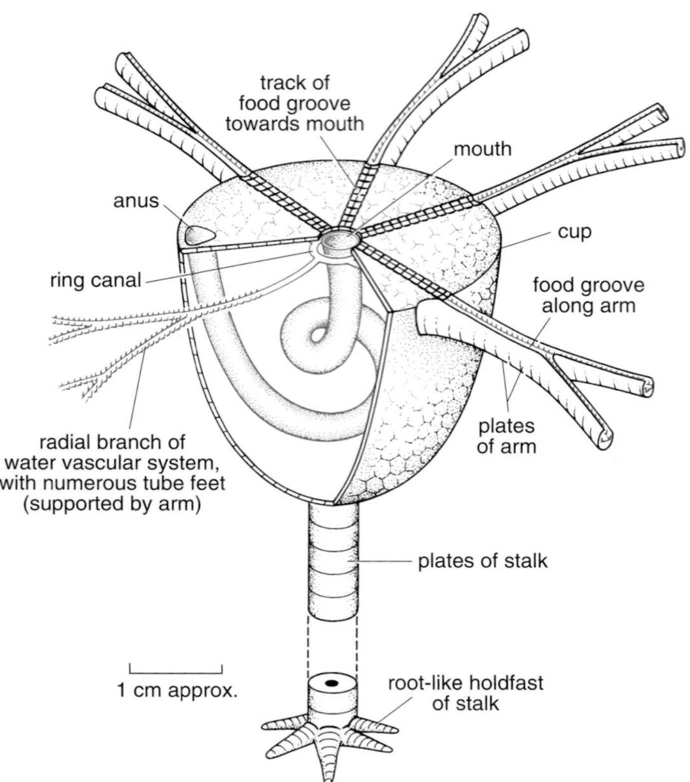

FS H shows the cup to have a jigsaw-like skeleton of plates, and the stalk and arms consist of stacks of plates looking like piles of coins. As with the echinoids (Section 9.3.2), these plates are of calcite, with a sponge-like three-dimensional network and they enclose the bulk of the soft parts; but again, the plates are endoskeletal and are secreted within soft tissue, which coats the animal's outer surface. The stacked plates of the arms support a muscular coating that controls arm movement, as well as radial branches of the water vascular system with the tube feet, which run along their upper surfaces. As in the echinoids, again, these branches radiate from a ring canal around the mouth, though in crinoids there is no perforated plate connecting the system with seawater. The exposure of the branches of the system on the arms also contrasts with their internal position in echinoids.

Underwater photographs of living crinoids show that feeding is accomplished by spreading the arms out to form an umbrella-like filtration fan (Figure 9.21). Microscopic food particles are trapped on the tube feet arrayed along the arms. From there they are passed to the food grooves overlying the radial branches of the water vascular system (Figure 9.20), and then towards the mouth.

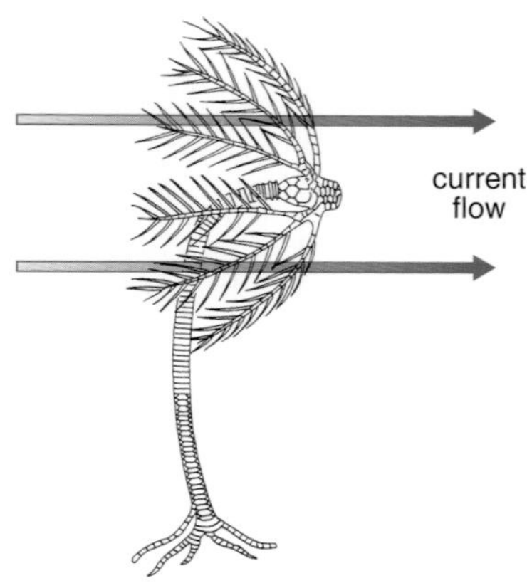

Figure 9.21 Feeding posture of a crinoid in a current.

Question 9.10 FS H is preserved with its arms bunched together. What geometrical aspect of their construction suggests that they nevertheless spread out like those of living crinoids during feeding?

FS H is unusual in being preserved with its assembly of skeletal plates intact. As with the echinoids (Section 9.3.2), the plates were connected only by organic material (collagen) in life. Many of these connections were flexible, as was clearly the case in the arms of this specimen. Thus, upon death and decay of the organic material, crinoid skeletons tend to disintegrate rather rapidly. For example, most of the crinoid remains in RS 24 are individual plates from stalks that have become disarticulated (separated).

❑ What does the mode of preservation of FS H suggest, therefore, about how this individual died?

- For the plates to have been preserved intact, without falling apart, this crinoid must have been entirely buried, either alive or shortly after death, by a rapid influx of sediment; the way that the arms are bunched inwards indeed suggests that it may well have been killed by this burial event.

Inferences such as this, derived from the mode of preservation of fossils, can provide valuable clues to conditions in ancient environments – in this case, for example, testifying to occasional, probably storm-driven, fluxes of sediment. We will be exploring this topic in greater detail in Section 10.

Question 9.11 Summarize what you have learned about the life habits of FS H by assigning it to the appropriate box in Table 9.2 (p. 80).

9.3.7 Corals (fossil specimens E and M)

There are three major groups of corals which probably evolved independently (making 'corals' a ragbag grouping, from an evolutionary perspective). The first two groups are confined to the Palaeozoic, while the third arose in the Triassic and is still flourishing today. There are also soft-bodied, close relatives, such as the sea-anemones, but these are rare in the fossil record. FS E and M both belong to one of the Palaeozoic groups. Throughout their history, corals (of all groups) appear to have been marine, and certain forms have played a major role in the growth of reefs at various stages in geological history.

A coral individual, or **polyp**, is essentially similar to a sea-anemone in bodyplan, though, unlike the latter, it builds up a cup of calcium carbonate beneath itself as a sort of platform (Figure 9.22). As the outer wall of the cup grows upwards, so too do the septa (Section 9.1.1) supporting the base of the polyp. In FS E, much of the outer wall has been lost, revealing the outer edges of the septa. Successive 'false floors' are also formed beneath the polyp (Figure 9.22), but these are usually not visible from the outside. Look closely inside the cup in FS E (with the hand lens) and you might see some thin, sometimes draped, partitions between the septa, especially towards their inner ends. However, they are not clearly visible within the many cups in FS M.

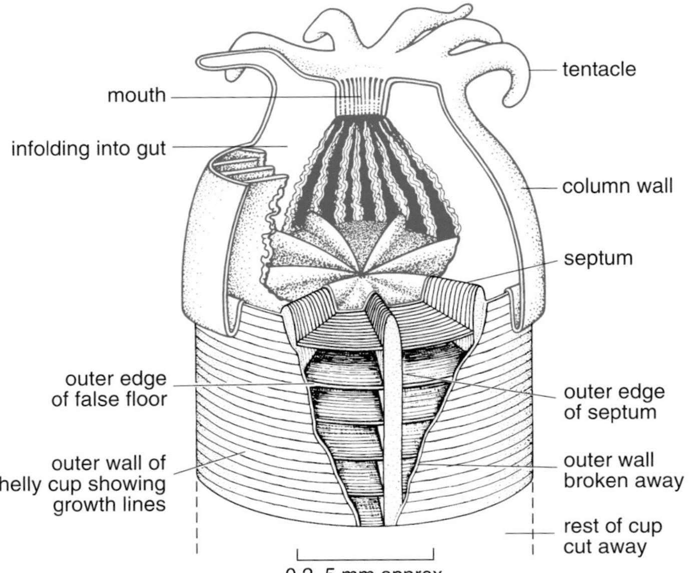

Figure 9.22 A single coral polyp showing 'false floors' beneath it.

The polyp is of simple construction, comprising a thick-walled sac with an upward-facing opening surrounded by a ring of tentacles. The internal cavity is where it digests its food and the single opening serves as both entrance and exit (for undigested matter). The sidewalls of the cavity have numerous infoldings, separated by the septa, which greatly increase the surface area of the cavity, facilitating absorption of the products of digestion. The tentacles are armed with stinging cells, which paralyse the prey (mainly microscopic planktonic animals) as it is passed to the opening of the internal cavity.

In the groups we have considered so far, reproduction is almost always sexual. In the case of corals, however, new individuals can arise both by sexual reproduction or by asexual reproduction (budding). In some species, separation of the buds is incomplete, resulting in a more or less densely connected cluster or **colony** of polyps. The mode of budding (which is characteristic for each species) affects the way the colony grows, although the overall colony form is also subject to environmental influences. In FS M, which represents part of a colonial coral, new buds arose by sideways splitting of polyps, so that new corallites (Section 9.1.1) developed between existing corallites. The pattern of budding in FS M produced a massive globular colony, whose upper surface became increasingly domed as more and more buds were packed into it.

The colonial habit in corals, with its potential for generating robust interlocking skeletal frameworks on the sea-floor, has been a major factor in their history of reef-building (something we will pursue in Section 16.5). Another factor, seen in living corals, is that many have great numbers of single-celled algae dwelling within their soft tissues, and photosynthesizing with the benefit of the breakdown products of their hosts' own digestive processes. This symbiotic association ensures a tight recycling of nutrients within the corals, allowing them to thrive especially in waters with low nutrient levels, where the growth of fleshy algae (which elsewhere can swamp the slower-growing corals) is kept in check by grazers. Today, such corals are prominent constituents of tropical shallow-water reefs in areas that might otherwise be described as 'nutrient deserts'. However, there is much debate as to which fossil corals did, or did not, have such single-celled algae in their tissues in life. One clue suggested by living corals is that the symbiotic forms tend to have smaller, more closely integrated corallites, though there is considerable overlap with non-symbiotic forms in these respects. Thus, it would be merely speculative at this stage to attribute this particular ecological trait to unrelated Palaeozoic forms such as FS E and M (remember the warning in Section 8 about over-interpreting from living relatives, in relation to the example of mammoths).

Question 9.12 Enter the information on the life habits of FS E and M in the appropriate box of Table 9.2 (p. 80).

9.3.8 TRILOBITES (FOSSIL SPECIMEN G)

Trilobites are extinct, and are entirely restricted to Palaeozoic strata, being particularly abundant in the Lower Palaeozoic. They are only remotely related to groups alive today. Thus one cannot so readily reconstruct the form and function of their soft-part anatomy on the basis of homology as it was possible to do to some extent with the ammonoids by using *Nautilus*.

The basic features of the dorsal exoskeleton of the trilobite were described in Section 9.1.3. Like most invertebrates, trilobites grew continuously, albeit more slowly during adulthood. The exoskeleton, however, was inflexible and so periodic shedding, and secretion of a new skeleton was necessary to permit growth, just as in crabs and lobsters today. **Moulting** involved breakage of the skeleton, along lines of built-in weakness, into distinct components.

❑ Whereabouts can you see any such lines of weakness on the head-shield of FS G?

■ A fine line is visible on each side of the head-shield, running outwards from immediately behind each eye.

In fact, the line on each side continues in towards the central portion of the head-shield in a W-shape, with its centre running around the back of each eye, though this is not clearly visible in this specimen. The flanks of the head-shield (including the eye surfaces) would thus have broken off along these lines of weakness. Moulting means that during growth each individual left behind a succession of potentially fossilizable skeletal elements. Widespread scattering of these ensured an abundance of trilobite remains in a wide variety of deposits.

❑ Does FS G represent a moulted skeleton or an individual that died?

■ Since the exoskeleton is entire, and not broken up by the moulting process, it must represent an individual that died.

A distinctive feature on FS G is the pair of eyes on the head-shield. Trilobites are the oldest fossils which show evidence of a visual system. Inspect the eyes with the hand lens. You should see that the outward-facing surface of each eye consists of many rows of little low, rounded domes. Each dome represents a lens. They are thus compound eyes like those of flies (insects) and crayfish (crustaceans). However, the detailed structure of the compound eye differs in these groups. Compound eyes are particularly good at detecting motion of nearby objects because the same changing pattern of light and shade is sensed many times over by the bank of lenses.

Question 9.13 Inspect the orientation of the wall of lenses on each eye. What was the trilobite's field of view? (*Note*: The right eye in this specimen is incomplete, but its form would have been the mirror image of the left eye, which is only slightly chipped.)

Preservation of the underside and appendages of trilobites is very rare, though sufficient exceptionally well-preserved fossils have been found to enable a reconstruction of the underside of a typical trilobite to be made (Figure 9.23). Each limb had two branches: a lower 'walking branch' and an upper, feather-like branch that may have acted as a gill. The front pair, however, formed simple antennae.

❑ What can you now infer about the mode of locomotion in FS G, and how does this match with your answer to Question 9.13?

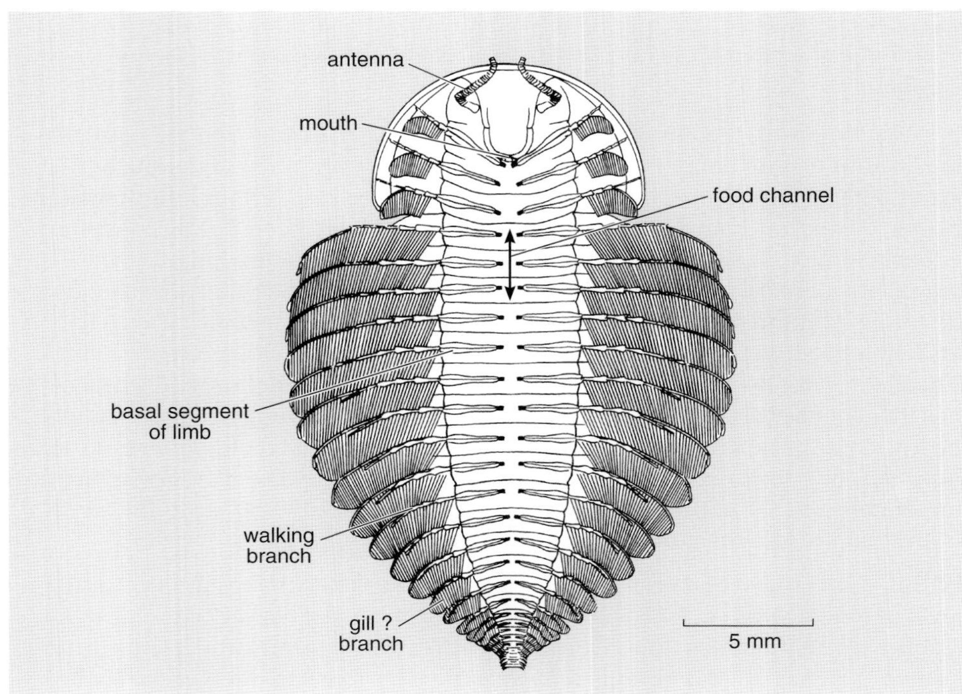

Figure 9.23 Ventral surface of a reconstructed trilobite showing appendages.

- If the limbs of FS G were like those in Figure 9.23, it is reasonable to suppose that the animal could scuttle around on the sea-floor. This agrees well with the answer to Question 9.13, for the trilobite's lower surface, out of the range of vision of the eyes, would usually have been applied to the sea-floor.

It is probable that such locomotion was typical of most trilobites, though some were evidently free-swimming, while others may have made shallow burrows. The turret-like projection of the eyes might also mean that FS G was capable of shallowly submerging itself in sediment, with only the eyes above the surface. Some trilobites are found fossilized in a curled-up position, like miniature startled armadillos. Notice how the articulation of the segments lying between the head-shield and the tailpiece would have allowed the animal from which FS G came to do this – the outer parts of each segment sliding under those of the one in front. This was probably a defence posture.

Finally, how did trilobites feed? Look at Figure 9.23 again. Running down the centre of the animal's underside was a sort of channel, the sides of which were formed by the basal segments of the limbs. These segments are known from very well-preserved specimens to have been furnished with bristles, which projected into the channel. This arrangement is analogous to that found in certain living primitive crustaceans, which feed on bottom detritus brushed into the central channel by scuffling movements of the appendages. There it is passed forward to the mouth by the paddling action of the bristles as the basal segments of the limbs move backwards and forwards. It is thus reasonable to suppose that many trilobites, including FS G, fed similarly, though certain species have also been interpreted as predators and/or scavengers on account of the large size (and presumably power) of the limb bases, and of their sharp bristles. Some other trilobites are considered more likely to have been suspension feeders.

Question 9.14 Summarize the data on the life habits of FS G in Table 9.2 (p. 80).

9.4 Relationships and classification

We started this Section with a simple but practical key system for classifying the specimens in the Home Kit. Now that you have learned something about the biology of the original organisms, we shall take a brief look at an hierarchical biological classification.

It is easy to group the familiar large multicellular organisms that we regularly see around us into animals, plants and fungi. These are three of the major groupings, or **kingdoms**, into which life can be classified. Other kingdoms can be detected among single-celled organisms, though we need not be concerned with them here. It has also long been realized that, within the kingdoms, organisms naturally fall into broad groups, each of which is characterized by a distinct bodyplan differing from that of other such groups. Such a group is termed a **phylum** (plural: phyla). For example, humans belong to the Phylum Chordata. Within each phylum there are subgroups called **classes**. Each class has its own characteristic modification of the basic phylum bodyplan. As with phyla, classes are thus recognized by their morphological distinctiveness. Note, however, that they need not contain large numbers of species (though many do); for example, some of the short-lived classes of animals that arose in the early Cambrian are known from only a few species.

Within classes, a nested series of smaller subgroups may be recognized until the level of single **species**, which is the natural basic unit of classification, is reached. From the class downwards, these are: **order**, **family** and **genus** (plural: **genera**) (Table 9.3), and various intermediate divisions are denoted by such prefixes as 'sub-', as in 'subclass' or 'suborder'. The successive groupings, generally known as **taxa** (singular: taxon), thus form a nested series, or **'taxonomic hierarchy'**. The taxonomic hierarchy has many uses. It provides an efficient means of identifying a specimen. For example, knowing that FS I is a bivalve means that we can narrow the search to reference books on the Class Bivalvia (to give it its formal name, which, by convention, is always latinized). We can then try to decide to which order it belongs, and so on, until the appropriate species description is found. The available information on the species can then be used to determine, say, the likely age of a fossil specimen (for stratigraphic correlation), or to infer its life habits and hence what the original environment was like. Perhaps the most important aspect of the taxonomic hierarchy is that, ideally, it should reflect evolutionary relationships.

Table 9.3 The taxonomic hierarchy.

Taxon*	Example	Contents
Kingdom	Animalia	animals
Phylum	Chordata	vertebrates plus some minor groups
Class	Mammalia	mammals
Order	Carnivora	dogs, cats, badgers, seals etc.
Family	Felidae	cat family
Genus	*Felis*	wild and domestic cats
Species	*catus**	domestic cat

*Formally, the names of all taxa should begin with a capital letter, except for the species (or 'trivial') name, which begins with a lower-case letter. Moreover, the species name is always given together with the genus name, in italics, so the domestic cat is therefore referred to as *Felis catus*.

9.4.1 Recognizing bodyplans and phyla

The classes Echinoidea (Section 9.3.2) and Crinoidea (Section 9.3.6) are united in the Phylum Echinodermata.

Question 9.15 What are the three principal features in common between these two classes?

So, although echinoids and crinoids look quite different, they have the same bodyplan and are constructed from the same basic components. The fundamental similarities (homologies) place them in the same phylum whilst the differences (relating to radically different general life habits) dictate that they belong to distinct classes.

Apart from the Echinodermata, only one other phylum, the **Mollusca**, (informally called 'molluscs') is represented in the Home Kit by more than one class. The following features are shared by practically all molluscs: a dorsally attached calcareous shell is grown incrementally by a lining of mantle tissue. Beneath, next to the body, there is a mantle cavity, containing gills. The body itself has a dorsally situated visceral mass, beneath which is a muscular organ of locomotion. All but one of the molluscan classes are also characterized by the possession of a radula inside the mouth.

Question 9.16 Considering this list of characters, which of the groups represented by specimens in the Home Kit belong to the Phylum Mollusca?

Some of these features mentioned above may be greatly modified, or even lost in particular forms. The octopus, for example, lacks a shell, but it is still recognizable as a cephalopod mollusc because it retains other typical features of that class (beak-like jaws, radula, tentacles etc.).

You may have been tempted to include brachiopods in your answer to Question 9.16, because they too have an incrementally grown calcareous shell, mantle tissue and a mantle cavity. Brachiopods, however, are quite separate with a different orientation of the shell (which is not dorsally situated, for example), possession of a distinctive tentaculate feeding organ and absence of both a foot and radula. They belong to a distinct phylum (the Brachiopoda). This means that the features similar to those found in the Mollusca, such as the mantle cavity, must be analogous rather than homologous.

Corals (Section 9.3.7), together with anemones, are placed in the Class **Anthozoa** (informally called 'anthozoans'). Together with various other creatures such as jellyfish, sea-whips and sea-fans, they belong to the Phylum **Cnidaria** (pronounced 'nigh-dare-rhea' and informally called 'cnidarians'). The cnidarian bodyplan is simpler than those of the other phyla considered here, and tends to contain an abundance of gelatinous material, especially in the jellyfish. The body usually has radial symmetry, expressed in the cylindrical body and ring of tentacles around the opening of the internal cavity. The tentacles bear characteristic stinging cells that are a unique cnidarian feature. The gut (internal cavity) is one-ended, combining the functions of mouth and anus.

The final group to consider is the Trilobita. These comprise a class within the Phylum **Arthropoda** (informally called 'arthropods'), an enormous assemblage that today consists of over three-quarters of all known living animal species (mostly insects, but also including crustaceans such as crabs and shrimps). The characteristics of arthropods include a segmented body with paired limbs. The body and limbs are covered with a stiff, jointed exoskeleton that is periodically moulted. Compound eyes are often present.

The taxonomic information given in this Section can now be summarized as follows (Table 9.4):

Table 9.4

Phylum	Class	Specimen FS
Echinodermata	Echinoidea	C, J
	Crinoidea	H
Mollusca	Bivalvia	B, I
	Gastropoda	A, K
	Cephalopoda (subclass Ammonoidea)	D
Brachiopoda	Articulata	F, L
Cnidaria	Anthozoa (corals)	E, M
Arthropoda	Trilobita	G

9.5 Body fossils of other organisms

So far we have concentrated on the most commonly encountered kinds of marine invertebrate body fossils. However, macrofossils from other groups of organisms, including other invertebrates as well as vertebrates and plants, can be abundant and/or of great interpretative value in certain deposits, while microfossils from many different groups of organisms are widely distributed, though the somewhat specialized techniques required for studying them preclude treating them in any detail here.

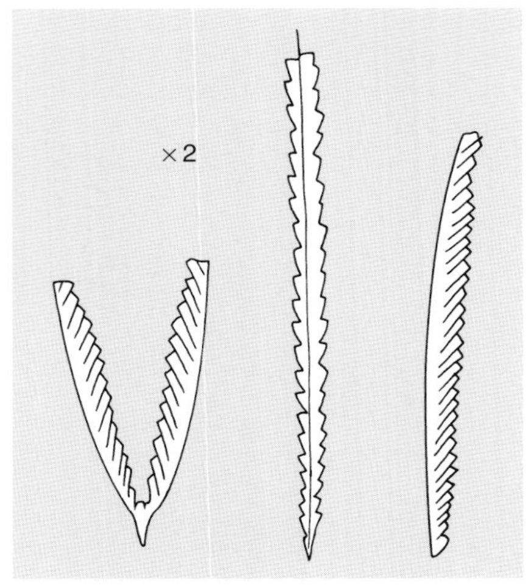

Figure 9.24 Three varieties of planktonic graptolites. Only the resistant rows of cups, shown here, are normally preserved as fossils. Forms similar to the two-branched 'tuning fork' graptolite on the left are characteristic of the Lower to Middle Ordovician. Single-branch varieties with cups on both sides (centre) are typical of the Middle Ordovician to Lower Silurian. Single branch varieties with cups on one side only (right) are limited to, though widespread in, Silurian to Devonian strata.

Among other invertebrate taxa, some are of particular value for stratigraphical correlation. **Graptolites**, for example, were colonial animals, probably suspension feeders, distantly related to chordates. Tiny individual members of the colony lived in comb-like rows of cups made of readily fossilizable resistant organic material. Primitive forms, dating from early Ordovician times, but persisting until the Carboniferous had colonies comprised of numerous branches. Graptolites with only a few colonial branches (Figure 9.24) – known more specifically as *graptoloids* – were limited to Ordovician to early Devonian times. Graptoloids were pelagic, part of the plankton, and so are found widely distributed in marine sediments. This, combined with their relatively rapid evolution, makes them very useful for precise correlation of strata deposited during that interval (like the ammonites in the Mesozoic).

Vertebrates – animals with backbones – constitute the vast majority of the Phylum Chordata, comprising such familiar animals as fish, amphibians, reptiles, birds and mammals. Scattered scales and teeth, coated by highly resistant enamel, provide the most common examples of body fossils, in both marine and non-marine environments. In favourable circumstances, however, bones, entire skeletons (Plate 7.1a), or even soft tissues in exceptional cases, may also be preserved.

Because of their possession of relatively resistant supporting tissues, land plants, especially woody forms, are commonly fossilized, especially in settings associated with rapid burial (as on delta tops). Ironically, being burnt first (e.g. in forest fires) can enhance the chances of fossilization of the remains, because charcoal provides no sustenance to bacteria and fungi and so does not rot. Carbonaceous fossils of marine algae are rare, by contrast, although a number of algal taxa secrete calcareous skeletons, which may furnish abundant fossils in certain shallow marine deposits, e.g. fossil reefs.

Also important in marine limestones are **stromatolites** (Figure 5.3), which are finely layered, often crinkled or lumpy-looking accumulations of carbonate muds or sands, trapped by surface-dwelling mats of filamentous photosynthetic bacteria or algae (shown later in Figures 16.4 and 16.5). They were widespread and common in the Proterozoic, during which diverse and intricate forms of stromatolites were produced. In the Phanerozoic, by contrast, they became largely confined to restricted environments, such as intertidal to supratidal flats in hot and arid areas, excessively salty bays and lagoons, and hidden cavities in reefs, for example. This restriction seems to have been largely due to the destruction of the mats by grazing animals in most normal marine environments.

Microfossils are derived from all sorts of organisms, both large and small. The former kind include plant spores and minute teeth and scales from animals, including the fine siliceous or calcareous spicules from the skeletal frameworks of sponges, while the latter comprise whole microscopic organisms, most of which are single-celled, such as certain algae and skeletalized single-celled organisms (e.g. **foraminifers**, Figure 9.25). Microfossils also come from all sorts of settings, including the land (e.g. pollen), the plankton (e.g. many single-celled algae and free-floating foraminifers) and the benthos (e.g. other foraminifers

Figure 9.25 Fossil foraminifers. These are planktonic species, of Neogene age, from the tropical west Pacific, and are shown at different magnifications, diameters ranging from *c*. 200–600 μm.

and various tiny crustaceans). Chalk is mostly composed of microfossils, chiefly the tiny plates, or coccoliths, from certain single-celled planktonic algae, as well as foraminifers and other forms.

9.6 Trace fossils

You will need FS N–P and the hand lens from the Home Kit for your study of this Section.

Palaeontologists have been studying body fossils systematically for well over 200 years, but it is only in the last few decades that they have been analysing trace fossils in any great detail. Indeed, for a considerable time some branching trace fossils (Figure 9.26) were identified as fossil seaweeds.

Figure 9.26 The trace fossil (branching burrow system) *Chondrites* (pronounced 'Kon-dry-tees'), from the Jurassic of Germany, which superficially resembles a plant.

Typically, trace fossils consist of burrows in soft sediment, borings into hard surfaces, tracks and trails across the sea-floor and terrestrial footprints. Trace fossils are an important source of palaeontological information, since they record the activity of organisms and so tell us something about their behaviour.

Behaviour and the mode of production of traces are in turn closely linked with ambient conditions. Moreover, most trace fossils are simply rearrangements of the sediment particles and cannot be transported without being destroyed (exceptions are borings in shells and pebbles which may survive transport and reworking), unlike many body fossils. Hence trace fossils have much to offer in the reconstruction of past environments.

Some sessile epifaunal animals can leave traces. The stalks of certain brachiopods, for example, penetrate hard surfaces (e.g. pebbles), leaving distinctive marks. Many trace fossils, however, reflect the activities of soft-bodied organisms, such as burrowing worms, which are themselves likely to be preserved as body fossils only in exceptional circumstances. Thus the body, and trace, fossil records are to some extent complementary.

Most of the preserved trace fossils in sedimentary rocks that were deposited underwater were made by infaunal animals, especially deposit feeders. This leads to problems when we try to find modern equivalents of ancient traces. Trails made today on sediment surfaces are easy to observe, especially in shallow water and intertidal areas, but these are liable to erosion and only rarely are fossilized. Infaunal traces have a far higher potential for preservation, though modern examples tend to be inaccessible. Various techniques have been developed to overcome this problem. One method is to inject quick-setting resin into burrow systems, but this has the disadvantage of only filling spaces: any filled areas, perhaps stuffed with faecal pellets, will remain undetected. Another method is to X-ray a block of sediment. For a considerable number of trace fossils, however, modern equivalents are not known. The material contrast between the trace and surrounding sediment, often slight in newly formed traces, is usually accentuated during fossilization. For this and other reasons, it is often easier and more informative to study fossil rather than modern traces.

Trace fossils are given generic and specific names for ease of reference, but it is essential to realize that they are not equivalent to genera and species of living organisms or their body fossils. An animal might produce different traces at different times, in which case each type of trace will be given its own name. Furthermore, two unrelated animals might have the same type of behaviour and so produce the same genus of trace fossil.

Finally, since many traces are originally formed as hollow burrows, bores or tracks excavated into sediment, the trace fossils that result are commonly moulds made in opposite relief by the sediment that later filled the hollows. Thus, when trails are cut into a muddy sea-floor and then buried by sand, it is the positive mould standing proud on the base of the resulting sandstone bed that is most commonly found, because the mudstone beneath tends to erode away.

You are now about to look at some distinct, easily identifiable examples of trace fossils. It is important to realize, however, that many trace fossils lack such diagnostic and identifiable features, and can only be classed as a 'burrow' or 'trail'. Furthermore, in some cases the sediment may be completely reworked by burrowers which destroy the primary sedimentary structures like cross-stratification. This churning of the sediment, known as **bioturbation**, leaves few clear individual traces. Nevertheless, even this effect is informative for interpreting ancient environments. Strongly bioturbated sediments indicate both well-oxygenated conditions, capable of supporting a rich infauna, and either relatively slow rates of deposition that allow thorough mixing of the sediments or a pause in deposition. Where there is a thriving infauna, but also more or less frequent current activity – e.g. as in many shallow marine settings – the changing balance of preservation of sedimentary lamination versus bioturbation can provide a detailed record of subtle variations in, say, depth of deposition, sediment influx or storm activity.

9.6.1 MINING BURIED FOOD IN QUIET ENVIRONMENTS

Some of the most interesting traces come from sediments deposited in fairly deep water, below the **photic zone**, which is the layer of water penetrated by sunlight. At these depths, there is no photosynthesis by algae, and food, in the form of dead plankton and other detritus raining down from shallower levels, is usually in comparatively short supply. Many of the traces, therefore, show regular patterns formed by the systematic search by deposit feeders through sediment for the food that has accumulated there, relatively undisturbed by currents. It is possible to replicate such regular traces by programming a computer with a series of simple commands, such as 'turn after a given distance', 'avoid crossing an earlier trace' and so forth (Figure 9.27). Note, however, that the original trace-makers were not rigidly programmed like a computer, and if a patch of sea-floor proved unproductive they stopped the systematic search and moved elsewhere.

Figure 9.27 Two trace fossils (left) and analogous computer-generated simulations (right).

Figure 9.28a shows a complex horizontal burrow system known as *Nereites* (pronounced 'Nerry-eye-tees'). This is a Palaeozoic trace, probably formed on or near the sediment surface as the animal meandered through the sediment. In this example, you are looking at the actual trail of disturbed sediment itself (rather than a mould) as seen on the top of a bed. It consists of a median zone with lobate edges, and each lobe contains fine curved lines. The lobate areas are interpreted as patches of sediment searched for food particles and then back-filled, while the central groove appears to have been filled with a faecal string of sediment that has passed through the body (Figure 9.28b). *Nereites* evidently represents a trail produced by a deposit feeder. But what was the original animal like? As with many trace fossils, the nature of the original animal remains mysterious. In the case of *Nereites*, it has been suggested tentatively that some form of worm or mollusc was responsible but no firm evidence is available.

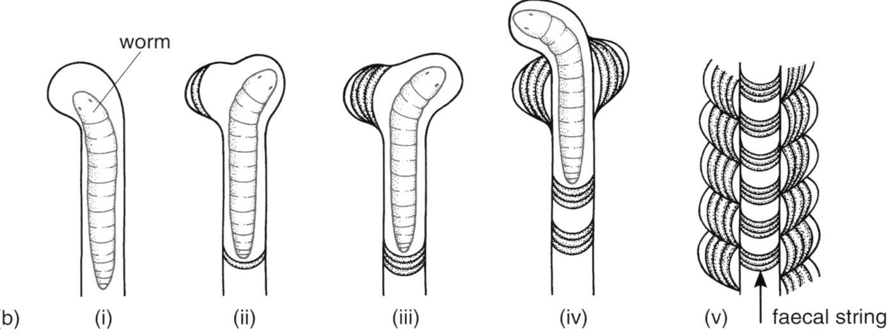

Figure 9.28 (a) The trace fossil *Nereites* (specimen is *c*. 7 cm long). (b) Possible mode of formation of *Nereites*, seen from above. A soft-bodied animal (?worm) excavates areas on each side of its anterior end, sorts sediment for food and replaces the processed sediment in the lateral excavations. As the animal crawls forward and repeats this process, a faecal string fills the burrow behind.

For an example of extreme regularity in a trace fossil, turn to FS N. It is the trace fossil *Paleodictyon*, which forms a characteristic honeycomb-like network (the finer network visible on the slab – the larger and more irregular network represents the activities of another animal). Unlike *Nereites,* in which the trace was formed by being packed with sediment, *Paleodictyon* was originally a system of hollow tubes beneath the muddy sediment surface (Figure 9.29). However, it is now preserved in positive relief on the base of an overlying bed of sandstone, with the tubes filled with sediment from above. How did this filling occur? To understand the process, recall what you learned about turbidity currents in Section 4.3.3. One interpretation is that *Paleodictyon* was formed in mud just beneath the sea-floor, with short vertical tubes connecting the

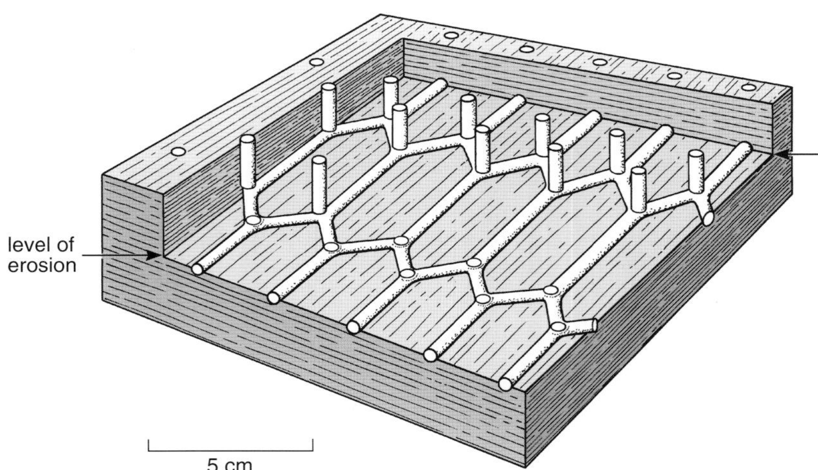

Figure 9.29 Interpretation of the trace fossil *Paleodictyon*. Erosion by a turbidity current to the arrowed level destroyed the vertical tubes leading to the sediment surface and exposed the hexagonal network which was then filled with sediment.

hexagonal network with the overlying seawater. A turbidity current flowed across the sea-floor and scoured away the roofing mud and vertical tubes, laying open the trace. Coarser-grained sediment from the turbidity flow then filled the tubes to give an internal mould of the original trace on the underside of the turbidite bed.

Question 9.17 Inspect FS N with this model in mind. Notice how one end of the *Paleodictyon* network stands proud on a slight bump on the surface of the slab, ending along a well-defined boundary, while the other end is more subdued, seemingly fading into the slab. How might you explain these differences in relation to the model described above? (*Hint*: remember that you are looking at relief on the *underside* of a turbidite; you might find it helpful to hold the slab with the trace fossils facing down, and to look up at them from beneath, in order to imagine the processes of scour and fill that took place.)

As with *Nereites*, the maker of *Paleodictyon* is not known. The net-like pattern is often interpreted as the result of a systematic search by a deposit feeder for food. More recently, however, it has been suggested that although the net-like form originally arose from the methodical search of sediment, it was later exploited as a sort of microbial 'garden'. The food material that drifts down into the deep sea has already been stripped of much of its available organic content and consequently the remaining material is difficult for deposit feeders to digest. Bacteria and other microbes living in the *Paleodictyon* network, however, may have been able to accelerate breakdown of such material to the benefit of the deposit feeder. This is not a far-fetched idea, as some modern marine burrowing crustaceans are also known to be 'composters'.

Both the *Nereites* and *Paleodictyon* specimens were formed in deep-water deposits of predominantly fine sediments with influxes of coarser-grained material in turbidity flows. Their morphology reflects the need for efficient exploitation of scarce food supplies.

9.6.2 Feeding in current-swept environments

The remaining trace fossils to be studied were probably formed in shallower water than *Nereites* and *Paleodictyon*, and furthermore can be attributed reasonably to known organisms. Look first at FS O. This specimen again shows the underside of a sandstone bed, so that the features projecting from its surface are the fills of depressions in the underlying mud. In this case, however, we are not looking at the sand fills of pre-existing traces exposed by scour, but instead those formed at the interface between mud and sand layers by animals that had burrowed down through the sand to this level.

❑ Study the flatter areas around some of the protruding trace fossils on FS O. Can you detect any evidence for the overlying sand layer itself having been disturbed by the burrowers? Again, it may help to look up at the slab from beneath to answer this question.

■ Here and there, little oval depressions can be seen on the surface of the slab, similar in size and outline to the pod-like projections. These depressions would appear to represent spots where the projections have broken off, implying connection of the latter with tubes of sediment of somewhat different consistency (i.e. burrows) passing through the sandstone, and now weathered out as shallow pits with well-defined edges.

The most obvious feature in FS O is the large five-rayed trace.

❑ Casting your mind back to the discussion of bodyplans in Section 9.4, of which phylum is this trace-maker likely to have been a member?

■ The five-fold symmetry of the trace strongly hints at an echinoderm.

There can be little doubt that this trace was formed by the digging movements of a starfish – a member of another echinoderm class, the **Asteroidea**. In some cases, successions of these traces are found one above the other; they represent the escape of the starfish from an influx of sediment. The coarse striations arising from the central zone along each arm were formed by the digging action of the tube feet, which ran along the underside of each arm.

The other traces on this slab are the numerous pod-like projections mentioned above, where the sand has filled excavations made by infaunal animals.

❑ Recalling what you learned in Section 9.3 about the shape, locomotion and habits of various burrowing animals, suggest what sort of creature might have made these pod-like traces?

■ Bivalves are the most likely candidates. In burrowing, infaunal bivalves form a pod-shaped chamber in the sediment (Figure 9.8b). When the bivalve leaves this, it is filled with sediment from above to produce the trace fossil.

These pod-like trace fossils are called *Lockeia*. Every bivalve trace cannot have been simultaneously occupied, as some traces cut across others. Some may have been formed at the same time as the starfish trace, but others must have been made earlier or later than the starfish trace, as they either are cut by, or cut across it. Whether the bivalves were suspension feeders (most likely), or surface detritus collectors, their food was taken directly from above the sediment surface, unlike that mined by the makers of *Nereites* and *Paleodictyon*. Given the numbers of traces here, evidently made over a period of time, there must have been ample replenishment of food particles at the surface, and this in turn suggests some current activity, either transporting material in or resuspending detritus from the sediment itself.

The association of traces formed by starfish and bivalves is interesting because many starfish prey on bivalves either by ingesting them whole, or by wrapping their arms around them and pulling the valves apart using the tube feet as suckers to grip them with. Fossil assemblages comprising distinct clusters of bivalves and starfish, with some of the latter still wrapped over the former, have also been found.

FS P is a Cambrian example of a trace fossil known as *Cruziana*. It represents a crawling trace. The original trace was a furrow in the muddy sea-bed, but the trace fossil is again preserved as a mould of somewhat coarser-grained sediment filling the hollow. Much of the trace consists of pairs of scratch or rake-like marks that come together to form a chevron pattern.

❑ Given this information, which kind of organism represented in the Home Kit is most likely to have produced this trace?

■ A trilobite (FS G), which possessed numerous pairs of jointed legs (Figure 9.23) well-suited for scratching and raking as the animal ploughed through the topmost layer of sediment.

The filling of the original furrow by coarse-grained sediment, which must have been carried in a relatively fast current, raises the question of why the trace itself was not eroded by the current. By looking at Figure 4.8 in Section 4.2, you will see that fine-grained sediment is cohesive, and needs high speed currents to erode it. Thus it may still be resistant to erosion, even in a current that may carry coarse-grained sand. Furthermore, experiments have shown that if the mud loses water and becomes even more consolidated, its resistance to erosion further increases. Thus, the mud over which the animal moved must already have been somewhat consolidated, perhaps through compaction beneath a subsequently eroded layer of sand.

In forming *Cruziana*, the walking legs of the trilobite swept backwards and inwards so that the chevrons point in the opposite direction to that of movement. The angle formed by the two sets of striations is not constant; it may vary from about 50° to 140°. The variation in the angle is probably related to the speeds of locomotion, smaller angles being produced at somewhat higher speeds. Much of this specimen of *Cruziana* is occupied by the scratch marks of the walking limbs. There are, however, two sets of fine 'wispy' structures in a few areas along the edges of the chevron pattern, just within the sharp ridges that run down its sides.

❑ What part of the trilobite body could have made these markings?

■ Possibly the gill branches (Section 9.3.8) as they dragged over the sediment.

In other specimens of *Cruziana*, the marks made by the gills form a well-defined zone on each side of the scratch marks. The other notable feature in FS P is the ridge running along each side.

❑ Remembering that each ridge marks an original furrow in the underlying mud, which parts of the trilobite body could have produced the grooves (taking FS G as a model)?

■ The most likely parts are the spines that trail from the posterior corners of the head-shield.

Note, however, that FS G is a Silurian species of trilobite and the trace is Cambrian. The original trilobite that produced the *Cruziana* in FS G would certainly have been from a different species.

While studying *Cruziana*, you may have noticed at one end of the specimen a ribbed oval area with a coarse unfurrowed texture. This is also a trace fossil. The same animal made both this trace and *Cruziana*, but because the traces differ, the oval trace is given a separate name. It is called *Rusophycus* and was made while the trilobite rested on the sea-floor. Imagine that the walking legs, instead of striking backwards and inwards to produce the raking marks of *Cruziana*, scooped together to excavate an oval hollow, into which the trilobite settled (Figure 9.30). FS P is highly unusual in that *Rusophycus* and *Cruziana* are directly associated, whereas in nearly all other cases they occur separately.

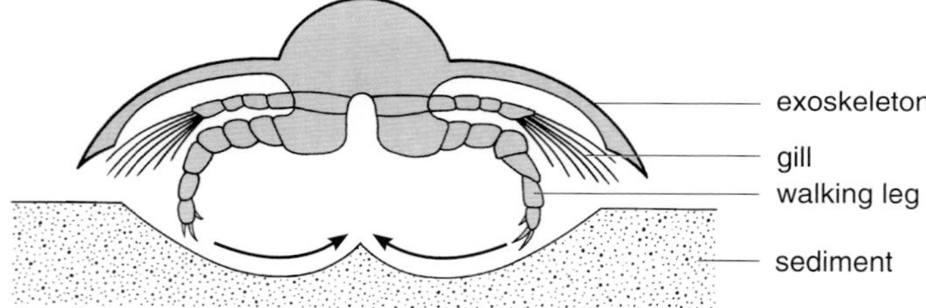

Figure 9.30 Diagrammatic section across a trilobite excavating a resting hollow (*Rusophycus*).

Question 9.18 What sequence of activity would you now deduce from FS P?

The *Cruziana* in FS P was clearly made by a trilobite that had its body close to the sediment surface, with the limbs excavating a pronounced trough and the head spines being dragged through the sediment, leaving a pair of grooves. One possible explanation is that it was ploughing through the overlying layer of sand that we postulated earlier, with its body pressed down onto the top surface of the mud, and that the sand was later washed away. In contrast, there is also evidence that some trilobites walked with only their legs on the sea-floor and the body held well clear, so producing a trace that consisted of a paired array of scratch marks where only the ends of the legs had come into contact with the sediment.

We have stressed how *Cruziana* and *Rusophycus* were apparently made by trilobites. However, the body fossil record shows that trilobites declined in numbers during the late Palaeozoic and the class finally became extinct during the Permian. Well-preserved *Cruziana* have, however, been described from the Triassic of Greenland. It is thus likely that not all *Cruziana* were made by trilobites, although the apparent abundance of trilobites in the Lower Palaeozoic suggests that most *Cruziana* specimens there were formed by trilobites. This example emphasizes again that almost identical traces can be made by entirely different animals.

Some trace fossils reflect the migration of burrow systems through the sediment. Figure 9.31a shows a U-shaped burrow known as *Rhizocorallium*. The burrow lay almost parallel to the sea-floor and was connected to the overlying seawater via two more steeply inclined galleries (Figure 9.31b).

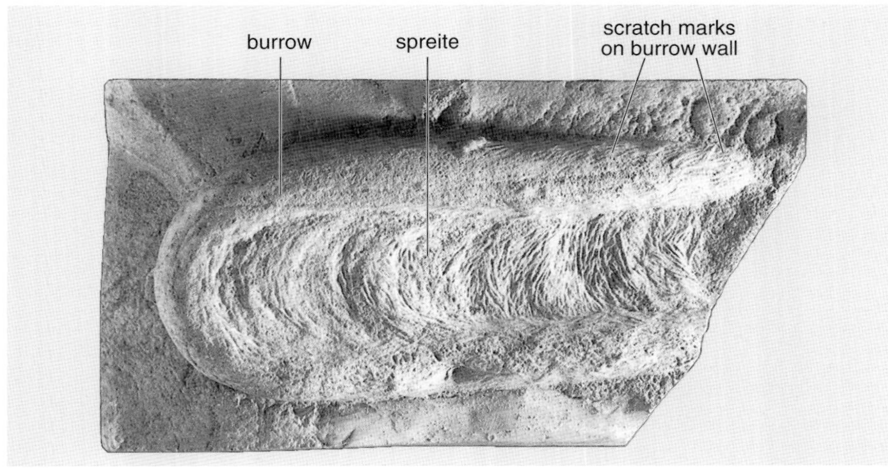

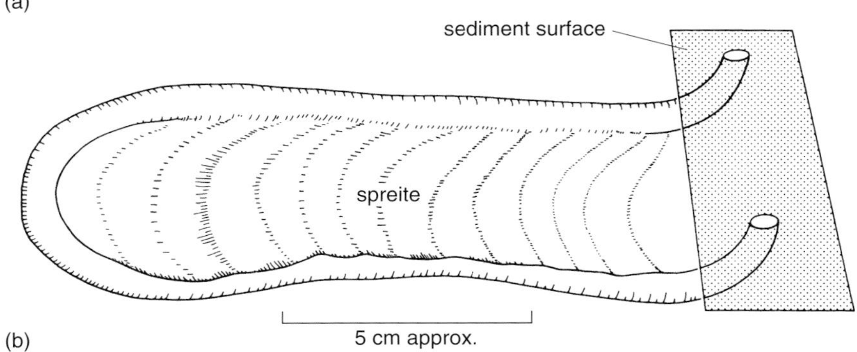

Figure 9.31 (a) The trace fossil *Rhizocorallium*. (b) Diagram showing the relation of the burrow to the sea-bed, seen obliquely from above.

The U-tube was originally hollow, but has subsequently been filled. What sort of animal constructed this trace? A clue comes again from the scratch marks along the tube walls, where the burrow-maker had evidently scrabbled material away. As with *Cruziana*, such scratches are likely to have been made by another stiff-legged arthropod, and in the case of *Rhizocorallium* probably by a shrimp-like creature.

Between the sides of the tube is a web-like area known as a **spreite** (pronounced 'spryter'; plural: spreiten, from the German for the web between the toes of a duck), filled with a series of curved markings. These markings formed as the burrow was extended, with the animal scratching away the outer surface of the U and plastering layers of sediment against its inner surface. The animal that built *Rhizocorallium* probably occupied the burrow for some time. It has been suggested that the animal was a deposit feeder while it excavated the sediment to form the burrow and, having exhausted this food supply, was thereafter a suspension feeder drawing in water through one of the openings.

Another spreite-bearing trace fossil is *Diplocraterion*. This trace fossil is similar to *Rhizocorallium*, but here the U-tube is vertical. The spreiten of *Diplocraterion* are believed to reflect the animals' responses to sediment deposition or erosion rather than their feeding behaviour. Downward movement, in response to erosion of the overlying sediment, resulted in a spreite between the two branches of the U-tube, whereas upward movement following sediment deposition resulted in a spreite beneath the tube (Figure 9.32). Such structures must have been associated with fairly extensive current-driven fluxes of sediment as may occur, for example, in tidal deposits. They also tend to show some alignment with presumed currents. By studying sequences of the two kinds of *Diplocraterion*, it is possible to reconstruct the history of sediment influxes and erosion.

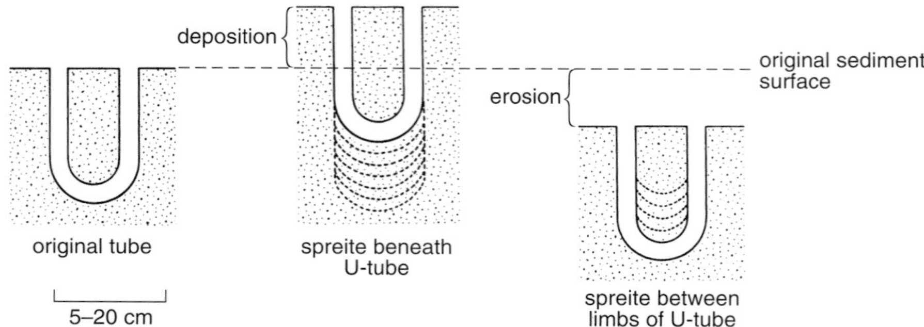

Figure 9.32 Spreiten in the trace fossil *Diplocraterion* formed in response to deposition and erosion.

9.6.3 TRACES AND WATER DEPTH

You have just studied some traces made in relatively deep water (*Nereites*, *Paleodictyon*) and in shallow water (*Cruziana*, *Lockeia*, *Rhizocorallium*, *Diplocraterion*). The relationship with water depth arises not because of any direct response by the trace-makers to depth *per se*, but as a consequence of their responses to environmental conditions that tend to vary with depth. The nature of the food supply is a major factor, though this in turn is affected by, for example, nutrient supply, light levels and current activity. Because of variations according to local circumstances, it is not usually possible to put absolute values on the water depths inferred. By looking at Figure 9.33, however, you can assign some of the trace fossils you have studied to broadly defined zones of water

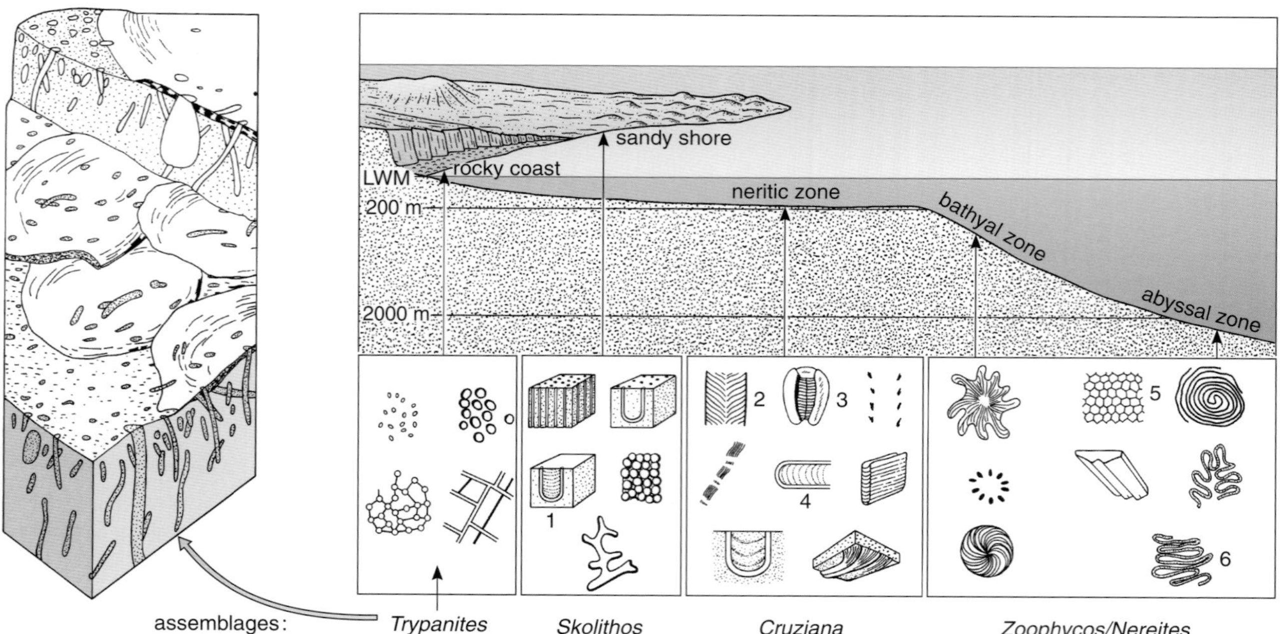

Figure 9.33 Assemblages of trace fossils and their interpreted relation to environment, especially relative water depth. Each assemblage is named after a characteristic trace fossil. Trace fossil examples that you should be able to recognize are: 1, *Diplocraterion*; 2, *Cruziana*; 3, *Rusophycus*; 4, *Rhizocorallium*; 5, *Paleodictyon*; and 6, *Nereites*. Arrow at left points to detail of *Trypanites* assemblage borings into rock surfaces.

depth. Each suite is named after a characteristic trace fossil, but note that the actual genus, e.g. *Cruziana*, does not have to be present for the depth zone to be recognized, nor is it necessarily confined to the depth zone; the name is just a convenient label for a given commonly observed association. The differences between the suites of trace fossils are largely connected with feeding habits, as discussed above. In shallow water, where plankton and detritus re-suspended by frequent current activity are abundant, suspension feeders commonly predominate, often producing vertically oriented burrows. In the quieter and muddier sediments of deep water, by contrast, particulate edible material is mostly incorporated in the bottom sediments, where thriving bacteria re-process it, often enhancing its food value to consumers. Here, deposit feeders tend to dominate. In the deepest water, many traces reflect a methodical search for food, usually with predominantly horizontal burrow systems.

The foregoing discussion of trace fossils and water depths is generalized. *Not all deep-water traces are complex, nor are all shallow-water traces simple.* Lagoonal muds deposited in shallow but quiet conditions may be dominated by deposit feeders capable of making relatively complex traces. *Nereites*, for example, has been recorded in such deposits associated with coals (formed from peats). In contrast, localized areas in the deep sea (as you will see in Section 17, when you read about submarine fans) may be subject to relatively rapid deposition and reworking of sediment, and 'shallow-water'-type suites of trace fossils have been found in ancient deposits from such environments. Therefore, when studying trace fossils it is essential to study entire assemblages rather than particular specimens, and to test your interpretations for consilience with other depth-related (e.g. sedimentological) criteria.

9.7 SUMMARY OF SECTION 9

Shelly marine invertebrates yield the commonest fossils. Eight frequently encountered classes, in five phyla, are represented by fossil replicas (FS) A–M in the Home Kit:

(1) Molluscs (including bivalves, gastropods and cephalopods) are characterized by a dorsal calcareous shell (providing protection) and a ventral muscular foot (assisting locomotion), with a mantle cavity between the body and the shell, housing gills.

In bivalves (FS B and I), the shell is divided into two enclosing lateral valves that are opened by a dorsal ligament and closed by adductor muscles (with shear between the valves prevented by hinge teeth and sockets). Most species have externally symmetrical valves, and burrow with a blade-shaped foot, circulating water via siphons from above the sediment surface, for suspension feeding on expanded gills. There are also epifaunal and boring forms, as well as deposit feeders.

Gastropods (FS A and K) have a broad creeping foot, which can be retracted into the shell. The shell is typically coiled asymmetrically over the back of the foot, as a consequence of torsion. Some forms retain a pair of gills, and water enters on either side and is exhaled between them, via an indentation of the outer shell margin. But most have one gill, and water enters, and exits, at opposite ends of the aperture; the inhalent site may then be extended as a siphon, housed in a shelly scroll. Most gastropods are grazers, though there are also detritus collectors and scavengers or predators.

In cephalopods (FS D), the foot is wrapped forward under the body, to form (in part) a funnel for water squirted from the mantle cavity, enabling swimming by jet propulsion. The shell, where retained externally (e.g. in *Nautilus* and the extinct ammonoids), is chambered, serving as a gas-filled buoyancy device.

Grasping tentacles around the mouth allow the capture of prey (large and/or microscopic), and scavenging.

(2) Echinoderms (including echinoids and crinoids) have an enclosing endoskeleton consisting of numerous calcite plates, each of which is a single crystal with a microscopic spongy texture. A water vascular system has radial arms that support exposed flexible tube feet, used in feeding and/or locomotion. Most echinoderms show a five-fold arrangement (if not symmetry).

Echinoids (FS C and J) have a globular test, consisting of interlocking plates, covered in spines that are articulated on tubercles on its outer surface. Tube feet project from rows of pore pairs on five radial sectors (ambulacra) of the test. Regular echinoids show five-fold symmetry, with the mouth centred on the lower surface and the anus on top. They are epifaunal grazers to predators. Irregular echinoids show bilateral symmetry, with relocation of the mouth and anus, and are detritus collectors.

In crinoids (FS H), the mouth faces upwards, with the anus alongside, surrounded by a ring of branching arms supporting the tube feet. The latter trap suspended food particles, and pass them in to the mouth. Fossil forms were commonly attached to the sea-floor by a stalk, though most living forms are stalkless.

(3) Brachiopods (FS F and L) have a two-valved calcite shell, but differ from bivalves in growing them dorsally and ventrally, with a plane of bilateral symmetry bisecting each valve. Muscles control both opening and closure. Articulate brachiopods are hinged posteriorly by teeth (in the ventral valve) and sockets (in the dorsal valve), features lacking in inarticulate brachiopods. Most are (or were) attached to the sea-floor by a horny stalk emerging from a hole in the umbo of the ventral valve, though unattached and cemented forms are also known. Suspension feeding is by means of a tentaculate feeding organ housed within the mantle cavity.

(4) Cnidarians most notably include corals (FS E and M). Each individual (polyp) builds a cup-like calcareous shell, divided internally by radial walls (septa), which lie between infoldings of the polyp's internal cavity. The polyp (like an anemone) has a ring of stinging tentacles, which trap microscopic prey and pass them to the single orifice of the internal cavity. Many living examples have symbiotic single-celled algae in their tissues, which allow the hosts to thrive in waters with low nutrient levels. Corals may be solitary, or, as a result of budding, colonial.

(5) Arthropods are the largest animal phylum, mainly through the inclusion of insects, though also including the extinct trilobites (FS G). The bilaterally symmetrical exoskeleton of the latter consists of numerous articulated segments, and was periodically moulted. Its calcified dorsal part comprises a head-shield (bearing compound eyes), a central trunk region and a tailpiece, all divided longitudinally into three lobes. The unmineralized underside of the animal included numerous paired jointed limbs, flanking a central channel along which food particles (detritus or bits of prey) were passed forward to the mouth.

The feeding/habitat classification of the specimens from the Home Kit (the completed version of Table 9.2) is shown in Table 9.5.

Other important kinds of body fossils include the resistant remains of graptolites (extinct colonial suspension feeders, mainly planktonic), vertebrates, land plants, calcareous marine algae, sponges and microscopic foraminifers as well as accumulations of laminated carbonate sediment trapped by microbial mats (stromatolites).

Trace fossils are usually named to indicate particular types of behaviour, not the organism that made them. They reflect environmental conditions, and inferences from marine examples can thus be linked with water depth.

Table 9.5 Combined feeding/habitat classification for marine animals (completed Table 9.2).

Habitat		Feeding type					
					deposit		
		predator	scavenger	grazer	swallower	collector	suspension
pelagic	nekton	?D					?D
	plankton						
epifaunal	vagrant	(?C)		CK		G	
	sessile	EM					FLH
infaunal	vagrant	A				J	(I) (B) ↑ ↑
	sessile						↓ ↓

Meandering (usually more or less horizontal) trails/burrows, reflecting systematic deposit feeding and/or 'microbial gardening', tend to characterize deeper, more offshore deposits (e.g. *Nereites* and *Paleodictyon* (FS N)). Simpler traces (often vertically extended), reflecting a variety of behaviours including responses to rapid sediment gain or loss, tend to characterize shallower waters (e.g. starfish excavations and *Lockeia* (FS O), *Cruziana* and *Rusophycus* (FS G), *Rhizocorallium* and *Diplocraterion*). Trace fossils are not foolproof evidence of water depth, however, as exceptions occur. They must be interpreted in conjunction with other sedimentological data.

Activity 9.5

To test your understanding of Section 9, you should now do Activity 9.5, which asks you to observe and interpret fossils and related material within the Sedimentary Environments Panorama on DVD 3.

9.8 Objectives for Section 9

Now you have completed this Section, you should be able to:

9.1 Make simple but accurate drawings to record your observations of given fossil specimens.

9.2 Relate recorded observations of fossil specimens to information supplied on the biology of selected organisms, so as to interpret aspects of the palaeobiology of the specimens in question (especially their habitats and mode of feeding).

9.3 Compare and contrast body fossil specimens from the major groups of shelly marine invertebrates, and classify them to the level of class, at least, according to the homologies of their bodyplans.

9.4 Interpret, in broad terms, the likely modes of formation of given trace fossil specimens, and the conditions in which they probably formed.

10 FOSSILIZATION

Consideration of how the processes of fossilization affect the record of past life is crucial in reconstructing ancient environments. You will occasionally need to refer to some of the Home Kit fossil replicas while reading through this Section, so you should have them somewhere easily to hand.

How do fossils form? This is a deceptively simple question. In principle, fossilization involves eventual burial, either in sediment, or more rarely by other means, for example through igneous processes such as volcanic ash falls or even lava flows. But the routes to fossilization normally entail many complex, and in some cases incompletely understood, processes. Normally, the first thing that happens is that the soft parts decay or are eaten, except in very rare cases where extraordinary conditions permit their preservation. The remaining hard parts may themselves undergo change, perhaps involving recrystallization or dissolution of the original skeletal material, replacement by a new mineral, deformation or even complete destruction.

The various processes that lead to the formation of fossils fall under the general title of **taphonomy** (which means 'tomb law' in classical Greek). This is a field in which enormous strides in understanding have been achieved in recent years, including work on the highly publicized issue of 'fossil DNA'. An understanding of taphonomy not only identifies the kinds of biases that must be allowed for in interpreting the fossil record, but it can also make its own direct contribution to the business of reconstructing past environments, as you saw earlier with the example of the mode of preservation of the Permian fish fossils discussed in Section 7.

The processes that occur between the death of an organism and its discovery as a fossil will be considered here in two stages: (i) those that occur between death and final burial; and (ii) those that occur between final burial and before discovery, due to diagenesis, which embraces all the physical and chemical processes occurring in a sediment after its deposition, but excluding metamorphic processes. The emphasis on *final* burial is important, since the remains of organisms may be buried and exhumed several times before eventual long-term burial in a sedimentary succession.

10.1 FROM DEATH TO FINAL BURIAL

10.1.1 DEATH

Although the great majority of body fossils do record the deaths of organisms, there are exceptions.

❑ List some instances.

■ Arthropod moults, shed from live individuals, were discussed in Section 9.3.8. Other examples you may have thought of include teeth, scales, feathers and other such expendable items shed by animals, as well as leaves, flowers, and even branches, from plants.

Biological causes of death include predation, disease, shortage of food or simply old age. For some species, predation wields by far the largest scythe. Many common bivalves, for example, are favourites on the menu of numerous predators, some of whom leave distinctive traces on the shells of their victims, allowing the palaeontological sleuth to detect 'whodunit'.

❑ Cite an example illustrated earlier.

- ■ The distinctive drill hole of a gastropod predator on a bivalve was illustrated in Figure 9.16b.

Physical causes of death include severe climatic conditions (drought, flood, frost etc.), sudden burial under a thick layer of sediment, or changes in salinity, oxygen level and/or other vitally important environmental variables. Such events may affect different species unequally, though occasionally an entire fauna may be virtually annihilated in a **mass mortality**, which refers to the simultaneous killing of huge numbers of individuals through some common cause. **Catastrophic burial** by sediment (or sometimes volcanic ash or lava) is one important mechanism of mass mortality. Other mechanisms include poisoning by volcanic gases, upwelling of seawater depleted in oxygen and often also containing toxic hydrogen sulfide, the formation of 'red tides', where blooms of a certain kind of single-celled alga produce toxins, and, more recently, man-made pollution events such as oil spills and atomic bomb tests.

Catastrophic burial is of especial interest to palaeontologists. Sessile animals such as brachiopods and crinoids are almost certainly doomed to be smothered in such a circumstance and will thus be preserved where they lived. If the covering of sediment is not too thick, some mobile animals, especially burrowers, may be able to escape.

- ❑ What record of their presence would nevertheless probably remain?
- ■ Their escape behaviour would be likely to leave trace fossils behind (Section 9.6.2).

Greater thicknesses of deposited sediment (say, half a metre or more), perhaps the result of a major storm or turbidity current, may prevent animals escaping altogether. Some may be preserved in the position in which they lived, while some make futile movements before expiring. The catastrophic burial of a fauna is of particular palaeontological interest for three reasons:

1 It is geologically instantaneous, preserving an assemblage of organisms that lived together at the same time, whereas in sediments that accumulate gradually, thousands of generations may stock even a thin bed. Catastrophic burial is thus of great value in reconstructing past ecological relationships.

2 Some, especially sessile, individuals, may be preserved exactly where they lived, if not in their living orientation. Examples include crinoids and other echinoderms attached to the sea-floor (Figure 10.1), and bivalves still in their burrows.

3 Rapid burial will help to keep the remains of organisms intact, before they rot and fall apart, especially in the case of those animals with many skeletal elements.

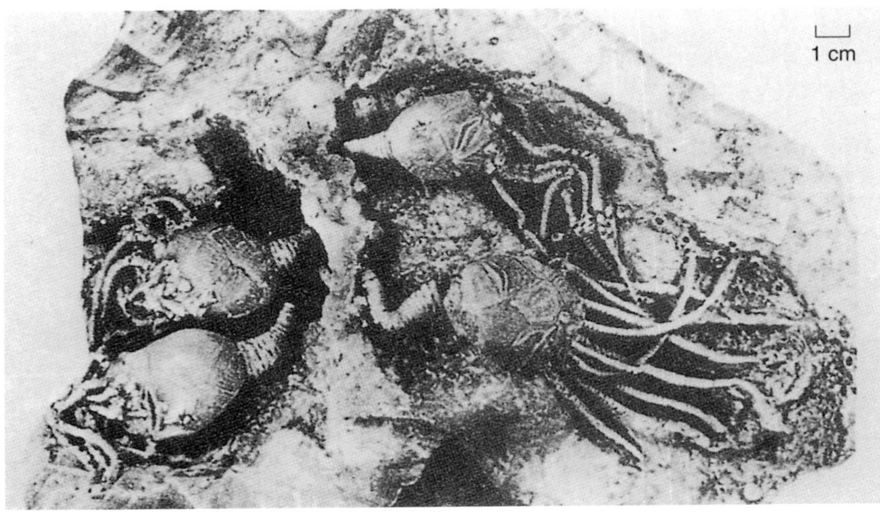

Figure 10.1 Complete fossil primitive stalked echinoderms (called cystoids, distantly related to crinoids), attached on an ancient erosion surface, preserved by catastrophic burial.

Question 10.1 Which of the common fossil marine invertebrates that you met in Section 9 would have been prone to such break-up, unless catastrophically buried?

Infaunal organisms might, of course, die in their burrows and hence stand a chance of intact preservation without the need for catastrophic burial. This is probably the main reason for the relatively better fossil record of the infaunal irregular echinoids than of the epifaunal regular echinoids.

Although rapid deposition of sediment is a frequent cause of death, rapid erosion is another important factor. Once again, sessile animals are particularly vulnerable; they can be ripped from their moorings. Even burrowers may not be able to escape. Downward-digging bivalves, for instance, may still be exposed if they meet an impenetrable layer. Such erosion frequently goes hand in hand with rapid burial in the eventual preservation of specimens. Thus, animals may be washed out of their burrows, transported with, and then reburied in, the originally surrounding sediment.

❑ How would you expect bivalves to be preserved in such a circumstance, in terms of completeness and orientation?

■ If still alive when buried, their shells would be tightly closed, and would be preserved thus. Unless they attempted to re-establish themselves, however, they would not be preserved in their characteristic living position.

All these scenarios involving erosion and/or sudden burial typically accompany storm activity on shallow marine shelves.

10.1.2 Decay and disintegration

Death, when not involving complete consumption by predators, is followed by bacterial decay and sometimes disintegration caused by various scavengers. Bacteria that cause decay can be found in all marine environments, but are most numerous in the topmost layers (cm to tens of cm) of the sediment. Deeper down, their numbers decline, so organisms buried under an appreciable thickness of sediment may decay more slowly. Decay may be either **aerobic** involving the consumption of oxygen in the breakdown of organic molecules (analogous to burning in air), or **anaerobic** when the breakdown of organic molecules is carried out without the consumption of oxygen. Different bacteria specialize in each. Indeed, aerobic bacteria cannot survive in the absence of oxygen, which is, by contrast, toxic to anaerobic bacteria. In anaerobic decay, some bacteria generate hydrogen sulfide (H_2S) and methane (CH_4).

The distinction between aerobic and anaerobic decay is important since these processes have different consequences. Aerobic decay is rapid, yielding readily lost products (mainly CO_2 and H_2O), and in warm temperatures the soft parts may entirely disappear in a few days. At lower temperatures and/or conditions of high salinity (or desiccation), however, aerobic decay may be slowed. Anaerobic decay is usually not as rapid as aerobic decay, but more importantly it may sometimes yield finely crystalline products that can permeate the decaying remains and even lead to the fossilization of soft tissues. We will return to such exceptional preservation in the discussion of diagenesis (Section 10.2). Conditions of anaerobic decay may arise in several ways. Some bodies of water, such as the Black Sea, are permanently anoxic for a variable distance above the sea-floor, because of restricted circulation. Organisms that sink or are transported into this anoxic zone are often well preserved.

❑ Recall an instance of such a style of preservation from what you read earlier.

■ The environment in which the Permian fish discussed in Section 7 became fossilized seems to have been like this.

Even in oxygenated conditions, burial by sediment may lead to anaerobic decay as the oxygen is rapidly used up by the bacteria. If you turn over a spadeful of mud on a sheltered stretch of coastline, you will commonly find that, although it is rust-brown to yellow at the surface, the sediment is dark grey to black just a few centimetres (or less) down. The blackness is due to finely dispersed pyrite (FeS_2) and related chemical compounds produced by the reaction of ferrous (Fe^{2+}) ions and hydrogen sulfide (H_2S). The hydrogen sulfide itself is a product of anaerobic respiration involving the reduction of sulfate ions (SO_4^{2-}, Equation 10.1). The blackness thus signals the presence of anoxic water in the sediment. A similar process may follow a mass mortality, when oxygen is soon depleted by the initially aerobic decay of the excessive concentration of organic matter. Thus in many cases anaerobic decay may follow aerobic decay:

$$SO_4^{2-} + \underset{\text{organic matter}}{2CH_2O} + 2H^+ \longrightarrow H_2S + 2CO_2 + 2H_2O \qquad (10.1)$$

Another important aspect of decay, especially in oxygenated conditions, is the production of gas bubbles which, if trapped within the body, may allow it to float. Decay continues during flotation and parts may be scattered over a wide area of the sea-floor before the remains sink to the bottom. The destruction of soft parts is also hastened by scavengers, which often congregate around a corpse, such as crustaceans that tend to tear the corpse and scatter the remains. This is a particularly important mode of spreading remains around in quiet-water environments where there is little or no dispersal by water currents.

There are distinct stages in the disintegration of a corpse, some of which have recognizable equivalents in the fossil record that indicate how far decay went before burial. For example, the decay sequence in a modern echinoid has been observed to include the shedding of the spines by day 7 after death and the dropping out of the small plates around the anus by day 12. Such taphonomic clues can of course be turned around to provide insights on rates of sediment flux on the sea-floor.

So far, we have concentrated on the effects of decay on soft tissues, but skeletal materials contain subsidiary amounts of organic material, too, which can eventually rot. Unless diagenetic mineralization comes to the rescue, so to speak, to bond the crystalline components together, the hard parts will themselves crumble.

Plant material is also subject to decay. The principal structural materials in plants include cellulose, which surrounds all plant cells as cell walls, and lignin (in woody tissues). Despite their apparent toughness, they are readily attacked by a variety of organisms, especially fungi and some bacteria. The striking exception to this is a substance known as sporopollenin, which forms the coat of spores and pollen. Sporopollenin is remarkably resistant to decay and this explains why fossil plant spores may be found so abundantly. Certain kinds of coal, for example, consist to a large extent of flattened spore cases.

10.1.3 FOSSILIZATION POTENTIAL

From the above, it follows that different organisms have different chances of leaving fossil evidence of their existence – i.e. different **fossilization potentials** – depending not only on their anatomy and way of life but also on where they lived and died. You may recall from Section 9.1 that the fossil record is dominated by marine organisms (a bias reflected in the Home Kit specimens) because the chances of permanent burial are greater in the sea than on land.

Clearly, in most circumstances soft-bodied animals are much less likely to be fossilized than those with hard parts. Nevertheless, even the fossilization potential of hard parts is far from uniform. In the case of marine organisms, two of the most important factors are their position relative to the sea-floor and whether the skeleton is in a single piece or made of many elements (Question 10.1).

Question 10.2 Refer back to the information you have recorded in Table 9.2 (or the completed version in Table 9.5) on the living positions of the Home Kit specimens, and consider the life habits of the organisms represented by FS B, G, H, K and J, together with the construction of their skeletons. Comment on the relative potential for the intact preservation of their hard parts under normal circumstances (i.e. *without* catastrophic burial).

Having considered the likely fates of the bodies and skeletal components of different organisms, we must now turn our attention to the biological, physical and chemical processes that affect the skeletal material itself.

10.1.4 Biological attack

Much of the breakdown of skeletal material may be caused by predators chipping or entirely smashing shells. This usually results in angular shell fragments (Figure 10.2a), unlike the sustained battering meted out by current transport, which tends to round them (Figure 10.2b). An easy trap to fall into, when in the field, is to attribute all scattering and breakage of shells to current activity; remember always to be cautious of such a supposition and to consider the alternative possibilities of predation and scavenging.

Predation, however, does not necessarily result in shell destruction. Some predators swallow whole animals, digest their soft parts and then eject the empty shells, whereas others extract the soft parts without damaging the shell.

Another important form of biological attack is the invasion of the mineral material of the shell by boring organisms (Figure 10.2c). Although this mainly affects the shells of dead organisms lying on the sea-floor, it can also seriously damage the shells of living epifaunal organisms. The most common borers include sponges, algae, and fungi. Some of these borers may derive

Figure 10.2 Taphonomic factors contributing to the breakdown of shells: (a) predation and scavenging (in this case, of mussels, by seabirds); (b) shattering and rounding by currents; (c) boring; (d) dissolution by acidic groundwaters (here the shells were incorporated in a coastal soil). Field of view a few cm across in each case.

nourishment from the organic matrices of the shell material, but others appear to use their borings simply as protective refuges. Extensive boring weakens the host shell and hastens its destruction, both through physical abrasion, and through further attack by grazers (such as many regular echinoids) rasping away to gain access to the boring organisms.

10.1.5 Physical effects

The principal physical effects prior to final burial result from the transport of skeletal remains over the sea-bed and their fragmentation. Animal skeletons that separate into two or more pieces, such as the shell of a bivalve or the plates of a crinoid, often show variable behaviour during transport. Moreover, bivalves with holes such as borings drilled by gastropod predators (as in Figure 9.16b) may respond to currents differently from unbored shells.

Transport of shells often also results in them coming to lie with a preferred orientation. This is best seen in very elongate shells, such as the Jurassic belemnites (an extinct group of squid-like cephalopods), where the degree of preferred orientation may be very pronounced (Figure 10.3). Usually, the shells end up with their long axes aligned more or less parallel with the main current direction.

Figure 10.3 Current-oriented belemnites from the Lower Jurassic of Yorkshire. Scale is shown by the compass which is *c*. 8 cm across.

Another important effect of transport on skeletal debris is abrasion, resulting in rounded shell fragments (Figure 10.2b). Experiments have been carried out where shells were tumbled in a barrel with an abrasive. Dense skeletons formed of many minute crystals separated by thin films of organic material (as in thick-shelled gastropods) were found to be more durable than low-density skeletons with more abundant organic matrix (as in echinoderms), or porous skeletons (as in pieces of coral colony). Only in high-energy beach environments where abrasion is more or less continuous is complete destruction of all shell material likely. In offshore areas, it seems unlikely that shells can be destroyed entirely by abrasion. Here, they are destroyed, if at all, by other mechanisms, such as by predation and scavenging, boring by other organisms, and to a limited extent (in deep cold water) by chemical solution.

As well as suffering abrasion, shells may be fractured by impact in high-energy environments such as a rocky shore where cobbles and stones pound against one another. In many organisms, the shell will fracture along preferential planes of weakness. In an echinoid, for example, the presence of irregular fractures cutting across plate boundaries would indicate that the animal was still alive when struck because all the plates are tightly bound by organic fibres in life, so that the entire skeleton acts as a single brittle shell. In dead echinoids, however, the fractures follow the plate boundaries because the connecting fibres have decayed.

Sometimes, current strengths are insufficient to transport the skeletal remains themselves, but winnow away the surrounding sediment, leaving a condensed lag deposit of shells, called a **shell bed**.

10.1.6 Chemical attack

By far the most common skeletal material in marine organisms is calcium carbonate ($CaCO_3$). Destruction of such shell material by chemical means usually involves dissolution (Figure 10.2d). In open shallow-marine waters, which are normally saturated with $CaCO_3$, dissolution is generally restricted. Old shells may thus survive for long periods on the shallow sea-floor. Some of the shells in the surface sediments flooring parts of the North Sea are reworked relics from older sediments exposed on the sea-bed.

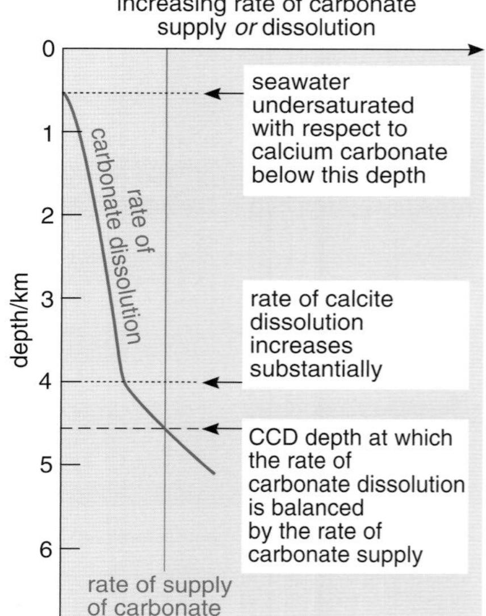

Figure 10.4 Graph to show the position of the carbonate compensation depth, in relation to the rate of carbonate dissolution and sediment supply. Note that the depths shown are typical for the tropical region of the world but that the absolute depth varies according to latitude and the local conditions in the ocean.

Chemical changes involving calcareous shells are, however, very important in deeper oceanic water where a combination of relatively high pressure and low temperature promotes the dissolution of calcium carbonate (Section 5.2). The depth at which the rate of solution of calcareous debris balances the rate of supply of material sinking down from shallower levels is known as the **carbonate compensation depth** (**CCD**) (Figure 10.4). Below this depth, no carbonate sediment accumulates. The depth of the CCD is greater for calcite than for the more soluble aragonite and varies from ocean to ocean. In the Atlantic, it is about 4000 m for calcite, whereas in the colder waters around Antarctica it is about 500 m beneath sea-level. The CCD affects principally pelagic calcareous organisms that on death sink towards the sea-floor. Those sinking to depths below the CCD are prone to dissolution, but those settling on areas of high relief, such as mid-oceanic ridges, which project above the CCD, will accumulate as calcareous sediments.

In addition, some researchers think that, at certain times in the past, calcium carbonate has been rather more prone to dissolution in seawater than today, because seawater chemistry has varied over geological time. Part of the evidence for this is that in some ancient shallow marine deposits, aragonite shells left lying on the sea-floor apparently started to dissolve, effectively meaning that the CCD for aragonite was much shallower than at present.

The mineral component of bone, teeth and fish scales is calcium hydroxyapatite [$Ca_{10}(PO_4)_6.2OH$], which is a good deal less soluble than calcium carbonate. Such remains may accumulate preferentially where calcareous shells are subject to dissolution.

10.1.7 Burial and final burial: pre-fossilization

So far, we have considered pre-burial factors in isolation. However, detailed study of some fossils often reveals a complicated history prior to final burial, involving one or more episodes of partial or complete burial followed by exhumation. This is especially common in shallow marine sediments. In some cases, diagenetic alteration of the skeleton occurs during preliminary burial so that upon exhumation its resistance to destruction is different. This process is known as **pre-fossilization**. For example, pre-fossilization of bones often involves filling of pore spaces by more phosphate, increasing their resistance to abrasion if temporarily exhumed. In times of reduced sedimentation, a resistant deposit of pre-fossilized bone fragments may gradually accumulate to form a bone bed.

10.2 DIAGENESIS FOLLOWING FINAL BURIAL

When a potential fossil is finally buried, it is then subject to diagenetic changes that frequently lead to marked alteration in the microstructure and/or composition of the skeleton. Some diagenetic processes may indeed destroy the remains. Although diagenetic processes are grouped under separate headings below, the processes described do not act in isolation, but may grade into one another.

10.2.1 MOULDS

The decay of soft parts inside a shell leaves an internal space into which sediment may seep, forming a core that replicates features of the inner surface of the shell.

❑ Recall, from your earlier reading, what such a sedimentary infill of a shell is called. Which specimen in the Home Kit is an example?

■ It is termed an *internal mould* (Section 9.1). FS D is an example of an internal mould – in this case, of an ammonite.

Question 10.3 What features would you expect an internal mould of the bivalve *Circomphalus* (FS I) to show?

Usually, the internal mould becomes visible because the enclosing skeleton is dissolved naturally (Figure 10.5; see also Figure 10.2d). In some cases, however, a palaeontologist might deliberately remove skeletal material in order to reveal the internal mould. For example, in fossil vertebrates a mould of the skull interior may give important information on brain shape and size.

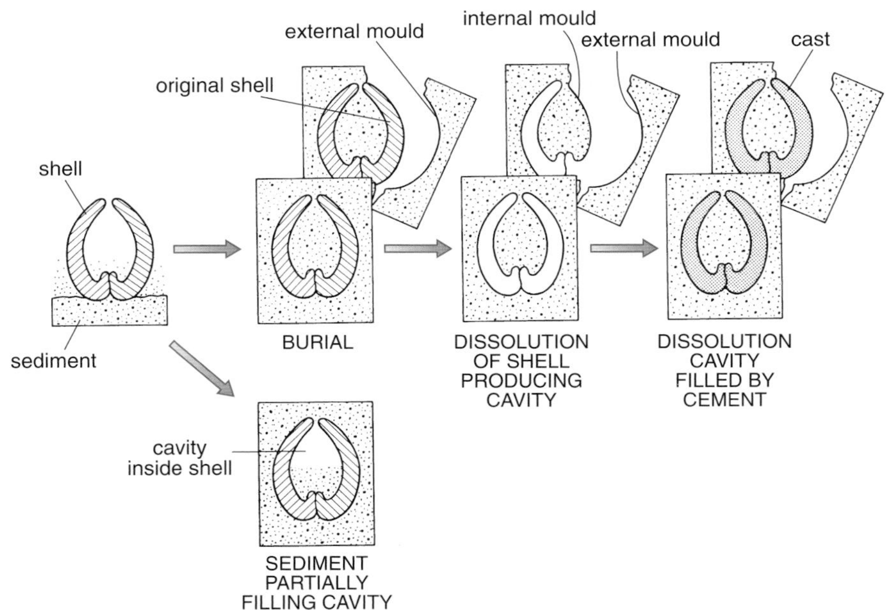

Figure 10.5 Diagram illustrating the formation of moulds and casts. Along the centre is shown the sequence of events that may follow burial of a shell – dissolution of the shell and filling of the resulting cavity by cement crystals. The row of diagrams depicted above and behind show what is revealed if the rock is broken open at the different stages. The diagram at the bottom shows how a partial sedimentary fill can act as a sort of 'spirit level'.

In some cases, sediment fails to fill the shell interior entirely. The remaining space may either be left as a void, which may subsequently be crushed flat during sediment compaction, or become lined or even filled by cement crystals, often of calcite. Incomplete fillings by sediment can be very useful for the interpretation of geological structures in the field, because they can indicate the horizontal at the time of deposition, like a fossilized spirit level (Figure 10.5, bottom diagram). Examples can be seen in RS 21/TS R. Figure 10.5 also shows that if the sediment surrounding a shell becomes sufficiently consolidated it may preserve an impression of the external surface of the skeleton. If the skeleton is dissolved or otherwise removed, an **external mould** is left.

In case you have been wondering how the 'plaster replicas' in the Home Kit relate to these terms, they are artificial casts taken from rubber impressions, themselves artificially prepared from the original fossil specimens. The preparation of artificial casts (usually of rubber) from natural external moulds is, incidentally, a valuable technique for studying the fossil content of rocks from which all shell material has been leached away, leaving only dissolution cavities in the rock (as in the centre of Figure 10.5). Frequently, however, the casting has already occurred in nature, a process to which we now turn.

10.2.2 Casts and skeletal alteration

Many body fossils that you see in museums and most of the originals represented by the plaster specimens in the Home Kit are natural replicas of skeletal components, distinguishable because their texture and/or composition have been altered. The greatest alteration is to be found in a **cast**. This is formed when the cavity between an external and an internal mould becomes filled by another material, most commonly a coarse mosaic of cement crystals, precipitated from pore-waters (Figure 10.5). Amongst calcareous shells, this form of preservation is much more common in those originally consisting of aragonite (such as most gastropod and bivalve shells) than in those that were of the less readily soluble calcite (such as brachiopods).

Such wholesale replacement via an open cavity stage usually means that no vestige of the original skeletal microstructure is preserved. Although most casts of originally calcareous shells are themselves calcareous, numerous examples of replacement by exotic mineral cements are known, including fossils preserved as opal (SiO_2), baryte ($BaSO_4$), cassiterite (SnO_2), pyrite (FeS_2) and haematite (Fe_2O_3).

In contrast, the crystalline microstructure of skeletal materials may be altered without the involvement of open spaces, by the growth of new crystals at the direct expense of their neighbours. In other words, ions going into solution from one (dissolving) crystal face almost immediately become incorporated in the lattice of another, growing alongside. If the composition of the skeleton remains virtually the same, such alteration is termed **neomorphism** ('with a new form'). Neomorphic alteration to calcite is the usual fate of aragonite if it has not first been entirely removed by dissolution. Instances of neomorphism can be detected in thin section, because original microstructural features marked out by the organic matrix, such as growth lines, tend to be preserved as 'ghosts' within the new crystal mosaic. This form of preservation can be seen, for example, in the gastropod shells in RS 21/TS R. So the recrystallization you noted in Activity 6.1 (Section 6.4) is, more specifically, an instance of neomorphism.

Because of dissolution and neomorphism, the extent of aragonite preservation is broadly inversely proportional to age. Aragonitic fossils are common in the Cenozoic, but they are relatively uncommon in the Mesozoic and extremely rare in the Palaeozoic. However, preservation of aragonite is more common in some lithologies than in others. Mudstones, for example tend to provide better protection than sandstones, because their relative impermeability inhibits the flow of dissolving groundwater. The differences in the behaviour of aragonite and calcite lead to some important biases in the fossil record. In some rocks, e.g. the Chalk, only fossils with calcite skeletons are found in most beds, although originally there was also a diverse infauna with aragonite skeletons which failed to survive diagenesis. The principal line of evidence for the latter is where a sessile organism with a calcite skeleton, such as an oyster, has grown against an aragonite shell and so taken an imprint. At a few horizons in the Chalk, rapid lithification of the sediments also ensured the survival of some unfilled moulds of aragonite shells.

A process akin to neomorphism involves the gradual substitution of one mineral by another. This is termed **replacement**. The extent to which the original microstructure remains in ghosted form varies considerably, depending on the replacing mineral and the circumstances of replacement. The commonest replacing minerals for calcium carbonate are silica (SiO_2) and dolomite ($(Ca,Mg)(CO_3)_2$), particularly in limestones. Such replacement may be highly selective. If the limestone matrix of silicified fossils is dissolved, either naturally by rainwater or groundwater, or artificially by immersion in acid, specimens superbly preserved in three dimensions may be retrieved (Figure 10.6).

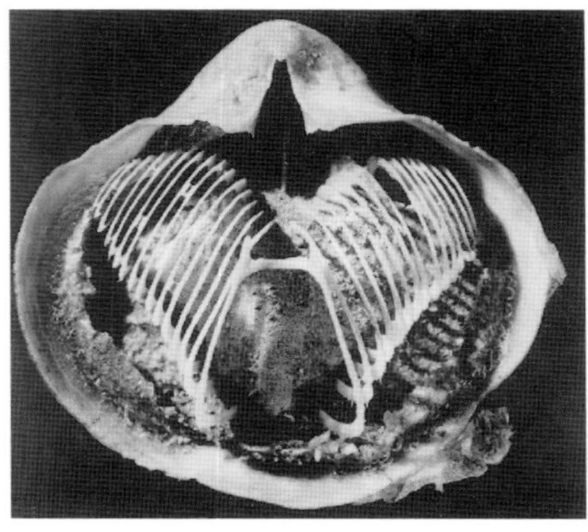

Figure 10.6 Silicified brachiopod shell revealed by dissolving away the limestone matrix with acid. The shell is about 2 cm across. Note that the support for the feeding tentacles has been preserved.

10.2.3 Exceptional preservation of soft tissues

In most circumstances, soft tissues entirely rot away before the remains of an organism can become subject to processes that might lead to eventual fossilization. In exceptional conditions, however, prolonged suppression of bacterial decay, and/or early diagenetic mineral growth, may allow such tissues to be preserved. Conditions suppressing decay include deep-freezing (as mentioned in Section 7), immersion in highly salty brines (effectively, pickling) and desiccation, either directly as mummies in arid settings, or, in the case of small organisms such as insects, through being trapped in resin (forming amber), which withdraws moisture from them. But suppression of decay alone provides no long-term guarantee of preservation: most of the instances cited above are relatively young (under a million years old), although beads of amber may last a good deal longer in sedimentary deposits. More important on geological time-scales is the growth of new diagenetic minerals within or around tissues, and here microbes such as bacteria and fungi can make a positive taphonomic contribution in contrast to the destructive role that has been emphasized so far.

Most sediments contain organic detritus that is decomposed by such microbes. Where there is an adequate oxygen supply, both the detritus and the decomposers are fed upon in turn by larger deposit feeders, which return excretory products and faecal material. This can lead to enrichment of the sediment by nutrients containing such essential elements as phosphorus, nitrogen and sulfur. The oxygen supply falls off rapidly with depth of burial, and where conditions become anoxic in the sediment, anaerobic bacteria start to reduce compounds to yield a cocktail of ions, some of which may then combine to form relatively insoluble minerals. For example, sulfide ions (S^{2-}), derived from the reduction of sulfate ions (SO_4^{2-}), are rapidly precipitated in contact with ferrous ions (Fe^{2+}) to form the insoluble FeS, which itself soon converts to pyrite (FeS_2). From a taphonomic perspective, what is important is that such new minerals may fill void spaces in biological tissues, or even replace them altogether, leaving a more enduring record of them when the organic material itself eventually disappears. Calcium phosphate is another mineral sometimes produced in such circumstances, and soft tissues can become fossilized with quite extraordinary fidelity, even showing details of cellular structures (Figure 10.7).

Figure 10.7 Scanning electron micrograph of muscle fibres replaced by apatite ($CaPO_4$), in a fish from the Lower Cretaceous of Brazil.

Environments where this felicitous overlap between the persistence of soft tissues and the early growth of diagenetic minerals are rare. Yet, numerous examples have accumulated through the immense spans of time represented by the geological record, giving us a spectacular series of detailed 'snapshots' of life through various stages in its history.

10.2.4 Sedimentary compaction

The amount by which sediments compact after deposition varies, but it is greatest in muds, which initially have high water contents and are composed of flaky minerals that reorientate with their long axes parallel to bedding and become compressed as the overburden increases (Section 6.1). Thus, the effects of compaction on fossils are usually most apparent in mudrocks, which have often been compacted to up to a tenth of their original thickness (yielding shales). Thin-shelled fossils may be crushed flat, or be distorted to give a variety of forms depending upon their orientation during compaction (Figure 10.8). If effects due to compaction go unrecognized, there is a danger that the various forms may be identified as separate species.

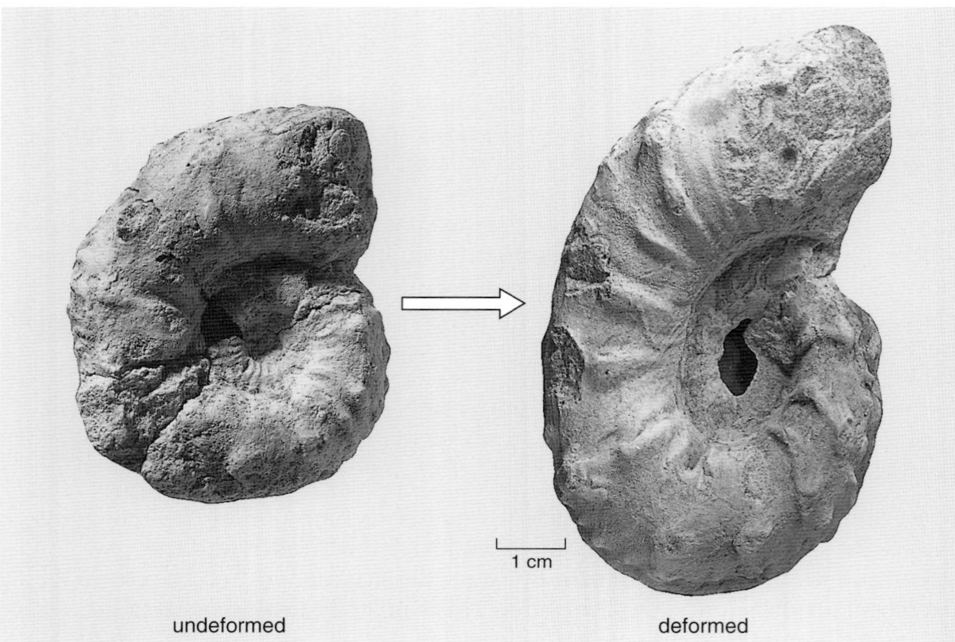

Figure 10.8 The effect of deformation by sedimentary compaction on a fossil ammonite shell.

10.3 Metamorphism of fossils

Under conditions of increasing temperature and pressure, fossils become distorted, experience change in mineral composition and are eventually destroyed. Fossils may be moderately abundant in some low-grade metamorphic rocks, and although generally extremely rare in medium and high-grade metamorphic rocks, have been described even from high-grade sillimanite granulites.

10.4 Summary of section 10

Fossils can be produced in many different ways, but eventual burial and replacive or protective mineral growth are nearly always involved. Taphonomy concerns the processes on the way to fossilization, including those causing death (by biological or physical means), decay (aerobic and anaerobic) and disintegration, as well as diagenetic changes, which are often crucial in the eventual preservation of a fossil.

From consideration of these factors, organisms of different construction (particularly composition and number of skeletal elements), and occupying different environments (especially in relation to sedimentation), can be inferred to have different fossilization potentials. Even skeletal components – the most likely candidates for eventual fossilization – may be subject to destructive biological (predation, scavenging and boring), physical (current transport) and chemical (dissolution) processes.

Following final burial, dissolution may leave only moulds (internal and/or external) of the remains, though new mineral growth in the moulds may provide casts. Internal structure is usually lost in such cases, though the simultaneous growth of new crystals at the expense of neighbouring ones, usually of the same composition (neomorphism), may replicate internal features such as growth lines.

In exceptional circumstances, diagenetic mineral growth may intervene before the decay of soft tissues has gone to completion, resulting in the fossilization of the latter. Such specimens can provide invaluable 'snapshots' from the history of life.

Subsequent processes affecting fossils include sedimentary compaction and metamorphism, involving deformation or even destruction.

10.5 Objectives for Section 10

Now you have completed this Section, you should be able to:

10.1 Outline the various taphonomic processes that may affect the remains of organisms (relating to manner of death or shedding, decay and disintegration, burial history and diagenesis, and subsequent deformation and metamorphism).

10.2 Determine which of these taphonomic processes are likely to have been experienced by given specimens, from observations of their modes of preservation, and so infer the conditions in which they became fossilized.

10.3 Make a reasoned estimate of the relative fossilization potentials of given organisms, based on their construction, life habits and habitats.

Now try the following question to test your understanding of Section 10:

Question 10.4

(a) Briefly describe the state of preservation of FS C (*Pseudodiadema*) from the Home Kit.

(b) In the light of your observations, explain which of the following taphonomic hypotheses is most likely:

(i) The animal was killed by a sudden influx of sediment, which permanently buried it.

(ii) The animal died by unknown means, became buried a few weeks later, but was re-exposed on the sea-floor for a period prior to final burial.

(iii) The animal died by unknown means and remained on the sea-floor for a few years before being buried.

(iv) The animal was killed by a predatory fish, discarded and buried a few days later.

11 Fossils as clues to past environments

For this Section, the only specimen you will be needing from the Home Kit is FS Q.

The study of how organisms interact with their environment and with each other is called ecology. Thus **palaeoecology** refers to the investigation of these relationships in the fossil record. While we can observe at first hand what eats what, or measure birth and death rates among living organisms, the biased sampling inherent to the fossil record clearly limits the scope of palaeoecology. For illustration, let us consider the palaeoecological work that might be done on a fairly ordinary assemblage of fossils (Figure 11.1).

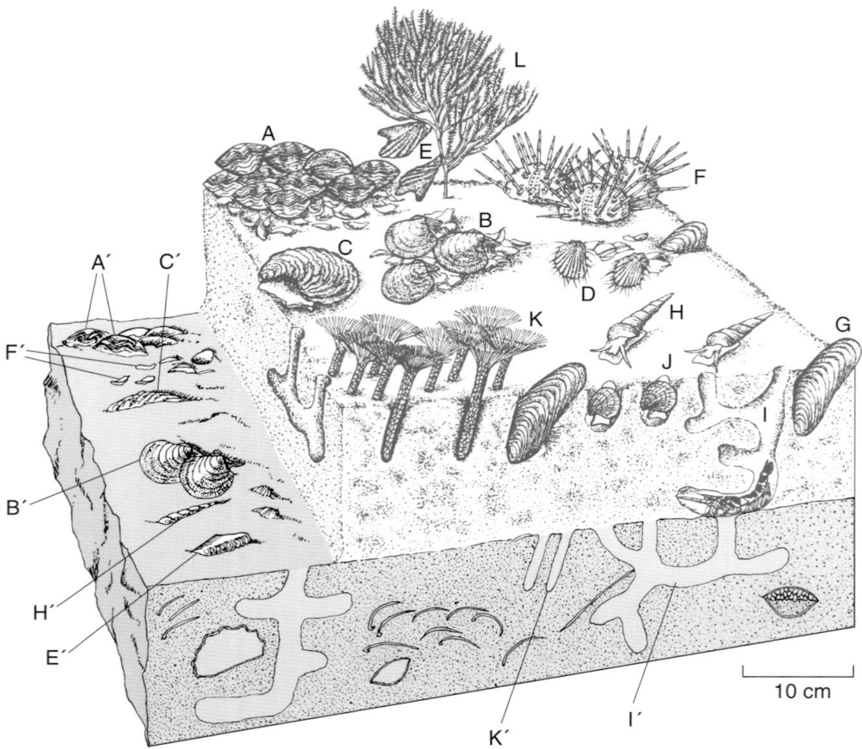

Figure 11.1 Reconstruction of the association of organisms that occupied a Jurassic shelly carbonate mud sea-floor. The lower half of the block depicts a slab of the bed of fossiliferous muddy limestone on which the reconstruction is based. Part of the upper surface (bedding plane) of the limestone is shown projecting to the left. The upper half is the reconstruction of a corresponding part of the original sea-floor and the association of organisms that dwelt there. The sediment surface was probably relatively firm, and deposition is likely to have taken place in a shallow but quiet marine bay comparable to modern examples in Florida and the Bahamas. Living organisms and their fossil counterparts are given the same letters, the fossils being marked by a prime mark (′); other features are explained in the text.

From Section 9, you have already seen how much information can be gleaned about the way of life and habitat of individual species represented by relatively intact body fossils.

- ❑ Which groups seem to be represented in Figure 11.1 by relatively intact body fossils?
- ■ The upper surface of the limestone block (at the left) shows complete fossil shells of molluscs (e.g. those of the bivalves B′ (scallop), C′ (oyster) and E′, and the gastropod H′) and brachiopods (e.g. A′). Note that all of these shells originally consisted of only one or two large components, with a high fossilization potential (Section 10.1.3).

Now imagine, for example, that you have recovered some fossil shells of the mussel (G). Certain living mussels closely similar in shape to this tend to live half-buried, and so the fossilized species might be inferred (by analogy) to have done likewise. This reconstruction might be verified by actually finding whole fossil mussel shells preserved where they lived in their excavations, i.e. in **life position**.

Reconstruction of organisms that had a lower fossilization potential depends more on lucky finds. You might have expected the echinoids (F), for example, to have been represented only by separated plates and spines – presenting quite a jigsaw puzzle for the palaeontologist. In this example, however, our knowledge of the species is based on the fortunate discovery of some 1500 specimens, in a quarry section, preserved intact with even their spines in place, and in life position.

- ❏ How might such preservation of so many specimens have come about?
- ■ Since surface-dwelling echinoids like these would have tended to start falling apart soon after death (Section 10.1.1), it is likely that these specimens were buried alive. Given the quiet marine bay setting inferred for the deposit, swamping of a population of the echinoids by a sudden influx of storm-driven sediment is the most probable explanation.

Such luck does not always attend the preservation of less readily fossilized species, so reconstruction of their forms and life habits is more tentative. Indeed, some organisms, or more particularly their life habits, can be inferred only from trace fossils. The branching burrows in the limestone (I′) commonly contain sediment-bearing faecal pellets, indicating that the occupant of the burrow was for at least some of the time a deposit feeder. The forms of the burrows and of the faecal pellets are similar to those associated with certain living infaunal crustaceans. We might, then, infer that the burrows belonged to such creatures (hence I), although (as so often with trace fossils) body fossils supporting this interpretation are very rare. The presence of the suspension-feeding fanworms (K) in the reconstruction is again based on distinctive trace fossils (K′), though simple vertical tubes lined with shell debris and fish scales leave a fair amount of latitude for interpretation.

The scope of palaeoecological work at the level of the single organism or species thus ranges from the precise to the speculative, depending upon the kinds of fossils left behind. But how complete a picture of the original community, or communities, of organisms that dwelt here is this reconstruction? Merely attempting to list the constituent species from the incomplete fossil record requires informed speculation, as you saw above. Most of the soft-bodied forms will have left no identifiable signature. As for the network of feeding relationships between organisms in communities, the evidence is yet more patchy. It is not difficult to interpret how various organisms fed, when considered in isolation, as you saw in Section 9, but it is quite another matter to reconstruct the intricate web of feeding interactions among the different species. Which, for example, were the main prey species for the predatory or grazing echinoid (F)? Did the suspension-feeding fanworms (K) compete for food with other suspension feeders such as the mussel (G) and brachiopod (A), or did they perhaps draw food particles from different levels in the water column, or specialize in food particles of a different size, and thus avoid competition with them? And was food (as opposed to, say, space for settlement of larvae) a limiting resource for them, in any case? Often, we can do little more than classify the organisms into broad feeding groups in a **trophic pyramid**, recognizing those that synthesized organic matter (e.g. by photosynthesis) – *producers* – as the basic food source for *consumers*. The latter in turn may be divided into those that ate the producers directly and those that ate other consumers.

Therefore, some questions are a lot easier to answer from the available evidence than others. Much of the skill of palaeoecological interpretation involves deciding what are answerable questions, given the partial nature of the evidence. In the early days of modern palaeoecology, in the 1960s and 1970s, there were many enthusiastic attempts to identify '*communities*' of organisms from fossil assemblages, like the living examples then being discussed by ecologists. At the time, many ecologists believed that the complex web of feeding and other interactions between organisms led to the establishment of highly regulated and predictable associations of species in given environments. So a search for these in the fossil record seemed a reasonable objective. Today, however, many ecologists are sceptical about the supposed extent of self-structuring of communities, especially those dependent on suspended or detrital food sources, and prefer to regard them as reflecting little more than the coincidental clumping of species sharing similar environmental preferences and constraints, with much variation, and overlap, of distributions. Moreover, progress in the understanding of taphonomy has emphasized the many and complex biases that intervene between living associations of species and fossil assemblages. Many of the earlier attempts to reconstruct communities from fossil assemblages can now be seen as naive, owing more to imagination than firm evidence. In this Section, therefore, we will not pursue the difficult objective of reconstructing entire ancient communities, nor discuss further the contentious issues raised by that exercise. Instead, we will focus upon some of the more straightforward palaeoecological inferences that can be drawn from fossils. Together with sedimentological information, these more modest insights can nevertheless make a useful contribution to palaeoenvironmental interpretation.

11.1 CLUES TO PHYSICAL CONDITIONS

Let us say you have seen a body fossil in a sedimentary formation, and you want to know what it can tell you about the original environment of deposition.

- ❑ What is the first question that you should ask yourself about the fossil?
- ■ The first thing you need to know is whether or not it lived in or near the place of burial.

If the fossil is preserved in life position, or shows little evidence of transport (Section 10.1.5), then you can use your knowledge of the biology of that organism to infer environmental conditions. If it had been transported from elsewhere, in contrast, then it could tell you little more than any other sedimentary grain about its eventual resting place. It is therefore worth having a simple scheme for classifying the various possibilities. An organism preserved more or less where it lived is said to be ***in situ***; it may be either in life position, as described above, or locally displaced. A sessile organism that normally lived implanted in the sediment (like G in Figure 11.1), for example, might have been washed out, but then buried in place as storm currents swept sediment over them. However, if remains have evidently been carried for some unknown distance from their source, by currents, scavengers or whatever, we could describe them as transported.

The possibility of transport can be a serious problem, especially in shallow-water deposits. The best kind of evidence for benthic organisms having been buried where they lived is if they are manifestly preserved in life position, such as cemented bivalves still attached to a bedding surface (Figure 11.2a), or burrowing bivalves preserved in their normal burrowing positions (Figure 11.2b). Alternatively, trace fossils may attest to the living activities of organisms at the site (e.g. the borings shown in Figure 11.2a).

If, on the other hand, you were to find some intact burrowing bivalves that were not preserved in the normal burrowing position, the problem of how far they might have been transported prior to being buried alive (as discussed in Section 10.1.1) would remain. So whenever you observe a fossil at an exposure, you should take careful note (preferably with sketches) of its orientation and position in the bed as well as its state of preservation. (You should, moreover, leave it there for others to see, unless you really need to collect it for further serious study.)

(a)

(b)

Figure 11.2 (a) A preserved cemented surface showing fossil evidence of both encrusting and boring bivalves. (b) Burrowing bivalves preserved in life position in a bed of fine-grained limestone. (Both from the Lower Cretaceous of Portugal.)

Following such an assessment, you are then better placed to make inferences about the ambient physical conditions from what is known about the biology of the original organisms.

Question 11.1 Suppose the only fossils that you could find in a finely laminated, black, pyrite- and organic-rich shale were a number of ammonoids with streamlined shells: what might you infer about the oxygen level (a) at the sea-floor, and (b) up in the overlying seawater?

Similar inferences can be made concerning salinity. Echinoderms today will not tolerate much salinity variation either side of 'normal marine salinity' of about 35 (*note*: salinity is often expressed as a number without units; however, it represents grammes of salt per kilogramme of seawater, i.e. 'parts per thousand' represented by the symbol ‰). Thus, if you found an intact fossil crinoid in a sedimentary formation, you could reasonably infer that deposition took place in normal marine water. Table 11.1 contains information about the present-day salinity tolerances of the major groups represented in the Home Kit.

Table 11.1 Some generalizations concerning the present-day salinity tolerances of the groups represented in the Home Kit.

Group	**Includes species restricted to water of normal marine salinity (35)**	**Includes species tolerant of water varying by as much as 10 above or below normal marine salinity**	**Includes species tolerant of (or even restricted to) freshwater**
echinoderms (echinoids and crinoids)	✓	×	×
molluscs:			
bivalves	✓	✓	✓
gastropods	✓	✓	✓
cephalopods	✓	×	×
arthropods*	✓	✓	✓
corals	✓	✓ (very few species)	×
brachiopods	✓	✓**	×

* Trilobites seem to have been restricted to water of normal marine salinity.

** Today, this only includes a few inarticulate brachiopods such as *Lingula*, and most of these survive for brief periods only by closing their valves tightly and waiting for conditions of normal salinity to return.

Fossils can also be used to investigate patterns of sedimentation. Intense bioturbation self-evidently indicates that the sediment remained unconsolidated near the surface for long enough to become infested with burrowers. This deduction in turn implies not only that conditions at the sea-floor fostered a thriving benthos, but that net rates of sediment accumulation were sufficiently low to have allowed the burrowers to get the upper hand over current activity, so to speak, in imposing their imprint on the accumulating sediment (Section 9.6). A sparser record of trace fossils accompanied by abundant current-generated sedimentary structures, by contrast, would suggest that current reworking and/or deposition of sediment 'had the upper hand' in this contest of influences between bioturbation and currents. Moreover, as you saw in Section 9.6.2, the spreiten in trace fossils such as *Diplocraterion* can yield further information concerning episodes of net sediment loss or gain.

This may all sound somewhat vague and qualitative, but the real value of such observations comes with the detection of vertical changes in such features in sedimentary successions. A shift from predominantly bioturbated to mainly wave cross-stratified sediment, for example, could be interpreted as reflecting the shallowing of the sea-floor into fairweather wave-base (see Section 4.2.5).

As another example, encrustation of a surface by a fauna like that in Figure 11.2a would indicate that the sediment surface had become hardened prior to the deposition of overlying beds. Borings present in the bedding surface (which cut through sediment grains and shells alike) add further weight to this interpretation.

The foregoing examples all illustrate simple inferences that can be made from the mere presence (or absence) of certain kinds of fossils. The approach can be refined by investigating the patterns of growth of the specimens. 'Abnormal' growth representing a response to adverse conditions, may be quite informative. A good example is provided by the sea shells of the Baltic Sea. These are markedly smaller than specimens at the same stage of growth found elsewhere on open marine coasts (Figure 11.3).

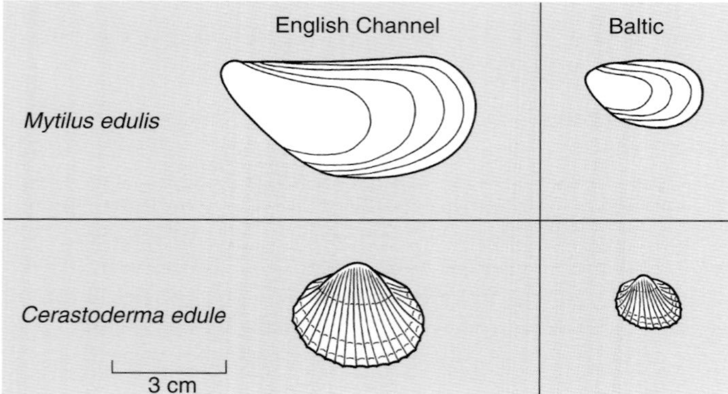

Figure 11.3 Shells of the common mussel (*Mytilus edulis*), above, and cockle (*Cerastoderma edule*), below, collected from the English Channel (left) and from the Baltic Sea (right).

The difference probably relates to the below-normal salinity of the Baltic Sea. In the fossil record, such stunting of growth might then be taken as signifying abnormal salinity, though care would have to be taken to ascertain that the shells were genuinely stunted with respect to others of their species. Otherwise, they might just have been juvenile specimens or a local species characterized by small body size, or alternatively a current-sorted assemblage.

11.2 PALAEOECOLOGICAL RELATIONSHIPS BETWEEN FOSSIL ORGANISMS

So far, we have been considering inferences from fossils only about the physical and chemical aspects of past environments. For each organism, however, the other organisms around it are just as much part of its environment.

The first task that has to be tackled in determining if species interacted is whether or not their fossilized representatives actually lived at the same time and place. This may sound simple – just find two fossil species in the same bed that have clearly been buried where they lived – but in practice it is not. Consider the Jurassic Kimmeridge Clay Formation of Dorset, which has a maximum thickness of between 500 and 600 m, and represents a period of time of approximately 10 Ma.

Question 11.2 What was the net rate of deposition of this formation in centimetres per 1000 years?

It is very unlikely that sedimentation took place at a regular rate. Rather, fluctuating deposition was probably interspersed with episodes of non-deposition or even erosion. This pattern is typical not only of shallow seas, due to storms, tides and so on, but also in deeper areas, as a consequence of fluctuations in sediment supply and in various deep currents. Nevertheless, using the net rate as a basis for calculation, a collection of shells made from one band only (say, a 3-cm-thick band in the thickest part of the Kimmeridge Clay) could represent a time-span of between 500 and 600 years – perhaps one or two orders of magnitude more than the maximum life-spans of most of the shelly organisms present. So a band containing several species of burrowing bivalves could be a record either of successive separate colonizations by each of the different species, or of all or some of the species living together there at the same time. Indeed, a shell-rich bed is quite likely to represent a relatively condensed lag deposit, from which sediment had been winnowed away (a '**shell bed**'). In that case, it would represent an even longer period of time than that calculated above. Assemblages of fossils that have accumulated in this way over long periods of time are said to be **time-averaged**. Attempting to unmix the 'cocktail' of organisms that lived at different times can be difficult, if not impossible.

The principal exceptions to this problem of demonstrating contemporaneity are assemblages preserved by catastrophic burial (Section 10.1.1), which consequently are of special value in palaeoecological reconstruction. In Figure 10.1, for example, we can be relatively certain from their common mode of preservation that the stalked echinoderms shown there (together with any associated organisms) did live – and die – together.

Establishing contemporaneity is only the first step. The precise relationships between organisms may still remain obscure. The only direct clues are likely to be traces left by one organism on another and examples of organisms that have clearly grown in intimate contact. Predation and grazing are examples of feeding relationships that might be revealed by trace fossils left on other body fossils. Many predators tackle their prey in a characteristic way and so tend to leave their 'fingerprints' on their victims.

- ❑ Recall an example of a predator that leaves recognizable marks on the shells of its prey.
- ■ In Section 9.3.3, you learned that some predatory gastropods penetrate the shells of their prey by drilling distinctive holes (Figure 9.16b).

Figure 11.4 shows some other examples of recognizable damage inflicted on modern molluscan shells by known predators. Examples of prey having survived such attacks, as indicated by repair and continued growth of the shell, testify to the attacks having taken place on the living prey.

(a)

(b)

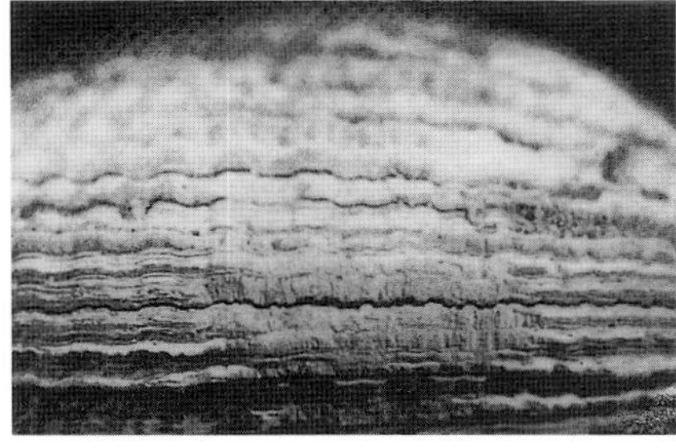
(c)

(d)

Figure 11.4 Mollusc shells showing characteristic damage inflicted by various predators: (a) cockle shell smashed by a lobster; (b) part of a bivalve's ornament plucked away by the tube feet of a starfish pulling the valves apart; (c) ventral valve rims of a bivalve, ground away by a whelk, to allow entry of its proboscis; (d) hole drilled in a gastropod shell by an octopus. (a) Actual size; (b–d) slightly magnified.

To illustrate the problems in interpreting closely attached organisms, turn now to FS Q from the Home Kit, which consists of a shell encrusted by a number of sinuous or coiled tubes.

❑ To what kind of organism did the large shell belong, and in what way is its shape exceptional for the group?

■ The large shell in FS Q is a brachiopod. It has two unequal valves, the umbonal parts of which, at least, show characteristic brachiopod symmetry (as in Figure 9.1b). The anterior side of the shell (opposite the umbones), however, lacks this symmetry, the shell margins being obliquely offset from one side to the other.

This distortion of the shell has nothing to do with the encrusting tubes, because, other members of the species, lacking the tubes, also show it. Some palaeontologists have suggested that it reflected a special modification of the suspension-feeding current system relative to normal brachiopods, but evidence to support this has not been found, and the functional significance of the distortion remains unclear.

There are three kinds of encrusting tube on the shell. The largest is in the form of a question mark near the anterior margin of the smaller, dorsal valve of the shell. On the opposite valve margin of the shell is a smaller, spirally coiled tube. The third kind is the smallest, and is represented by some long thin tubes running along the radial grooves on the outside of the shell. Three of these form a tight cluster in a single groove between two ribs, about one centimetre down from the spirally coiled tube (you will need the hand lens to distinguish these). All three kinds of tube can be attributed to sessile suspension-feeding worms (like those attached to the oral surface of the echinoid, FS C). From the open end of the tube in similar living examples, the worm projects a fan of tentacles which trap food particles from the surrounding water.

Having established what organisms were involved, we need to decide whether the worms encrusted the shell while the brachiopod was still alive, or after its death.

Question 11.3 Judging from the way in which the brachiopod shell itself has been preserved, can you detect any positive evidence that it lay exposed on the sea-floor after death and prior to final burial?

The taphonomic evidence for post-mortem exposure of the shell on the sea-floor, identified in Question 11.3, indicates that there was ample opportunity for encrustation by tube-building worms after, as well as during, the life of the brachiopod.

❑ Can you detect any preferred siting or orientation of the worm tubes with respect to the shell?

■ The large sinuous tube and the smaller spiral tube are both sited at the anterior margins of the shell, while the three small tubes near the spiral tube all have their apertures facing towards the anterior margin of the shell (you will need the hand lens to see this). The orientation of the other small tubes is less certain.

As suspension feeders, it is likely that the tube-dwelling worms would have grown in such a way as to keep their food-trapping tentacles well above the sea-floor, to avoid clogging by sedimentary particles.

❑ What, in the light of this assumption, was the likely orientation of the brachiopod shell on the sea-floor while it was being encrusted? Does this match the expected life orientation of the brachiopod (Section 9.3.5)?

■ Since the worm tubes are either sited near the anterior valve margins or have their aperture facing towards them, it is likely that the anterior part of the shell was facing upwards from the sea-floor while the shell was being encrusted. The umbones would thus have faced downwards. As the stalk in this species appears to have been reduced, or even lost, it is likely that the animal nestled in sediment with its umbones partially embedded (Section 9.3.5), as reconstructed in Figure 11.5.

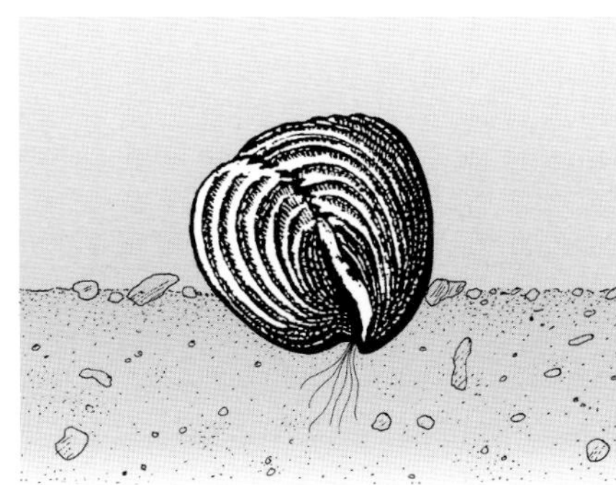

Figure 11.5 Reconstruction of the life position of FS Q.

Although this answer indicates probable encrustation of the shell while it was in life position, we still cannot be sure if the brachiopod was actually alive at the time. It is still possible that it had died, but remained in place and was only later encrusted by worm tubes preferentially growing on, or towards, its upper surface. Unless we can see evidence for a direct growth response by the brachiopod to the encrusters (not evident in this case), this is as far as we can go with this one specimen. Encrustation of live brachiopods might, however, be inferred if a survey of worm tubes on a large number of specimens showed them to be concentrated around those parts of the valve margins where feeding currents entered.

Your study of FS Q should have made it clear to you that there are difficulties even in establishing whether or not attached organisms were contemporaneous, and that single specimens often yield insufficient information to settle the issue. Yet only when such hurdles of interpretation have been successfully cleared are we entitled to go on to explore the possibilities of interactions such as parasitism or competition, for example.

Although the tone of this Section may sound a little discouraging, it is certainly not the case that all such investigations are in vain. The point we wish to stress is that inferences about palaeoecological relationships between organisms must be based on secure evidence for live interactions. Often these come only from the discovery of a one-in-a-hundred specimen that happens to show some serendipitous piece of evidence. Merely finding fossils in close association is not enough, and uncontrolled speculations based on them may be misleading.

Activity 11.1

To round off this and the previous Sections on fossils, you should now view the video sequence *Fossils as clues to past environments* on DVD 3, which illustrates the links between fossils and sedimentary environments.

11.3 Summary of Section 11

Palaeoecology and taphonomy together yield clues for interpreting past environments. Biases of preservation mean that fossil assemblages offer a limited ecological perspective, requiring critical evaluation of what constitute answerable questions. Preservation of a fossil in life position allows one to infer ambient conditions (such as salinity) from what is known about the biology of the organism concerned. Moreover, changes in the relative frequency (and style) of bioturbation and current-generated sedimentary structures may indicate deepening or shallowing trends in relation to wave-base. Abnormal growth patterns can also be informative.

Reconstructing interactions between fossil organisms is difficult. A primary task is to establish contemporaneity in view of the relatively condensed nature of much of the fossil record (time-averaging). Assemblages preserved by catastrophic burial greatly reduce this problem. Direct evidence for interactions includes recognizable traces on fossils, such as characteristic kinds of damage due to predation, especially if accompanied by examples of repair. Overgrowth (as illustrated by FS Q) still poses the problem of establishing contemporaneity, but this can be resolved by studying the distribution of the encrusters in relation to the life habits of the host in numerous examples.

11.4 Objective for Section 11

Now you have completed this Section, you should be able to:

11.1 Use your understanding of the palaeobiology of organisms now preserved as fossils to explain the conditions that existed in the environments in which they became fossilized.

Now try the following question to test your understanding of Section 11.

Question 11.4 Figure 11.6 shows a small part of an exposure of some Jurassic marine sedimentary rocks (the lens cap, for scale, is 5.5 cm across). Two layers of paler-coloured fine-grained sandstone are shown, each containing small-scale hummocky cross-stratification (see Section 4.2.5), sandwiched between darker silty mudstones of mottled appearance. Study the photograph closely and try to explain what you can see by answering the following questions. (a) What is likely to have caused the mottling of the darker units. (b) Describe the nature of the boundaries – sharp and simple, or irregular to gradational: (i) going up from the mottled units into the sandstone; and (ii) passing up from the sandstone back into mottled sediment, and interpret your observations in terms of the influences on deposition. (c) Decide whether or not the bottom waters were likely to have been anoxic where these sediments were being deposited. (d) At what depth, in relation to fairweather wave-base and/or storm wave-base, were these sediments probably deposited?

Figure 11.6 Photograph of an exposure, to accompany Question 11.4. Lens cap is *c*. 5 cm across.

12 Introduction to sedimentary environments

Sections 12–17 of this Block are about the subdivision and graphical representation of sedimentary deposits and their fossil content, together with the different environments in which these deposits form. There are eight environments that we want you to study (Figure 12.1).

Section 13 examines rivers and their flood plains, otherwise referred to as **alluvial** environments. We shall consider two types of alluvial environment based on the most common river channel morphologies. These are; alluvial, **meandering** where the river is confined to a single channel of sinuous form, and alluvial, **braided** systems where the channel is split into a number of smaller channels which may join or divide again. Section 14 describes the main features of **deserts**. These are regions of land where the rainfall is less than the amount of evaporation. In Section 15, three types of coastal environment will be discussed. These are: **strandplains** where there is a beach attached to the coastline; **deltas** where sediment-laden rivers reach the sea and build out a fan of sediment; and **barrier islands** which comprise a sand bar situated just offshore with a protected lagoon or lake behind. Section 16 considers marine areas of mostly shallow-water depth where mainly carbonates are deposited. These are referred to as **carbonate platforms**. Finally, Section 17 describes the processes and features in marine environments where the water depth is greater than about 500 m; such settings are here called **deep sea** environments.

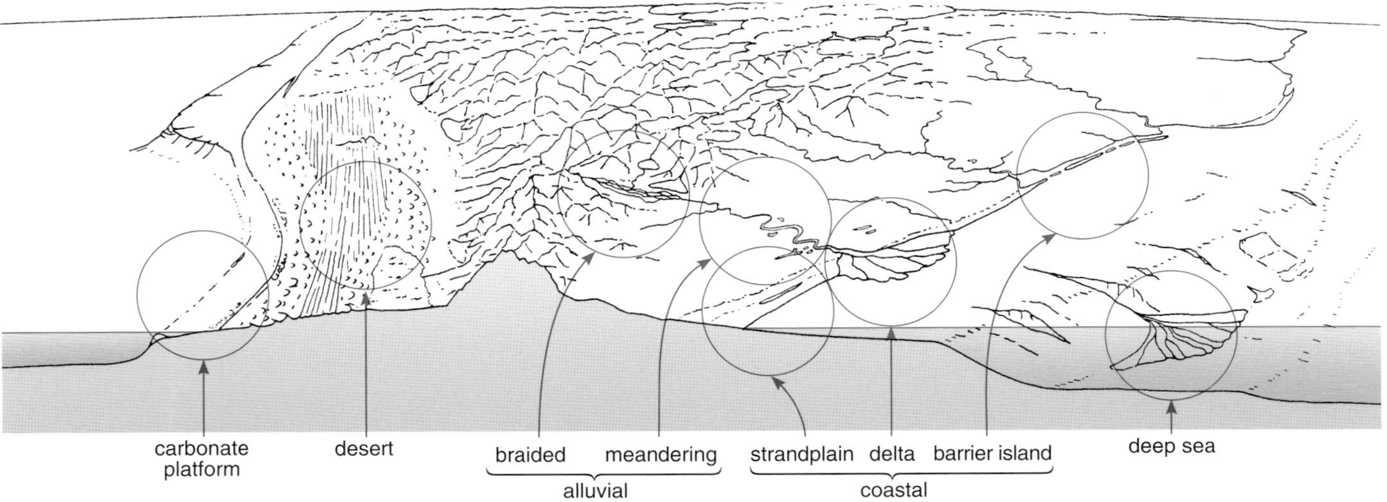

Figure 12.1 An imaginary panorama across a continent and adjacent oceans showing the eight principal sedimentary environments that we shall consider in this Block.

These environments are all illustrated in more detail on a larger version of Figure 12.1 on the Sedimentary Environments Poster and Sedimentary Environments Panorama (on DVD 3), to which we will refer you in several Activities. Although in the text, DVD and poster we have split our imaginary continent up into these separate 'environments', the boundaries are often gradational and a single sedimentary grain may visit many of the different environments during its life history. That is why we have shown you how the environments might connect together in the panorama. Imagine, for instance, chemical and physical weathering in the mountains leading to breakdown of rocks which may then have been transported down the mountainside. Initially, the breakdown of the rocks may produce a grain of quartz which then gets carried away in a braided river system; from there it is carried into a meandering river system. Eventually, it is swept through a delta and then carried across a shallow sea area by currents, finally being incorporated into sedimentary deposits in the deep sea. At any point, the sand grain could become trapped for a while in a particular environment before moving on again, or, alternatively, it may stay there and become preserved in the geological record as part of the depositional succession of rocks representing the particular environment in which it became trapped. However, it is more than just chance that leads to the preservation of sediments; exactly where they are deposited in an environment depends on the nature of the processes acting at that time and also what changes occur in terms of tectonics and climate. We will discuss these factors as we describe each sedimentary environment in the remainder of this Block.

Sections 13–17 consider the eight principal sedimentary environments introduced earlier so that we can teach you the features of each, but, of course, in the field a geologist has to work out the environment of deposition from observations on the rocks and their context. When we discuss the environments, processes, and source of material, we will use the terms **proximal** to mean the area nearest to the source of sediment in a river system or desert, or nearest to land if it is a marine environment, and conversely **distal** to mean the area furthest away from the source of sediment or furthest away from the land.

Glacial environments, dominated by processes particular to snow and ice, were important both globally and within the British Isles during certain periods of geological history but they are not discussed further in this Block.

12.1 FACTORS THAT CONTROL SEDIMENT DEPOSITION

The supply, reworking and deposition of sediments is controlled by several factors. For sediments to accumulate, there has to be: (1) a supply of sediment either from the weathering of igneous and metamorphic rocks, or the weathering and reworking of older sedimentary rocks; and (2) a space to put the sediments into. The tectonic setting and climatic conditions are the most important factors controlling both the supply of sediment and the creation of space to put the sediments into because they control:

1 The rate and type of weathering and erosion and hence the *amount* and *type of sediment supply*.

❑ Suppose a land-mass were to be uplifted. Do you think there would be more sediment produced or less, and why?

■ More sediment would be produced because increased elevation means that the rivers would have more energy and so cut down (erode) more rapidly to regain their equilibrium position.

❑ Climatic conditions can be either stable or changing. During which condition do you think the sediment supply will increase, and why?

■ In general, when the climate changes, more sediment is produced because of the fluctuating conditions. The sediment supply will stabilize when climatic equilibrium is again reached.

2 The conditions in the seawater, including the temperature, amount of dissolved oxygen and salinity. These parameters are particularly important in controlling where carbonates are deposited because a large proportion of them is composed of the remains of biota, the success of which is highly dependent on the seawater conditions.

❑ Suppose the oxygen content of the water is low. Do you think it will encourage or suppress the growth of carbonate-secreting organisms like corals, and why?

■ A low oxygen content would suppress the growth of carbonate-secreting organisms because they require oxygen to live; in fact, if there is not enough oxygen, they may not even exist in that environment.

❑ Do you think the majority of marine organisms live in normal salinity seawater or hypersaline (very salty) conditions?

■ The majority of marine organisms live in water of normal salinity; hypersaline conditions require special adaptation. Algal and bacterial mats (Section 9.5) are found in hypersaline water but the main reason for this is that the hypersaline conditions exclude other organisms that might otherwise graze on, and destroy, the mats.

3 The position of sea-level: this not only influences the marine environments, the type of marine sediments deposited and preserved, and how the whole sedimentary system moves around, but also the level of the water in the rivers and the water table beneath the land surface, and hence erosion levels in the non-marine environments of the alluvial systems and deserts.

❑ If sea-level falls, would the water level on land fall or rise?

■ If sea-level falls, then the water level on land would also fall.

Figure 12.2 shows the ideal stable longitudinal profile of a river in balance with sea-level. If sea-level falls (as shown by the blue line), then the river will respond by cutting down to a new equilibrium profile.

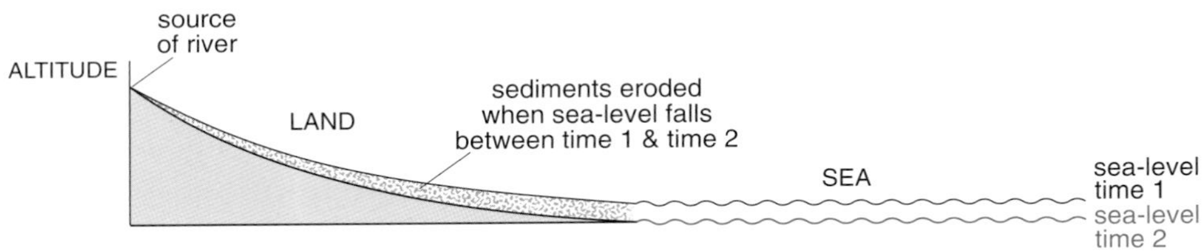

Figure 12.2 River profiles that are stable at two different positions of sea-level.

❑ If sea-level falls, will the beach move seaward (distally) or landward (proximally)?

■ The beach will move in a seaward (distal) direction as sea-level falls and a new coastline will be produced seaward of the previous one.

A further control that may influence some sediment deposition in both marine and non-marine environments is the type of biota. At different times in geological history, there has been a predominance of certain types of biota. For instance, during the late Mesozoic and Cenozoic, coccolithophores (a type of marine alga) were dominant, whereas there were none in the Palaeozoic. However, corals were important in the mid-Palaeozoic and mid-Mesozoic. This dominance of certain biota often exerts a particularly strong control over the type of carbonate sediments that have been produced at different times during geological history.

Organisms can also affect the deposition of siliciclastic and carbonate muds. Clay and silt particles would normally remain in suspension under many conditions because they are so fine-grained. However, many organisms create faecal pellets by the ingestion of many clay particles, along with organic matter upon which they feed; these are then redeposited as larger amalgamated particles.

Another method of depositing fine-grained sediments relies on clay particles naturally sticking together to make larger particles; this process is called **flocculation** and is the result of charge-based attractive forces between the particles. In freshwater, the clay particles are negatively charged and repel each other, but in saline waters these charges are neutralized by combination with other positively charged particles (such as organic matter from bacteria and other dissolved ions) and so the clay particles flocculate into dense masses and are deposited. As a result of flocculation, larger aggregated particles are formed which will then settle out at even higher current speeds.

❑ The cohesiveness of clay particles once they have stuck together is important in erosion of these sedimentary deposits. Can you recall and explain why this is so?

■ Because clay particles are cohesive, they are more difficult to erode, so a faster flow is required than you might expect for their grain size (Section 4.1.2).

12.2 Facies

In this Course so far, we have concentrated on the lithology of sedimentary rocks, in other words the composition and texture of a rock (e.g. mudstone, sandstone). We have also considered the body and trace fossils that you may find in a sedimentary rock, and the sedimentary structures. We can group the lithology, fossils and sedimentary structures together to describe a particular **(sedimentary) facies**. The word literally means the aspect or appearance of something, in this case a sedimentary rock. You were introduced to metamorphic facies in Section 8.4 of Block 3. The sedimentary facies concept is useful because it can be linked to a particular set of depositional processes. Facies are a more holistic way to describe sedimentary rocks.

For instance, imagine two lithologically identical sandstones, both being composed dominantly of quartz, medium-grained and well-sorted with moderately well-rounded grains. In other words, they are sandstones that are lithologically identical. However, one of these sandstones contains a few bivalves but shows no sedimentary structures and the other is planar-stratified but devoid of fossils. These differences define them as two distinct facies, deposited by slightly different processes. They may, of course, be closely related to each other; and could, for instance, both have been deposited on a beach. The planar-stratified sandstone might represent the main part of the beach deposited by waves during fairweather conditions; the sandstone with bivalves might represent a small bank on the beach deposited by waves during a storm, as illustrated in Figure 12.3a (other parts of this diagram are described later in the text). Facies may be identified by letters or numbers, e.g. 'Facies A', or else by brief descriptive names. The latter is more informative, so the two facies described ab ove could be called 'sandstone with bivalves facies' and 'planar-stratified sandstone facies'. Initially, this facies description is objective and descriptive but ultimately facies may be grouped together and given an environmental interpretation.

Question 12.1 Examine RS 21 and RS 24 and determine whether they represent different facies, giving reasons for your answer, and a brief description of each specimen. You may find it useful to look at the labelled pictures available on the Digital Kit.

12.2.1 Vertical successions of facies

So far, we have considered only facies that are horizontally adjacent and pass laterally into each other in plan view. We also need to consider how one facies ends up on top of another, as is found in the geological record, i.e. how facies are related to each other in time as well as space.

First, we need to define two more facies in our beach environment. Sandstones deposited by wind on the landward side of the beach will have large-scale cross-stratification resulting from the migration of dunes. We can call this facies

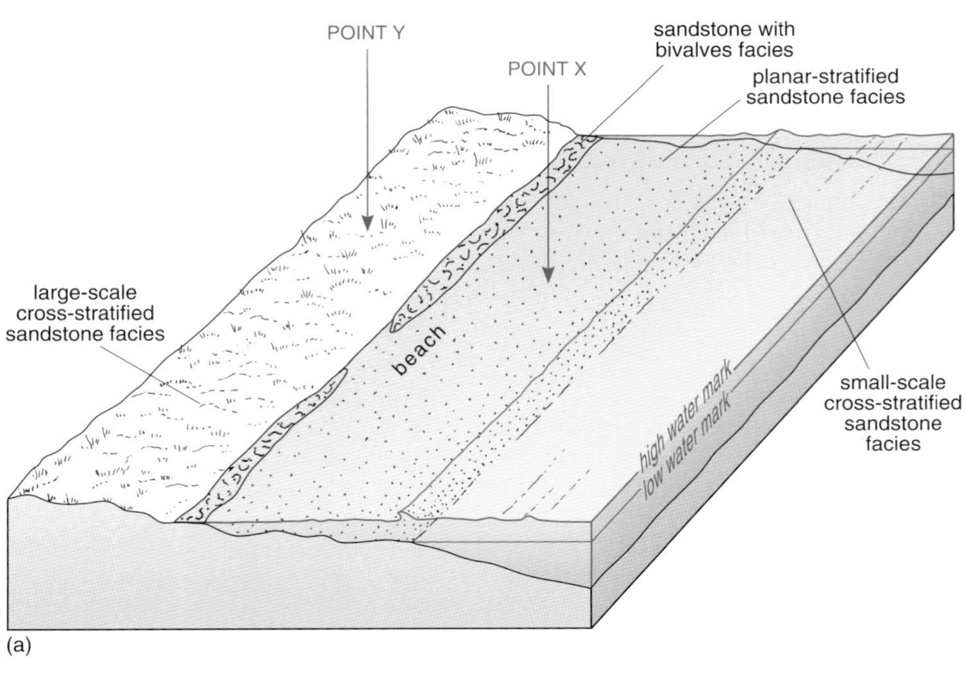

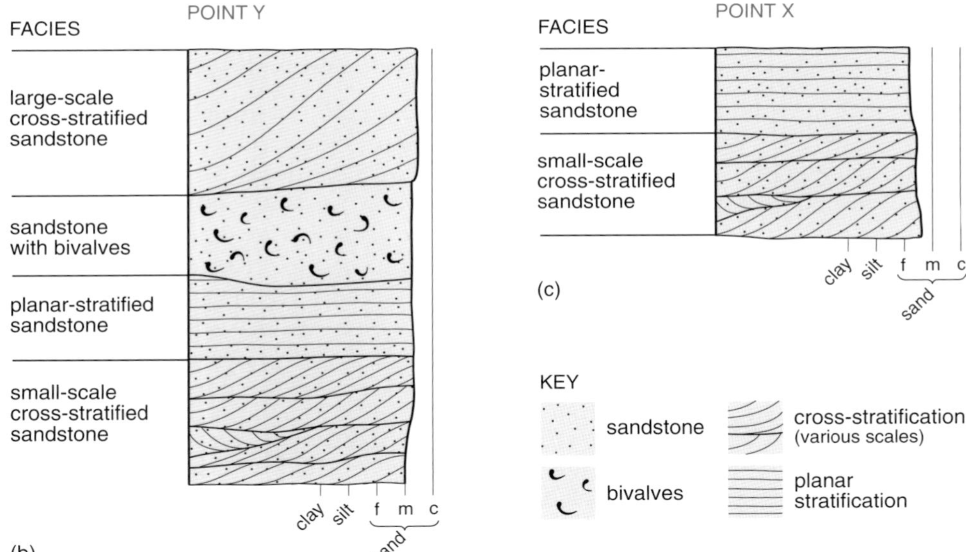

Figure 12.3 (a) Diagram to illustrate the plan view relationship of the various facies associated with a beach. (b) Vertical succession of facies at point Y assuming that the sedimentary deposits were building out in a distal direction. (c) Vertical succession of facies at point X assuming that the sedimentary deposits were building out in a distal direction.

'large-scale cross-stratified sandstone facies' (Figure 12.3). Below low water mark, the sand is constantly being reworked by the breaking waves so it will also be cross-stratified, but on a smaller scale than those formed by wind at the landward side of the beach. We can call this the 'small-scale cross-stratified sandstone facies' (Figure 12.3).

Let us now consider what will happen to these four facies which have been deposited adjacent to each other along a coastline. Imagine that over geological time either the beach was gently uplifted tectonically, or more sand was supplied, or sea-level slowly became lower. The end result of what happens to the facies is the same in each case and it is this result that we need to consider. What happens is that each facies will be deposited in a slightly more seaward (distal) position than it was before; we call such building of facies in a distal direction **progradation**.

- ❑ If the beach has prograded and you were to dig a hole at point X shown on Figure 12.3a, what vertical succession of facies would you find?

■ Planar-stratified sandstone facies overlying small-scale cross-stratified sandstone facies (Figure 12.3c).

❑ Assuming that progradation has been occurring continuously, if you were to dig a hole at point Y, what vertical succession of facies would you find from the top downwards?

■ At the top, large-scale cross-stratified sandstone overlying sandstone with bivalves overlying planar-stratified sandstone, overlying small-scale cross-stratified sandstone (Figure 12.3b).

12.2.2 WALTHER'S LAW

The relationship between facies that are originally laid down next to each other and then end up in a vertical succession was first noted in 1894 by Johannes Walther and so it is known as **Walther's Law**. Walther wrote that *'it is a basic statement of far-reaching significance that only those facies and facies areas can be superimposed primarily which can be observed beside each other at the present time.'* In clearer terms, this means that facies which were originally deposited laterally adjacent to each other can, given the correct conditions, be preserved as a vertical succession.

❑ Do you think that Walther's Law still applies if the contacts between the facies are sharp rather than gradational? Explain your reasoning.

■ No, it does not always apply because a sharp contact may indicate that some sediment is missing either through erosion or non-deposition.

So, the law must be applied carefully and if the contact between the facies is sharp, rather than gradational, then the facies cannot be assumed to have been laid down laterally adjacent to each other. Like all rules, there are some exceptions. Some facies have naturally sharp bases due to depositional processes, even though they are laid down next to each other, e.g. storm-deposited sandstone beds within an otherwise mudstone succession (Video Band 10 *Coastal processes*, which you studied in Activity 1.1). This indicates again how important it is to understand the processes that have led to sediment deposition. Figure 12.4 shows what a cross-section through a beach profile might look like and allows us to see how the facies relate to each other in time as well as in space.

Question 12.2 (a) Examine the bold black line labelled X–X´ in Figure 12.4. Explain whether it represents a time plane, i.e. a moment in time, or a facies boundary. (b) What do the thin black lines A–A´, B–B´, C–C´ and D–D´ represent? (c) What do lines W–W´, Y–Y´ and Z–Z´ represent?

Figure 12.4 is important because it explains how sedimentary rocks relate to each other in space, what a progradational vertical succession of rocks represents in terms of how the different beds originally related to each other in space, and how these all relate to the passage of time. Throughout the rest of the Block we will use 3D diagrams and vertical successions of rocks to summarize different features. You may find it useful to refer back to Figure 12.4 if you get confused.

12.3 VERTICAL SUCCESSIONS OF FACIES AND GRAPHIC LOGS

A **graphic log** is simply a visual representation of the *important* and *common* features of a vertical succession of sedimentary facies. Graphic logs show the vertical thickness of the individual beds, the nature of the contacts between each bed, and the lithology, sedimentary structures and fossils within the beds (i.e. the facies). Graphic logs are an excellent way of recording data because they

Plate 2.1 Rock debris on a mountain slope in the Glyders range, Snowdonia. (*Ruth and William Welsh*)

Plate 2.2 An accumulation of pebbles on a storm beach.

Plate 2.3 A Cambrian beach deposit from rocks in Pembrokeshire.

Plate 2.4 Glacial sediments.

Plate 2.5 Part of the tidal flat in the Dee estuary, Wirral.

(a)

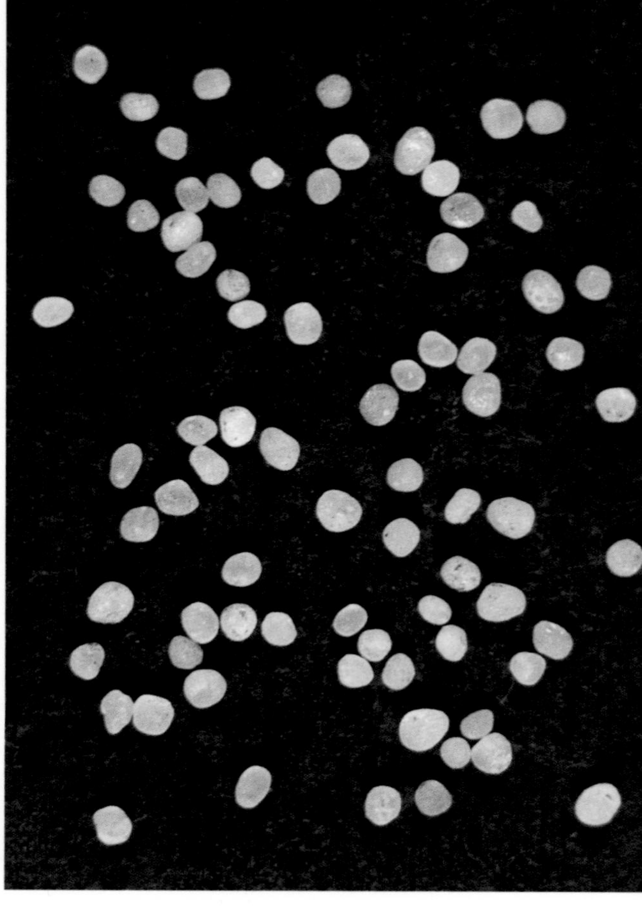

(b)

Plate 2.6 Sand grains. (a) Water-transported quartz sand grains about 1 mm in diameter. (b) Wind-blown quartz sand grains about 0.5 mm in diameter.

Plate 4.1 Cross-stratification in Triassic sandstones from Hilbre Island, off West Kirby, Wirral. The exposure shown is *c*. 3 m high.

Plate 4.2 Cross-stratification in Carboniferous sandstones, The Roaches, Staffordshire. The exposure shown is *c*. 80 cm high. (*Dave Rothery, Open University*)

Plate 4.3 Cross-stratification in an exposure of Triassic sandstone at Hoylake, Wirral. Car key shows scale.

Plate 4.4 Subaqueous dunes exposed in an estuary at Bantham, South Devon.

Plate 4.5 Herring-bone cross-stratification in Cretaceous sands, near Leighton Buzzard, Buckinghamshire. A single set of cross-stratification is labelled.

Plate 4.6 Mud drapes preserving ripples in cross-stratified Cretaceous sands. The mud drapes are the much darker laminae that alternate with the lighter-coloured sands. A particularly well-developed mud drape has been labelled.

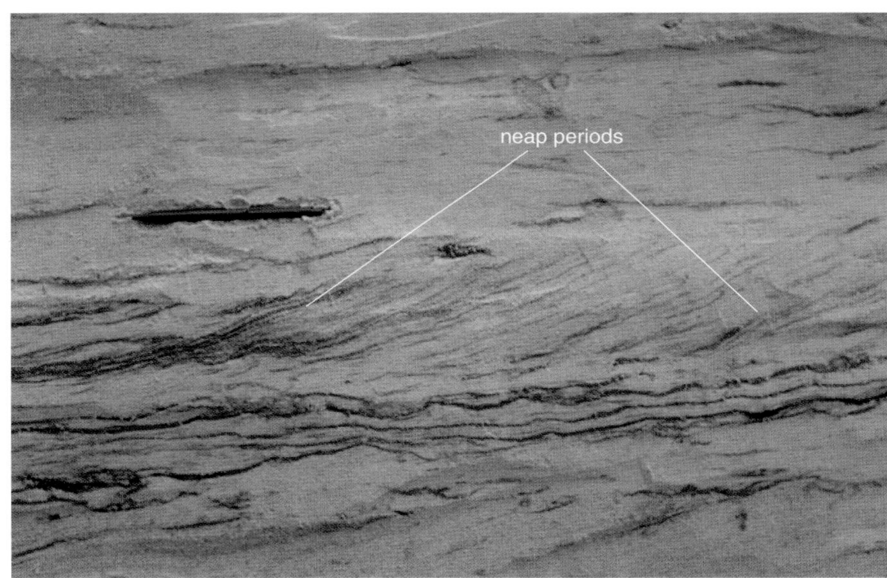

(a)

(b)

Plate 4.7 Tidal bundles. (a) The two areas of closely spaced mud drapes represent two periods of neap tides. The more widely spaced mud drapes represent spring tides (pen shows scale).
(b) For use with Question 4.8.

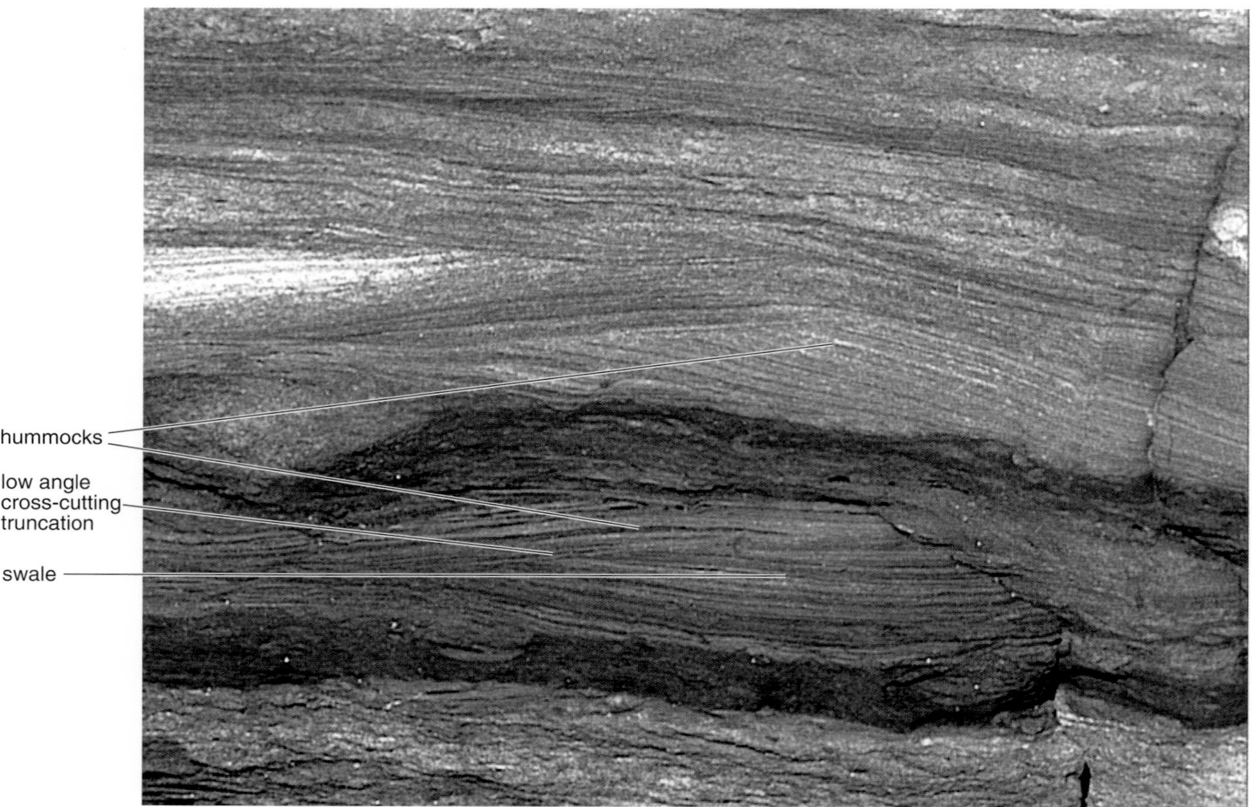

Plate 4.8 Hummocky cross-stratification in Carboniferous rocks from Northumberland, England. The exposure shown is *c*. 2 m high.

Plate 4.9 Cross-stratification in Jurassic aeolian sandstones, Utah, USA. The exposure shown is *c*. 12 m high. *(Jim Ogg, Purdue University, USA)*

(a)

(b)

(c)

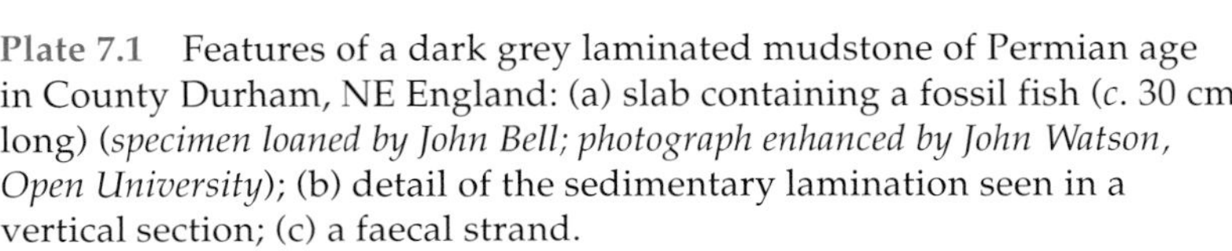

Plate 7.1 Features of a dark grey laminated mudstone of Permian age in County Durham, NE England: (a) slab containing a fossil fish (*c.* 30 cm long) (*specimen loaned by John Bell; photograph enhanced by John Watson, Open University*); (b) detail of the sedimentary lamination seen in a vertical section; (c) a faecal strand.

Plate 13.1 Aerial photograph of a meandering river system. (*Bob Spicer, Open University*)

Plate 13.2 Aerial photograph of a braided river system. (*Bob Spicer, Open University*)

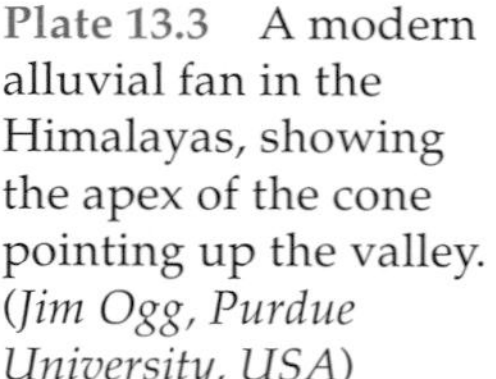

Plate 13.3 A modern alluvial fan in the Himalayas, showing the apex of the cone pointing up the valley. (*Jim Ogg, Purdue University, USA*)

Plate 14.1 A barchan dune in Oman. The steep slope is facing towards the right. (*Dave Rothery, Open University*)

(a)

(b)

Plate 14.2 (a) A steep-sided desert valley or wadi formed during heavy rainfall. This example is from Oman. (*Dave Rothery, Open University*)
(b) Wadi deposits in Oman showing crude stratification and grading. (*Dave Rothery, Open University*)

Plate 15.1 Gently seaward-dipping profile of a modern-day foreshore showing the planar-stratified sands in a trench in the foreground. The sea is towards the right.

Plate 15.2 Planar-stratified sandstones of ancient foreshore deposits (Carboniferous, NE England). Scale approximately 1 m from top to bottom.

Plate 15.3 Aeolian sand dunes stabilized by salt-tolerant grasses. The foreshore is to the left (Northumberland, England). (*Ros Todhunter, Open University*)

Plate 15.4 Jurassic organic-rich mudrocks from the Kimmeridge Clay Formation of Dorset, England. In this case, the darkest layer running across the middle of the photograph is the richest in organic matter.

Plate 15.5 A barrier island system called Kiawah Island, South Carolina, USA. The picture is taken at low tide. The barrier runs across the middle of the picture, the lagoon is to the left and the open sea to the right where the waves are breaking. Complex sandbodies are being deposited and channels cut where there is a break in the barrier about three-quarters of the way up the picture.

Plate 16.1 Mixed siliciclastic and carbonate succession, the Jurassic-age Corallian Beds, Dorset, England.

Plate 16.2 Cross-stratified oosparite from the Jurassic of the Dorset coast. Scale bar = 20 cm.

(a)

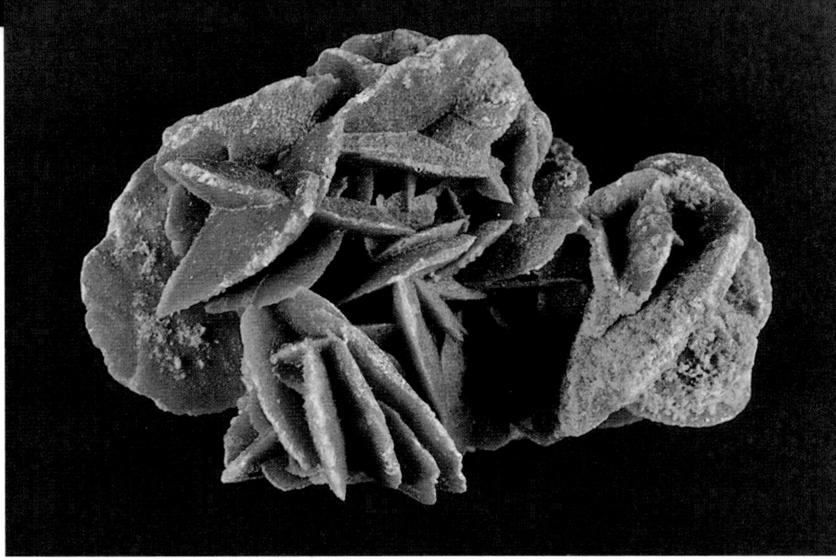

(b)

Plate 16.3 (a, b) The 'desert rose' form of gypsum. Both specimens are *c*. 8 cm across. (*Specimens loaned by Mike Henty, Open University*)

Plate 16.4 Chicken-wire texture, showing stringers of micrite between evaporite nodules (originally anhydrite, now gypsum because of contact on exposure with modern porewaters). These Jurassic deposits are from Mount Carmel, Utah, USA. Penknife shows scale. (*Jim Ogg, Purdue University, USA*)

Plate 16.5 An isolated carbonate platform, Majuro, Marshall Islands, Pacific. A barrier of coral reefs protects a lagoon area (bottom part of picture) from the breaking waves which can be seen on the open sea side of the barrier. The barrier is broken by restricted channels through which water and sediments may pass. Reworked carbonate sand derived from the reef supports vegetation in the subaerial part of the barrier. (*Jim Ogg, Purdue University, USA*)

Plate 16.6 The barrier reef of Majuro, Marshall Islands, Pacific is running diagonally across the picture. To the right of this is a subaqueous carbonate sandbody forming in the lagoon. The open sea is to the left of the barrier. (*Jim Ogg, Purdue University, USA*)

Plate 16.7 Intertidal and supratidal area of Florida Keys, USA. (*Jim Ogg, Purdue University, USA*)

Plate 17.1 A succession of turbidites and one debris flow from Portugal. Hammer shows scale.

Plate 17.2 A Neogene-age debris flow from Cyprus. The clast size varies from centimetres to large blocks several metres across, such as the white one on the left. The exposure shown is *c*. 12 m high.

Plate 17.3 Cretaceous-age pelagic chalks near Dover, Kent. Height of cliff shown is about 15 m.

Plate 17.4 Neogene-age pelagic chalks and interbedded marls from Cyprus. Height of cliff shown is about 10 m.

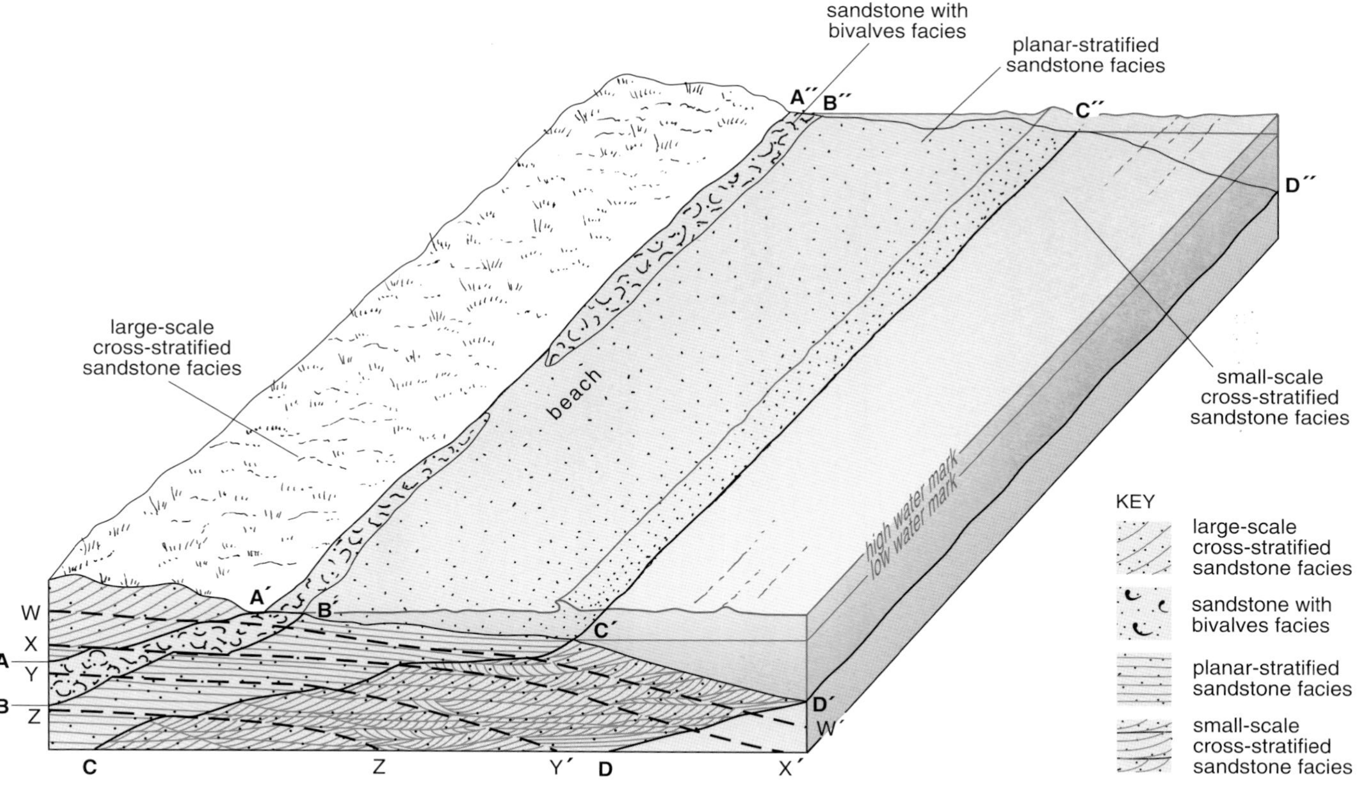

Figure 12.4 Plan view and cross-section showing the lateral and vertical relationship between adjacent facies associated with a beach. The boundaries between the facies are gradational and the facies have been prograded.

are a form of shorthand which shows the main features of the sedimentary rocks much more readily than lots of written notes. Figures 12.5 and 12.6 (overleaf) show examples of graphic logs. Figure 12.5 also shows a photograph of the exposure that was being recorded.

The different features of these sedimentary successions are shown in the following ways:

- *Bed thickness*: the vertical scale (Figure 12.5) on graphic logs is usually given in metres and shows the thickness of the individual beds. If the bed is variable in thickness, an *average* is usually shown.
- *Bed contacts*: these may be sharp or gradational and planar or undulose and are shown by the nature of the line drawn between the beds. See key to Figures 12.5 and 12.6 for examples.
- *Lithology*: the lithology of a bed is shown as a symbol within the bed, e.g. stippled for sandstones (as in bed 7, Figure 12.5), broken lines for claystones and mudstones (bed 4, Figure 12.5), alternating dashes and dots for siltstone (bed 2, Figure 12.5), and bricks for limestone (Figure 12.6). If it is a sandy limestone, then it would be shown as two overlapping symbols, in this case the brick ornament with some stipple (e.g. bed 4, Figure 12.6).
- *Grain size in siliciclastic rocks*: this is an important feature since it gives an indication of the energy involved in the depositional processes; it is shown as the horizontal scale on the graphic log. The scale used is basically a simplification of that shown in Table 6.1 with the finest grain size, 'clay', on the left-hand side and the coarsest grain size on the right-hand side. The graphic log is constructed to show the beds as wide as the volumetrically most abundant grain size of which they are composed. So, if the lithology is a fine-grained sandstone, then the width of the bed on the log would extend

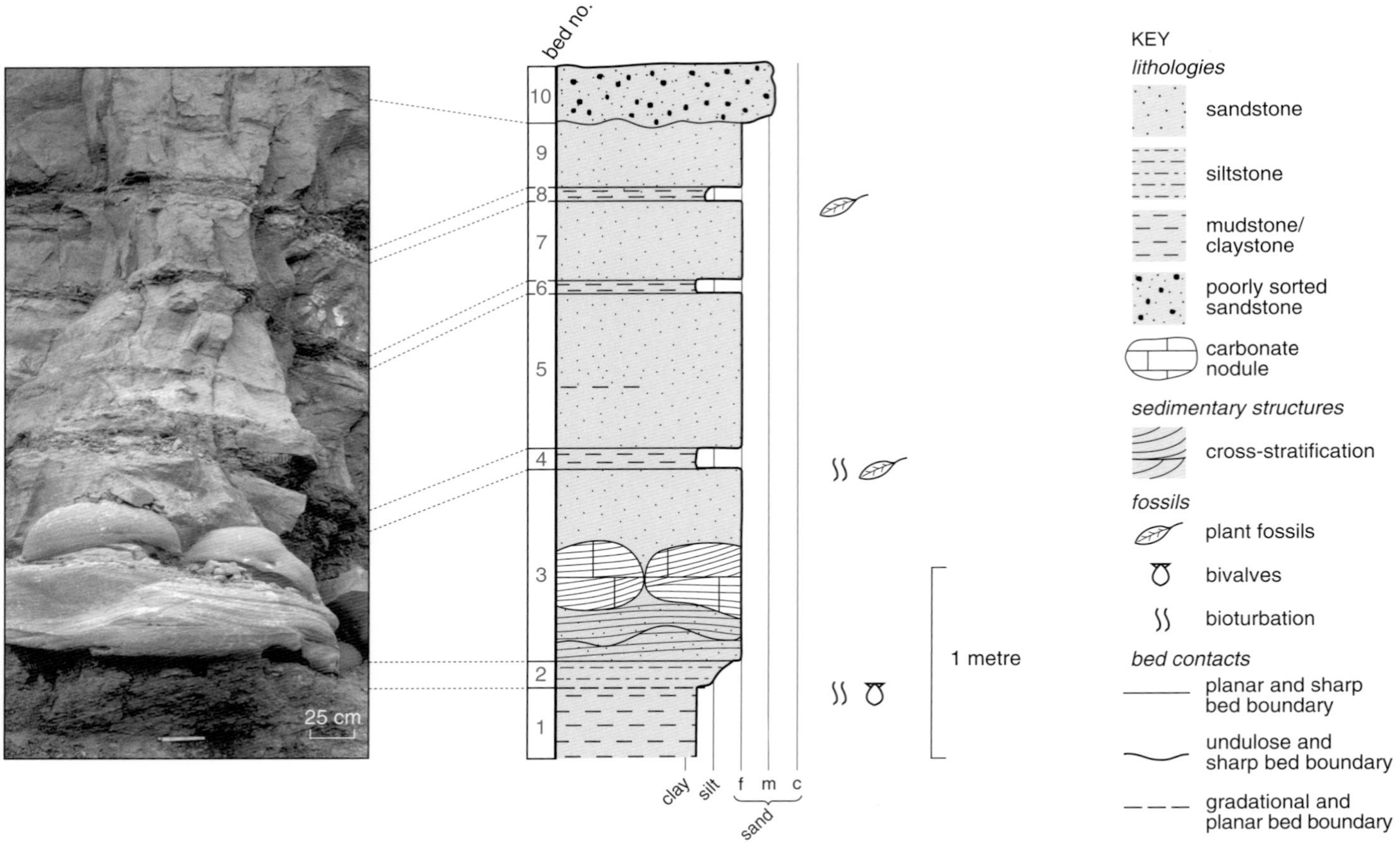

Figure 12.5 Example of a graphic log of a mainly siliciclastic succession and some commonly used symbols.

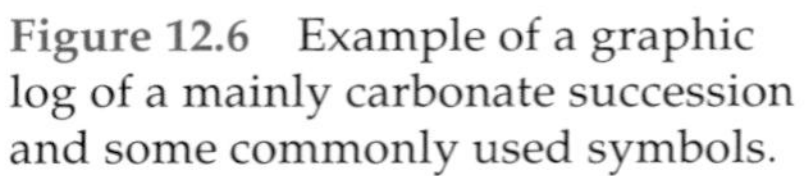

Figure 12.6 Example of a graphic log of a mainly carbonate succession and some commonly used symbols.

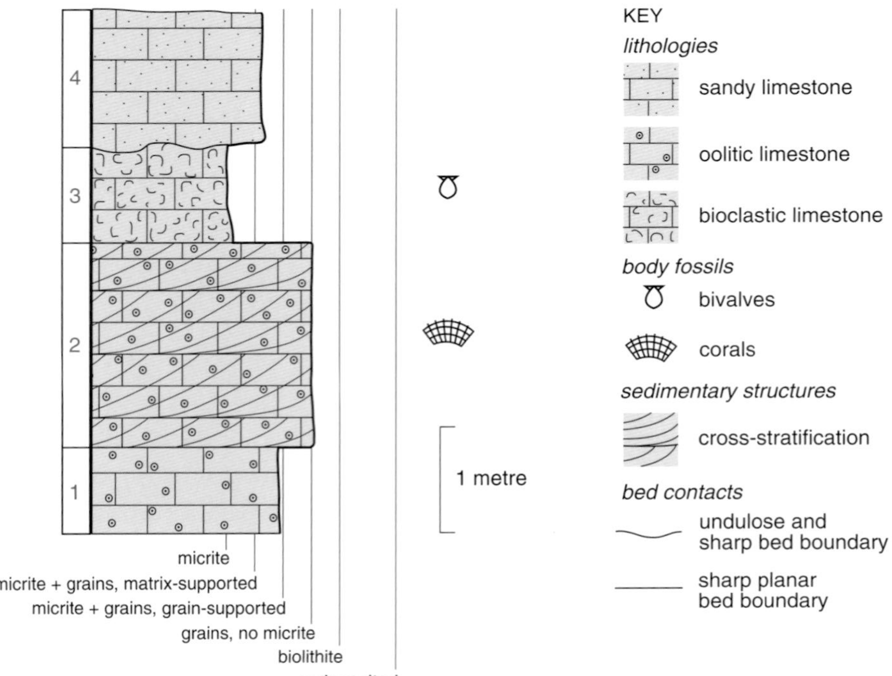

as far to the right as the fine-grained sandstone vertical line shown for bed 7 in Figure 12.5. If the rock is composed of a 50 : 50 mixture of fine-grained sand and coarse-grained sand, it would be shown as the volumetric average of these, i.e. a medium-grained sandstone (bed 10, Figure 12.5). A bed or group of beds that show a gradational coarsening upwards from say a siltstone at the base to a fine-grained sandstone at the top are shown by an inclined line that starts on the silt line and ends on the medium-grained sand line (e.g. bed 2, Figure 12.5).

❑ How do you think variations in grain sorting, such as poorly and well-sorted sandstones, might be indicated on a graphic log?

■ There are several ways that sorting could be indicated; the simplest is to show well-sorted sandstones with evenly sized stipple and poorly sorted with unevenly sized stipple (e.g. bed 10, Figure 12.5). Another way is to add an extra column to the graphic log to show this or to add a few notes to the right-hand side of the graphic log.

- *Grain size in limestones*: generally speaking, this is not as important for determining the energy of processes involved because many of the grains are biologically or chemically formed *in situ*, so large bioclasts are more a function of which organisms were living at the time the sediment was deposited rather than the result of transport by strong currents. The size of ooids depends on the amount of carbonate available, the time they have been forming and having the appropriate conditions and amount of agitation in the water. However, the amount of carbonate mud (micrite) present in a succession of carbonates is useful, because under high energy conditions the carbonate mud tends to be kept in suspension whereas under lower energy conditions it is deposited. This is similar to the behaviour of fine-grained particles in siliciclastic systems. So the amount of carbonate mud is indicated on the horizontal scale (Figure 12.6): the grain size lines from left to right would then become:
  - micrite (i.e. all carbonate mud);
  - micrite with grains, matrix-supported; where the grains are supported by the matrix (Figure 4.9);
  - micrite with grains, grain-supported; where the grains are in contact with each other (Figure 4.9);
  - grains, no micrite; grains only, usually cemented together with sparite.

  In this Course, we recognize two other distinct types of limestone, and because they are both high energy they are indicated by two further subdivisions to the right-hand side. These are:
  - biolithite (i.e. carbonates preserved *in situ* (Section 6.5.3));
  - redeposited (e.g. breccias and conglomerates composed of redeposited limestone clasts).
- *Fossils*: these are usually shown as symbols just to the right-hand side of the lithology column or if they are particularly abundant and are forming many of the grains within the rock, e.g. RS 21, they may be shown as part of the lithology symbol itself. A few of the common fossil symbols that we use in this Block are shown in Figures 12.5 and 12.6.
- *Sedimentary structures*: the best way to show these on a graphic log in terms of both their scale and geometry is to sketch a *simplified* but *accurate* version of the structure onto the bed. If the sedimentary structure is too small to be shown at the scale that the log is drawn then, again, like the fossils, a small symbol can be added to the right-hand side of the log. This is also the place to add other features not recorded elsewhere, e.g. pyrite nodules.

When a graphic log and its symbol key is complete, the author, or anyone else with a trained eye, should be able to recognize easily any of the individual beds in the field. We shall use graphic logs extensively throughout the rest of this

Block to illustrate the principal features of each environment that may be preserved in a vertical succession.

Question 12.3 Using Figure 12.5 and your grain-size scale, where on the horizontal scale would you plot beds of the same rock type as RS 10, RS 20 and RS 27?

Question 12.4 Using Figure 12.6 and a hand lens, where on the horizontal scale would you plot beds of the same rock type as RS 21 and RS 22?

Activity 12.1

You should now complete Activity 12.1 which reinforces the principles of graphic logging. It is based on the video sequence *Graphic logging* on DVD 3 which shows the process of graphic logging in the field using a Carboniferous succession exposed in north-east England.

Activity 12.2

You should now complete Activity 12.2 which will enable you to practise putting together a graphic log.

12.4 Facies and environmental models

Many facies may be grouped together into a model which summarizes the features of a sedimentary environment. However, it must be remembered that models represent an idealized situation, and no real life environment, either today or in the past, will be exactly like these models. Nevertheless, models are exceptionally useful for four reasons:

1 They act as the *'average'* for purposes of comparison.

2 The model provides a *framework* suitable for further observation on the real sedimentary successions.

3 They can help *predict* what is likely to occur in different geological situations.

4 They form a basis for an *integrated interpretation* of the environment they represent.

Figure 12.4 represents part of a model for a coastal strandplain environment. We will develop this model further in Section 15, and construct facies models for the other environments in the remaining Sections.

12.5 Summary of Section 12

- The Earth's surface can be split into a number of different environments; these include alluvial (meandering and braided), deserts, coastal (strandplains, deltas, barrier islands), carbonate platforms and the deep sea.
- For sediment deposition, there needs to be a supply of sediment and a space in which to put the sediments. These are controlled by tectonics and climate which, in turn, affect weathering, erosion, seawater conditions and sea-level. The type of biota present may also affect sediment deposition.
- Sedimentary facies are used to describe all aspects of sedimentary rocks, including their lithology, fossil content and sedimentary structures.
- Progradation of sediments occurs through continuous sediment supply, or tectonic uplift, or a lowering of sea-level.

- Walther's Law states that facies which were originally deposited adjacent to each other in space can, given the correct conditions, be preserved as a vertical succession. If the contacts between the facies are sharp, one or more facies may be absent and Walther's Law may not apply.
- Vertical successions of facies are conveniently represented by graphic logs. This is a shorthand pictorial way of representing sedimentary facies. The vertical axis represents the thickness of the beds and the horizontal axis represents the grain size. The lithology, fossils, sedimentary structures and nature of the bed contacts are all shown pictorially on graphic logs.
- Environmental models are useful for four reasons: (i) to act as the normal scenario for comparison, (ii) to provide a framework (iii) to predict what is likely to occur, and (iv) as a basis for integrated interpretation.

12.6 Objectives for Section 12

Now you have completed this Section, you should be able to:

12.1 Describe how climatic conditions, tectonic setting and the type of biota control the deposition of sediment.

12.2 Distinguish between the terms facies and lithology, and give examples of each.

12.3 Explain with the aid of a sketch how facies found near a prograding coastline may be deposited on top of each other.

12.4 Summarize Walther's Law and explain when it is not applicable.

12.5 Explain what different components are used to construct a graphic log.

12.6 Record your observations of sedimentary rocks (including their fossils and sedimentary structures) in the form of a graphic log and demonstrate how data can be assimilated from graphic logs.

12.7 State the four reasons why models of environments are useful in the interpretation of sedimentary rocks.

Now try the following questions to test your understanding of Section 12.

Question 12.5 If a land area changes from being tectonically stable to subsiding and sea-level is rising, would more or less sediment be likely to reach the sea, assuming that all other factors like climate remain stable?

Question 12.6 If the climate gradually becomes cooler and wetter, how might this affect; an area of (i) evaporite deposition; (ii) a siliciclastic coastline adjacent to a poorly vegetated land-mass?

Question 12.7

(a) Use the following notes to construct a graphic log summarizing the main depositional features of the following succession. Choose an appropriate scale by adding up the total thickness and deciding what size your paper is (we used 1 m = 2 cm). Draw a vertical line of the appropriate total length and then mark the grain size divisions along the base. Do not forget to include your vertical scale and a key to the symbols that you use. The succession from the bottom (oldest bed) upwards is as follows:

Bed A comprises a grey mudstone 2 m thick with 30 cm spheroidal nodules of calcium carbonate whose bases are 1.3 m above the base of the bed. The mudstone contains brachiopods and trilobites. There are no visible sedimentary structures. The bed is gradationally overlain by Bed B over about 20 cm. The contact is planar.

Bed B comprises 80 cm of pale-yellow siltstones with wave-formed ripples in the upper 20 cm. It also contains occasional brachiopods and is sharply overlain by Bed C. The contact between Bed B and Bed C is planar.

Bed C comprises 3 m of a red, coarse-grained, well-sorted sandstone. The sand grains are frosted on the surface and there is large-scale cross-stratification dipping at 25°. The bed sets are 1.5 m thick and there are no fossils in this bed.

(b) Which beds in the succession described in part (a) suggest a marine environment and which a non-marine? Give reasons for your answer.

(c) What geological term might be used to describe the contact between Bed B and Bed C?

13 ALLUVIAL ENVIRONMENTS

River systems dominate the surface processes that are seen in most of Europe today and so some features of them are probably familiar to you. They are important because they are the main routes for the transport of sediments to the sea from areas of continental weathering and erosion. River systems represent a complex system of erosion, sediment transport and deposition. Not only is the river channel important but also the land around the river which floods when the river swells, the catchment area for the rainwater, and the source of the sediments. The area that the river floods during high water levels, which is generally a flat area either side of the river channel, is called the **flood plain**.

❏ Describe the difference between the terms fluvial and alluvial.

■ The term 'fluvial' is used to describe the river channel itself, whereas the term 'alluvial' refers to the channel itself plus all the area under its influence, i.e. the river channel and its flood plain (Sections 4.2.1 and 12).

The pattern of river development, and the related sedimentation we see in western Europe, is only one example of a number of patterns that may be found in different parts of the world. Fluvial erosion and alluvial deposition operate to some extent in all types of continental environments, including deserts and glaciated regions, although the patterns of development vary.

The prime source of water for a river is precipitation in the area around the river. Rain or snow falling directly into a river channel make only a minor contribution. Most of the water arrives in the river channel by various indirect routes through the drainage basin in which the river is situated, as illustrated in Figure 13.1. Another important factor in alluvial environments is the position of the water table (Figure 13.1 and Section 6.1). Permanent rivers form only where the water table intersects the ground surface. In order to maintain a high water table, precipitation needs to exceed evaporation. So, in regions with humid climates, it is the vast amounts of water stored beneath the water table that permit continuous river flow. However, temporary rivers can form when the rate of influx of water exceeds the rate of infiltration into the ground and this is typical of what happens in arid or semi-arid regions. The rare, sporadic rainstorms create a lot of surface water and allow intermittent flow, but the results of such rainstorms are often catastrophic.

Question 13.1 Examine Figure 13.1 and explain whether you would expect the water table to remain at a constant depth below the ground surface (a) throughout the year, and (b) regardless of the underlying lithology.

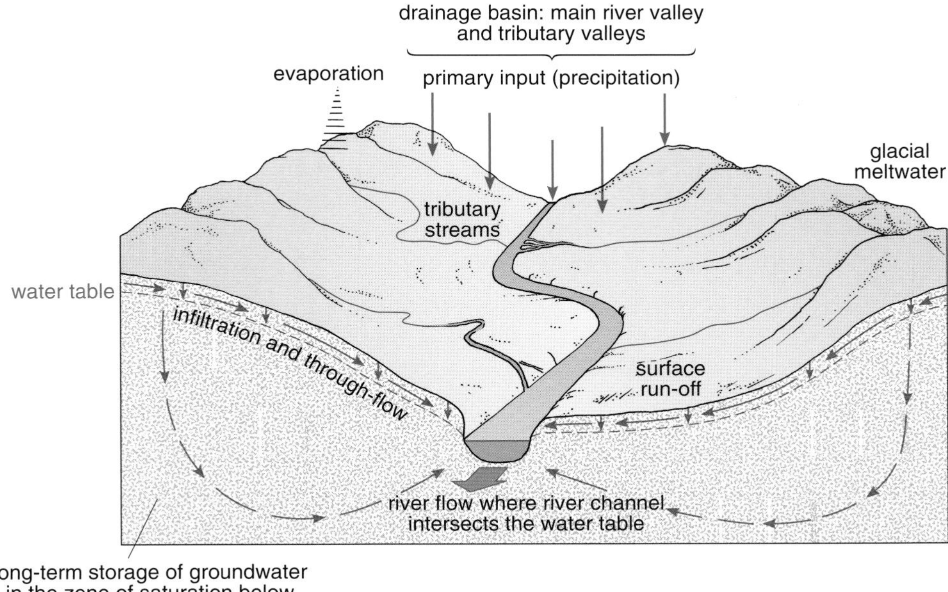

Figure 13.1 The different sources of water within a drainage basin that feed a river. A certain amount of water is lost by evaporation, but in humid climates the primary input from precipitation always exceeds losses due to evaporation.

The sediment derived from weathering is transported down the valley sides into a river channel by sediment gravity flow processes (Section 4.3) like those shown in Figure 13.2, and is supplemented by material eroded from the river banks by gradual attrition and more catastrophic river bank collapse.

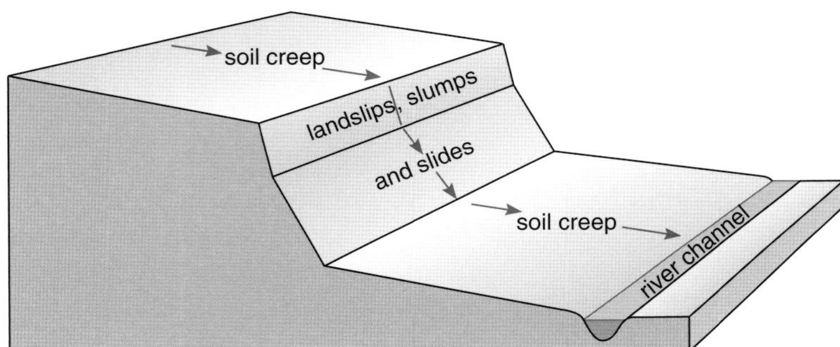

Figure 13.2 Cross-section to show the most important sediment gravity flow processes that lead to the development of river valley slopes, and transport weathered rock debris into a river channel.

The source of energy to move the water in a river and, indirectly the sediment, is the gravitational potential energy that results from the difference in height between the upstream and downstream sections of a river channel. As the water flows down-slope, much of this is converted to kinetic energy (the energy of the motion of the water) which transports sediments, erodes the river channel, and is capable of causing tremendous devastation at times of flood. The currents within the channel transport the sediment as dispersed grains.

❑ Recall the three ways in which the sediments move in water according to their size.

■ The largest grains (bedload) move by sliding or rolling, the medium-sized to smaller-sized grains (also part of the bedload) by bouncing along the bed (a movement called saltation), and the smallest grains (suspension load) are carried in suspension (Section 4.1.1).

13.1 THE RIVER VALLEY

In general, river valleys approximate to a 'V' shape in cross-section, particularly those in humid climates. They are wide at the top with the river occupying a channel at the narrow base. This is because mass movement of sediment, beginning with slumping of the channel sides, leads to the development of slopes on either side of a newly formed river channel. These slopes form new sites for water run-off and channel development, so that in time a drainage basin is developed (Figure 13.1) with its main river and many smaller feeders called tributaries.

If you were to follow a river downstream from the source to the mouth, you would see several changes. The volume of water flowing past in a given time, which is termed the **discharge**, increases exponentially (i.e. ever more rapidly), because water is being added continuously from tributaries and other sources like those shown in Figure 13.1. The valley and the river channel grow wider and deeper, which is not surprising because erosion is a continuous process related directly to the discharge. However, the gradient of the river bed *decreases* downstream as the discharge *increases*. This results in a longitudinal valley profile that is generally concave, with the highest gradient near the source but a gradient approaching zero near the mouth (Figure 12.2).

At the source of a river, the discharge is small, and consequently the channel is narrow and shallow. A lot of the energy is lost by friction at the base and sides of the channel, and by contact with the atmosphere, and so the discharge can only be maintained on a steep slope. Nearer the mouth of the river, the channel depth and width are greater, so the total area of water in contact with the base, sides and atmosphere in proportion to the volume of water in the channel is less. Energy losses due to friction are therefore less and so the discharge is able to be maintained on a shallow gradient. The fact that rivers flowing in narrow, shallow channels require a steeper gradient to maintain their discharge than rivers flowing in broad, deep channels is an important control on the pattern of river development, as you will see in Section 13.2.

Downcutting of a river channel should cease close to where it enters the sea because the gravitational potential energy available in the system approaches zero. However, as discussed in Section 12.1, sea-level has not remained constant through geological time and changes in sea-level affect the balance of erosion and deposition in rivers such that the ideal concave longitudinal profile (Figure 12.2) is rarely achieved. In fact, in some cases, what we see is a series of superimposed concave longitudinal profiles due to successive sea-level changes.

The river's flood plain is often very fertile due to moisture and nutrients from the river waters; it is therefore colonized by vegetation and soils develop. The type of soil formed depends on the climate and the composition of the bedrock and flood plain sediments. If the climate is humid, the area is almost continuously flooded, and the sediments are siliciclastic, abundant vegetation grows and peat may accumulate. On burial and compaction, the peat will form *coal*. The soil in which the coal-forming plants develop is often called a *seatearth*; these are particularly common in rocks of Carboniferous age in Britain. In hot arid climates, evaporation of the water between the sediment grains in the flood plain area can lead to the deposition of carbonates and rarely evaporites.

13.2 FLUVIAL CHANNEL STYLES AND THEIR TYPICAL SEDIMENTARY SUCCESSIONS

A river does not flow in the same direction over its entire length. It will change course if it meets an obstacle such as a resistant rock outcrop or a zone of more rapid erosion like an active fault zone. In Section 13.1 we discussed the fact that the gradient of a river changes along its length and that broad deep channels are found in areas of low gradient whereas narrow, shallow channels are found in areas of steep gradient. The amount of vegetation also governs the pattern of the channels: lush vegetation tends to bind the river banks together and help to prevent erosion. The climate governs the amount and periodicity of the rainfall and therefore the amount of water in the drainage basin; this in turn will influence the shape of the river channel. In summary, the factors that govern the morphology of the channel are: bedrock geology, tectonics, topography, vegetation and climate.

A wide range of different channel morphologies is possible but we shall consider just two. These two end members provide a good model for comparison of most ancient and modern river systems. The two features that distinguish these end members are:

- The **sinuosity** of the channels, which depends on the number of twists and turns in the channel and the length of the loops. In Figure 13.3, (a) has a low sinuosity whereas (b) has a high sinuosity.
- The number of times that the river channel splits into several channels which in turn rejoin and split again. This is called **braiding**. In Figure 13.3, (c) shows a river with only small amount of braiding whereas (d) shows a river with a lot of braiding.

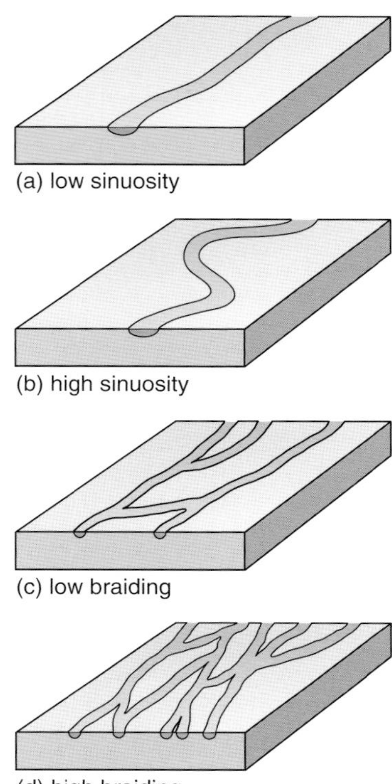

Figure 13.3 Sinuosity and braiding in river channels; (a) low sinuosity; (b) high sinuosity; (c) low braiding; (d) high braiding.

Rivers with high sinuosity but low braiding are called 'meandering' rivers. The pronounced bends in the river are **meanders**. Rivers with low sinuosity but high braiding are termed 'braided' rivers.

❑ Which part of Figure 13.3 would you call a meandering and which part a braided river?

■ Figure 13.3b is an example of a meandering and Figure 13.3d a braided river.

13.2.1 Meandering river systems

Components of the meandering river systems

Meandering river systems comprise a distinct sinuous channel and river banks (Figure 13.4, Plate 13.1). Meandering rivers are characterized by: (i) relatively low overall topographic slopes both within the channel and its flood plains, (ii) high suspension load to bedload ratio, and (iii) the area into which the channel has cut being made up of relatively cohesive deposits.

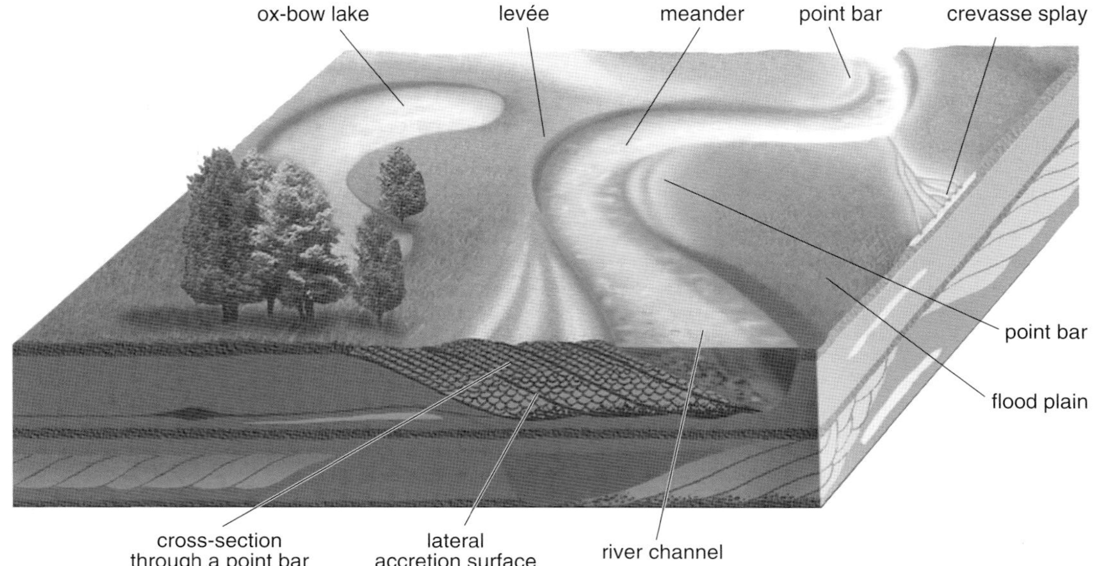

Figure 13.4 A block diagram of a meandering river with the various features shown. A coloured version of this Figure is shown on the Sedimentary Environments Poster and Sedimentary Environments Panorama (on DVD 3).

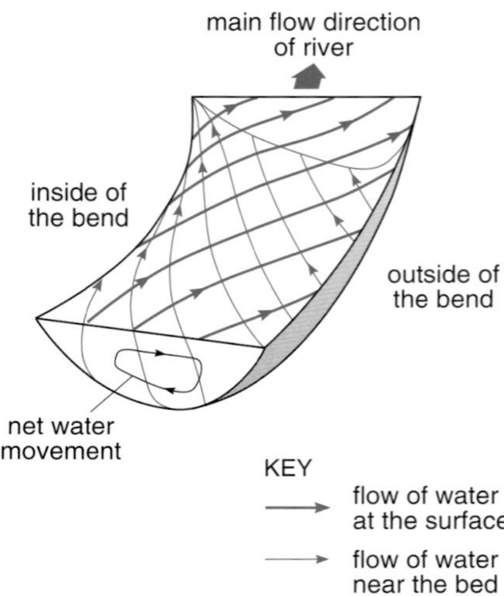

Figure 13.5 A plan view and lateral cross-section of a river meander to illustrate the helical flow of water. Water flows across the surface towards the outside of the bend and is returned by the flow across the river bed towards the inside bend. Consequently, the main current swings from side to side of the channel.

As a meandering river develops through time, the number and size of the meanders will increase, in effect lessening its slope compared to a river that flows straight down a slope. If you find this hard to visualize, then remember that roads on very steep hillsides are built in a zig-zag fashion with hairpin bends in order to lessen the gradient of the road. River meanders gradually increase in amplitude and migrate downstream, because a helical or corkscrew flow of water is superimposed on the overall downstream movement of water (Figure 13.5). This corkscrew flow is the movement of surface water across the meander towards the outside of the bend and its return across the base of the channel towards the inside of the bend. The result is that the main current does not flow straight down the river channel but impinges against each bank in turn causing erosion of the outside bend (Figure 13.6). See Box 13.1 for further explanation.

❑ Will the maximum erosion occur at the mid-point of the meander or somewhere else?

■ It will not occur at the mid-point, because the flow is helical. The point of maximum erosion actually occurs slightly downstream of the mid-point thus extending the development of the meander further downstream (Figure 13.6).

So how does the amplitude of the meander increase? On the outside of a bend, water flows at a faster speed than on the inside. So, while the outside bank of a meander is actively eroded, material is deposited on the inside bend of the next meander downstream where the flow is slower. The sediment is added slowly sideways on the inside of the bend and forms a bank of sediment called a **point bar** (Figure 13.4). At the same time, the outside bend is being eroded further. The overall result is that lateral erosion on the outside of the bends is compensated for by lateral deposition on the inside of the bends, and the amplitude of the meander increases (Figure 13.6).

Figure 13.6 The lateral migration (horizontal in the Figure) of a meander caused by erosion of the outside bend and point bar accretion on the inside bend. Downstream migration (upwards in the Figure) of the meander also occurs, because maximum erosion occurs downstream of the mid-point in the meander.

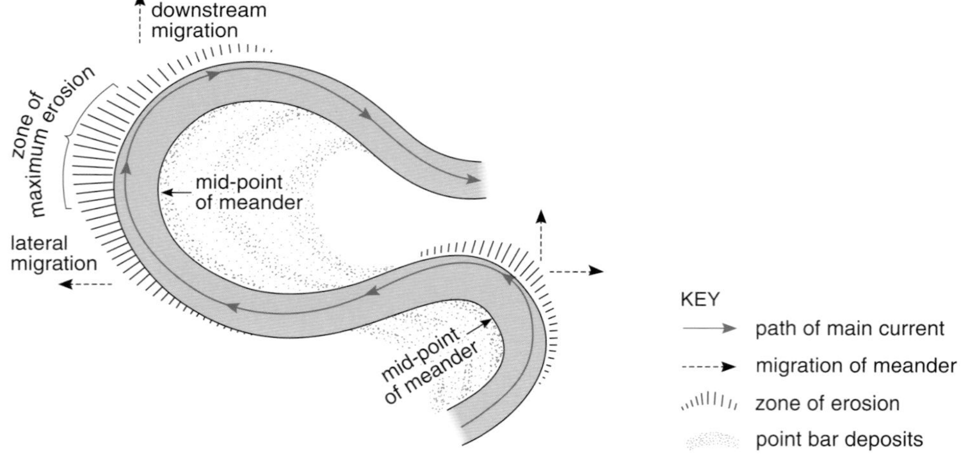

Box 13.1 Further explanation of helical flow

The simplified explanation of helical flow in this Box is given for interest; you are not expected to remember it. As water flows round a bend, centrifugal forces tend to push it towards the outside bank. However, the actual water velocity gradually decreases with depth, due to friction with the river bed, and so the pressure on the outside bank caused by the centrifugal effect also decreases gradually with depth. This leads to an excess of hydrostatic pressure at the surface which forces water to be moved towards the inner bank across the river bed. This, in turn, is compensated for by an outward movement of surface water and so helical flow develops.

A meander is usually only a temporary landform. If the rate of downstream migration of one meander is greater than the one further downstream, then the upstream meander may get closer to the downstream meander leaving only a narrow neck of land between. This neck of land is easily breached at periods of high discharge so that the river cuts a new channel (Figure 13.7a). A meander may also be breached at periods of normal discharge by progressive erosion through the point bar deposits (Figure 13.7b). In either case, the entrance and exit to an old meander soon become plugged with deposited sediment so that it is abandoned and forms a lake termed an **ox-bow lake**, so-called because of its curved shape (Figures 13.8 and 13.4). The ox-bow lake will gradually fill with fine-grained sediments brought in during flooding and with decayed vegetation.

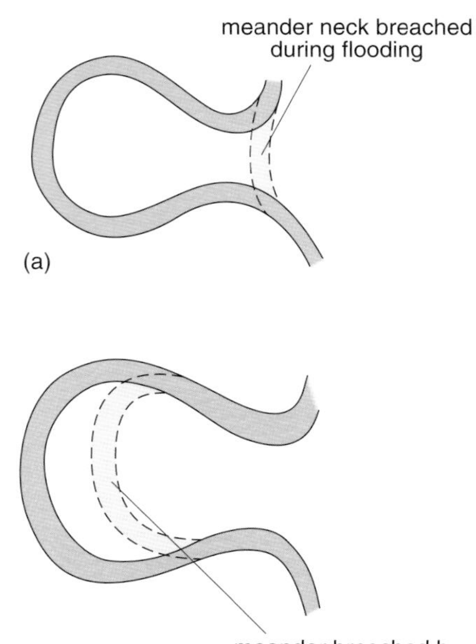

Figure 13.7 An illustration of the two ways in which an ox-bow lake may form: (a) the river may breach the meander neck during flooding, (b) alternatively, the river may gradually erode a new channel through former point bar deposits.

At times of increased water level in a river channel, water may flow over the top of the channel and any sediment carried in it will be deposited due to the sudden reduction in flow velocity outside of the channel. This sediment builds up to form **levées** (small ridges) on the edge of the river channel, analogous to the lava flow levées you met in Block 3 Section 6.1. If the water level rises a lot, the levées may be breached and widespread flooding occurs; this is called overbank flooding and sediment and water will be carried out over the flood plain. The strong river current during flooding may also breach the river bank at one particular point and deposit sediment on the edge of the flood plain in a fan shape called a **crevasse splay** (Figure 13.4). In humid areas, meandering river banks and their flood plains are usually areas of lush vegetation and good soil development. The periodic flooding from the river carries fresh nutrients over the flood plain encouraging plant growth. Plant debris from natural decay, flooding and fires often becomes incorporated into the sediments being deposited. A variety of animal life, both invertebrate and vertebrate, will live in the confines of the river system.

❑ How might evidence for animal life be preserved in alluvial sediments?

■ Animals which live in the sediment on the point bars, for instance bivalves, might easily be preserved in life position when they die. The point bar sediments may also preserve burrows made by various animals. The remains of vertebrates living on the river banks and flood plain will probably only be preserved during times of flooding when large volumes of sediment are deposited and they become buried. Flooding or natural causes could be the reason for their death.

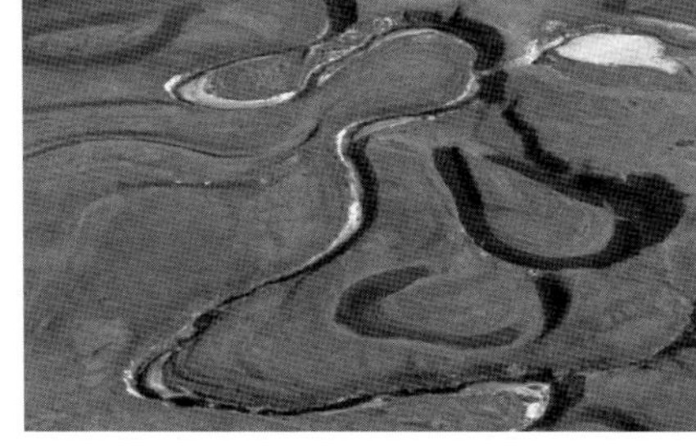

Figure 13.8 On the right-hand side of the photograph there are two distinct ox-bow lakes. The main river channel snakes across the middle of the photograph. Areas of active sediment deposition are light coloured.

Meandering river successions

Now that we have considered the three-dimensional nature of alluvial systems within a meandering river system, we are ready to construct an idealized vertical succession of sedimentary deposits that might be found and to examine the types of succession present in the geological record.

The most volumetrically significant deposits are those associated with the migration of the channel. Re-examine Figure 13.6 and remember that, as the current is eroding the bank on the outer side of the bend, sediment is being deposited on the inner bank of the meander and on the next meander downstream, forming point bars. The net effect, as shown in Figure 13.9, is that

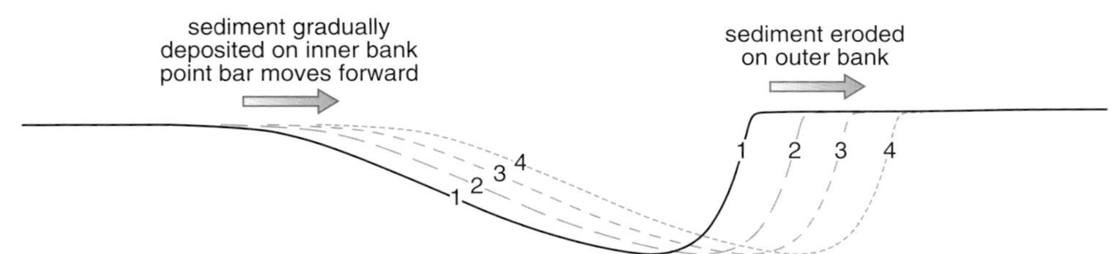

Figure 13.9 Cross-section of a meandering river channel showing the lateral movement of the channel through time. 1 = oldest river channel surface; 4 = youngest river channel surface.

as the outer eroding bank gradually migrates laterally, so the inner depositing bank gradually builds out laterally in the same direction as more and more sediment is deposited on the point bar. So, overall, in cross-section the channel maintains the same width and depth, but moves laterally.

The movement of the channel is usually slightly greater in some periods than in others due to variability in discharge and sediment content, resulting in slight pauses in deposition followed by more continuous deposition. We can see these various episodes in the point bar deposits as surfaces that each mimic the former shape and positions of half the channel; these are called **lateral accretion surfaces** (Figure 13.10).

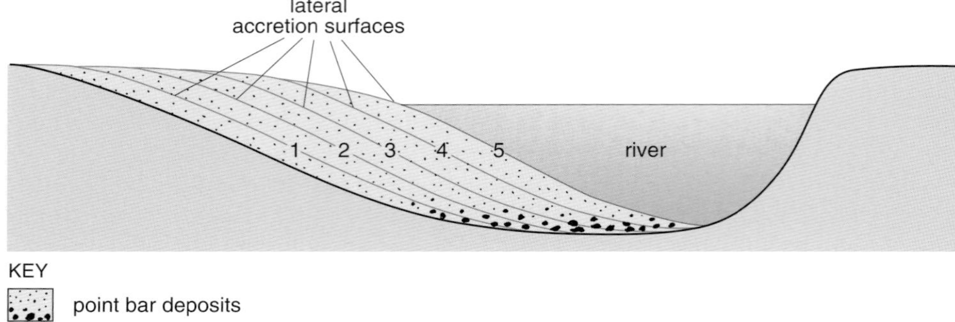

Figure 13.10 The formation of lateral accretion surfaces in point bars. 1 = oldest lateral accretion surface; 5 = youngest lateral accretion surface.

What about the sediments that make up the point bar? As the outer bank is eroded and collapses, fine-grained sediment is carried off in suspension, and the medium-grained sediment by saltation. The coarsest-grained sediment will only move by sliding or rolling at times of high discharge so most of it remains at the base of the channel, forming a lag deposit. The lag deposit may be added to by large clasts that have rolled down the river channel.

The currents in the channel and at the base of the point bar may build up subaqueous dune structures composed of the medium-grained sediments.

❑ What sedimentary structure will these dunes form?

■ Cross-stratification, and in this case, because of the type of currents in the river channel and the 3D beds that they form, usually trough cross-stratification (Figure 4.15a).

The finest-grained sediments will be deposited on the upper part of the point bar where the currents are slower and weaker. They often form current-formed ripples (Figure 4.10) which give rise to small-scale cross-stratified sandstones and siltstones. Thus, point bars are preserved as fining-upward successions with a lag of pebbles at the base overlain by cross-stratified sandstones, capped by small-scale cross-stratified sandstones and siltstones.

❑ In what orientation will the cross-stratification surfaces in the point bar deposits be in relation to the point bar lateral accretion surfaces?

■ The cross-stratification surfaces will be perpendicular to the slope of the lateral accretion surfaces, because the dunes and ripples have migrated downstream as the point bar has built out sideways into the river.

If the meander neck breaks to form an ox-bow lake, the ox-bow lake will gradually infill with fine-grained sediment and plant material during flooding. The levées will be composed mainly of the suspension load fraction of the river deposited at the edge of the river channel when the water level was high. Overbank flooding will carry the finest-grained sediments over the flood plain and as the flow slows down, a thin veneer of planar-stratified silts and clays is deposited. When the bank breaks at a confined point, sediments are deposited in a crevasse splay; these are usually composed of sediment with a mixture of grain sizes and are deposited quite rapidly so they show no systematic change in grain size. After the flood subsides the crevasse splay may become colonized by vegetation. New channels cut erosively through flood plain deposits and may cut through old crevasse splays resulting in a sharp contact between overbank and crevasse splay deposits and the overlying point bar deposits.

Overall, meandering river deposits typically comprise mudstones from the flood plain interbedded with thick sandstones that represent the point bars. An idealized graphic log summarizing each of these features is shown in Figure 13.11.

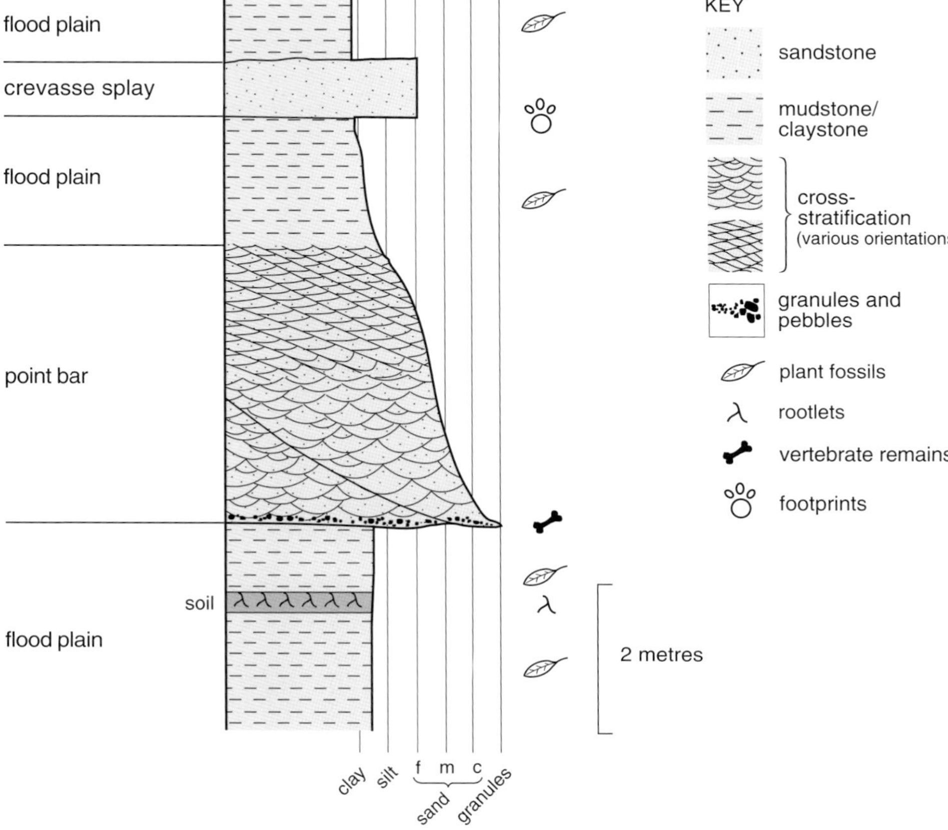

Figure 13.11 Idealized graphic log of the type of succession that might be produced by a meandering river. A coloured version of this Figure is shown on the Sedimentary Environments Poster and Sedimentary Environments Panorama (on DVD 3).

13.2.2 Braided river systems

Components of the braided river system

In contrast to meandering rivers, braided rivers (Plate 13.2) develop when (i) the discharge fluctuates widely, (ii) the volume of bedload transported is high and comprises coarse-grained, poorly sorted sediments, and (iii) where the slope over which the river flows is relatively steep. From what you know already about the erosion of sediment (Section 4.1.2) and river profiles (Section 13.1), you can begin to work out why these factors may be important.

❑ Are coarse- or very fine-grained sediments easier to erode?

■ Coarse-grained sediments are more easily eroded than very fine-grained sediments such as silts and clays because the latter are cohesive (Section 4.1.2).

At periods of high discharge, new channels are cut rapidly through the sediments that were deposited as the previous flood waters abated.

❑ How does discharge relate to the shape of river channel cross-sections and channel bed gradients?

- In a narrow, shallow channel, the discharge is maintained by a steep gradient. In a broad, deep channel, discharge is maintained by a gentle gradient.

Whereas a meandering river erodes meanders to decrease the overall gradient and thereby accommodate its discharge, a braided river is able to disperse the discharge through a number of smaller channels and so the gradient of each channel bed is not reduced. So the discharge is maintained by a network of small channels with relatively steep bed gradients instead of in one large sinuous channel with a gentle bed gradient.

The sorts of conditions that favour braiding are often found where heavily debris-laden rivers flow from a steeply dissected mountain range onto a lowland plain and where there are seasonally high discharge periods like in semi-arid and arid regions. The ancient braided river deposits in the sedimentary record are usually associated with such geographical features.

The braided river system thus comprises numerous braided channels all of which are within a larger channel-like feature usually with steep sides because of active erosion. Cones of sediment supplied (Section 13.3) from the mountain meet these steep valley sides and often the river channel dissects the end of the cone of sediment (Figure 13.12).

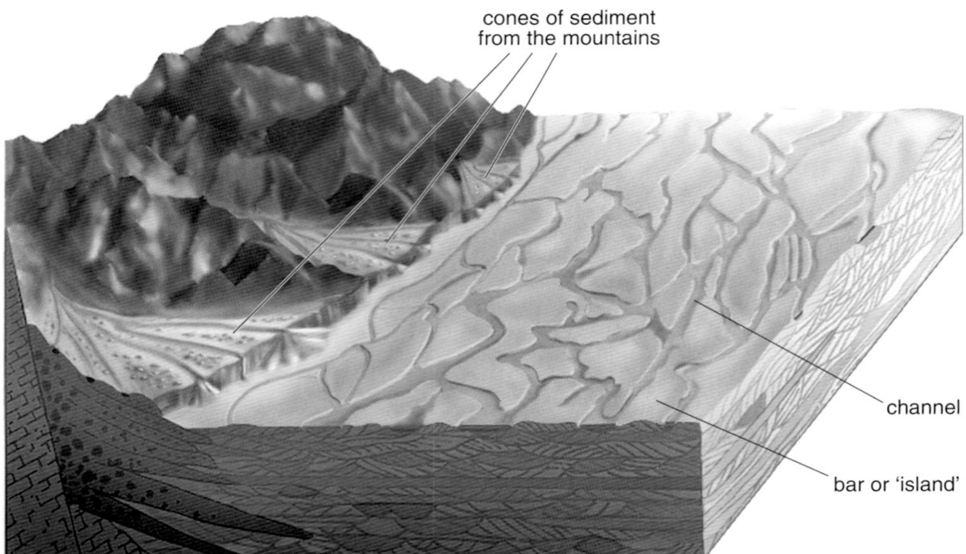

Figure 13.12 Block diagram of a braided river system showing the main features. This diagram only shows half of the larger channel-like feature in which all of the smaller braided channels lie. A coloured version of this Figure is shown on the Sedimentary Environments Poster and Sedimentary Environments Panorama (on DVD 3).

The individual channels usually contain coarse-grained lags at the base deposited when the discharge was high (Figure 13.12). These small channels are separated by 'islands' of sediment called **bars**, which build up in three ways. The first is similar to the way in which point bars in a meandering river system develop.

- ❑ Why do point bars in meandering river systems build up sideways, i.e., perpendicular to the flow of the river?
- ■ Because of the helical flow of the water, the channel is constantly migrating sideways as it erodes on the outer bank and deposits sediment on the inner bank to form the point bar.

The second way bars can be formed in braided rivers is because of the high flux of sediment and the variable discharge rate. This allows the bars to build vertically upwards because when the water level is high a sheet of sediment is deposited on the top of its formerly emergent surface. High water levels in the main river channel lead to the third way in which the bars can grow; this is by movement and deposition of sediment along the front of the bars as sediment is carried in a downstream direction. Dunes develop on the bars and build forward in a downstream direction. As the water level, and therefore the current strength, subsides, current-formed ripples may form on the bars and eventually the top surface of the bar may re-emerge. Flood plain sediments do not commonly accumulate in braided river systems because the continuous erosion and movement of the channels prevents this.

BRAIDED RIVER SUCCESSIONS

Braided river successions are dominated by the sediments that comprise the bars and channel infill.

❑ Braided river deposits mainly tend to comprise coarse-grained sand and gravels. Why do you think this is so?

■ Braided river systems tend to form where rocks are actively eroding, such as in mountain belts. The high discharge rate and steep gradient mean that the currents are strong and fast, and any fine-grained sediment tends to be carried downstream in suspension. However, the exact composition of the succession also depends on the nature of the rocks or sediments that are being eroded.

The migration of the bars will give rise to a wide variety of sedimentary structures including planar cross-stratification dipping downstream and some trough cross-stratification from the migration of bedforms dipping both downstream and perpendicular to the main flow. The number of channels and the constant migration of channels and bars means that there are many erosion surfaces and that individual beds are often not laterally extensive. Variation from high discharge rates during flooding to lower discharge rates during background conditions will result in a broadly fining-upwards succession in some of the channels.

So, in summary, braided river successions comprise channel and bar sand and gravel deposits with a variety of sedimentary structures and a high degree of lateral variability. The fine-grained sediments may be deposited on the top of the bars forming thin bar top deposits. A typical braided river succession is shown in Figure 13.13.

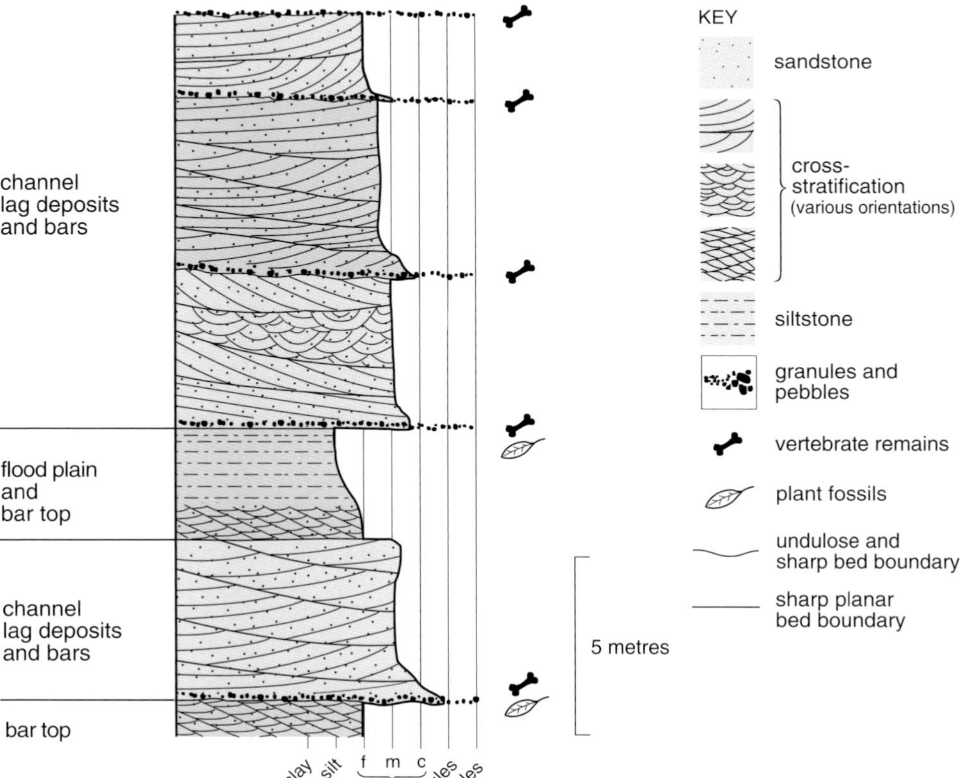

Figure 13.13 A typical succession deposited by a braided river. A coloured version of this Figure is shown on the Sedimentary Environments Poster and Sedimentary Environments Panorama (on DVD 3).

13.3 ALLUVIAL FANS

Alluvial fans are localized deposits that form where a river or stream loaded with sediment emerges from a confined mountain valley onto a topographically lower, flat lowland plain or major valley. Alluvial fans have a basic semi-conical shape. The apex of the cone points up the valley (Plate 13.3). As the water flows out from the steep valley and onto the flat plain, the sudden decrease in gradient is accompanied by extensive deposition from sediment gravity flows. Alluvial fans are most common in semi-arid and glacial climatic regimes where vegetation is sparse and run-off is seasonal, but alluvial fans can also form in humid climatic regimes.

Alluvial fan deposits are generally poorly sorted, matrix-supported and both texturally and compositionally immature, but they may show varying degrees of stratification or graded bedding depending on how much water remains in the flow. The actual degree of grain roundness is a function of the sediment source, how far the sediment gravity flow travelled and whether it was fairly viscous so that grains were cushioned. The fan is traversed by a network of rapidly shifting braided streams and so the sediments in the channels comprise gravels, sandy gravels and sands, which show grain support and sedimentary structures characteristic of those we described for braided rivers. The sands may be arkosic if the nearby mountains contain feldspar-bearing rocks and chemical weathering and grain transport are limited. A feature common to all alluvial fans is that the sediments become finer-grained with increasing distance from the apex of the fan. The flow of streams across the fan predominates where the supply of water is more abundant and occurs throughout the year. Alluvial fans are a prominent feature of most braided river systems in mountainous areas and where mountains meet flat desert areas. You will see some examples when you do Activity 14.1 in the next Section.

Activity 13.1

You should now complete Activity 13.1 to consolidate your knowledge of alluvial systems.

13.4 SUMMARY OF SECTION 13

- Sediment derived from weathering is transported into the river channel by sediment gravity flow processes. Material eroded from the banks of the river channel add to the sediment load.
- The water and sediment are moved downstream because of the gravitional potential energy caused by the river gradient, and the discharge increasing downstream. The river's bedload is moved by bouncing, rolling and saltation, while the suspension load is carried in suspension.
- Meandering river systems are characterized by a distinct meandering channel, low overall topographic gradient, high suspension load to bedload ratio, and the area into which the channel is cut being made up of cohesive sediments. The meanders increase in amplitude and migrate downstream by progradation of the point bar on the inside of the bend and erosion on the outside of the bend. Each phase of progradation is marked by a lateral accretion surface. Ox-bow lakes form where the meander is breached. Levées may form on the edge of the river channel following increased water level and these may, in turn, be breached during flooding to produce crevasse splays.

- Meandering river successions comprise sharp-based, cross-stratified, fining-upward units representing the point bar. These will be interbedded with fine-grained sedimentary rocks deposited during flooding of the flood plain. The flood plain deposits may contain sandstones representing crevasse splays.
- Braided river systems are characterized by high braiding and form where the discharge fluctuates widely, the bedload content is high, and where there is a steeply dipping slope over which the river flows. The bars which divide the channels accrete through point bar accretion, vertical build-up due to the high sediment flux, and deposition on the front of the bar.
- Braided river successions mainly comprise stacked channel fills, and bars. Occasional fine-grained bar top sediments are preserved. Braided river successions show a wide degree of vertical and lateral variability.
- Alluvial fans are localized deposits that form where a mountain river or stream emerges into a flat lowland plain or valley. They most commonly form under semi-arid and glacial climate conditions. Deposition from sediment gravity flows is the most important process, thus the deposits are generally texturally and compositionally immature. The sediments become finer grained away from the apex of the fan.

13.5 Objectives for Section 13

Now you have completed this Section, you should be able to:

13.1 Describe the physical, chemical and biological processes that are important in alluvial environments.

13.2 Sketch graphic logs of braided and meandering river successions.

13.3 Describe the main features and the type of sedimentary deposits and structures formed in alluvial environments.

13.4 Distinguish between meandering and braided river systems.

13.5 Describe the main features of alluvial fans.

Now try the following questions to test your understanding of Section 13.

Question 13.2 Explain whether or not you would expect to find: (a) thick flood plain suspension deposits; and (b) organic debris in a braided river succession.

Question 13.3 Figure 13.14 shows the progressive changes in the position of the channel of the River Mississippi during the periods 1765, 1820–30, 1881–1893 and 1930–32.

(a) Examine the pattern of the meanders and work out which of the channels A, B, C and D corresponds to which of these time periods. (You are given the direction of flow of the river.)

(b) Name what feature X indicated by the line would be in 1932.

(c) Describe the type of feature and sedimentary structures that would have been found at Y (indicated by the line) in 1932.

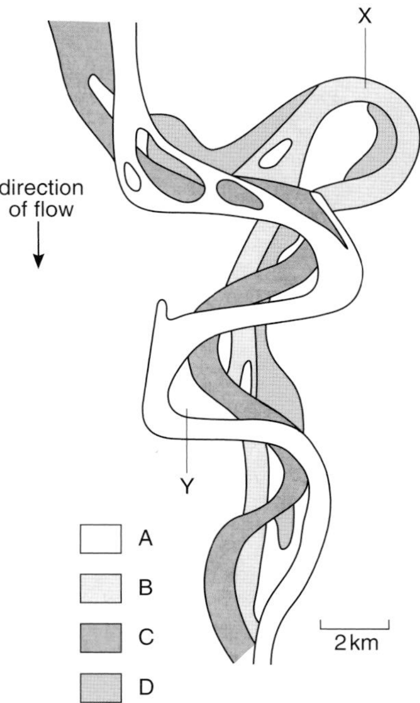

Figure 13.14 Shifts in part of the main channel of the River Mississippi between 1765 and 1932. (For use with Question 13.3.)

14 Deserts

To many of us, the word 'desert' conjures an image of a vast area of sand dunes, but in fact this is only one of several subenvironments found in deserts. Deserts are defined as areas where the average rate of evaporation exceeds the average rate of precipitation (rainfall, snow, etc.). This lack of water is the reason why they contain very little flora and fauna. Lack of vegetation and water also means deserts are dominated by physical weathering and that when precipitation does come erosion is rapid, thus sediments in deserts can be easily moved around. Deserts occur in both hot and cold climatic areas of the world. Hot deserts cover a larger area of the Earth today and we shall concentrate on these. Constant heating during the day and cooling at night in these deserts set up stresses within the rocks which leads to increased physical weathering.

The dominant agents of transport and deposition – wind and water – can be used to divide desert features and sedimentary deposits into two categories. The first category is those features and sedimentary deposits formed mainly by wind; these include aeolian sand dunes, extensive areas of which may be called sand seas, together with large areas of bare rock and rock debris. The second category is the features and sediments formed by the transient passage of water; these include alluvial fans (Section 13.3), river valleys and temporary lakes. We shall discuss the features of deserts in this Section by reference to these two categories.

Activity 14.1

As a general introduction to deserts, you should now do Activity 14.1, which is based on the video sequence *Deserts* on DVD 3.

14.1 Wind in the desert

Aeolian processes and the bedforms they produce were described in Section 4, particularly Section 4.2.6. You may find it useful to quickly re-read this Section now. Figure 14.1 shows a sketch cross-section through some ancient aeolian sand dunes and an associated sedimentary feature and deposit that formed in a desert environment. Examine the sketch carefully.

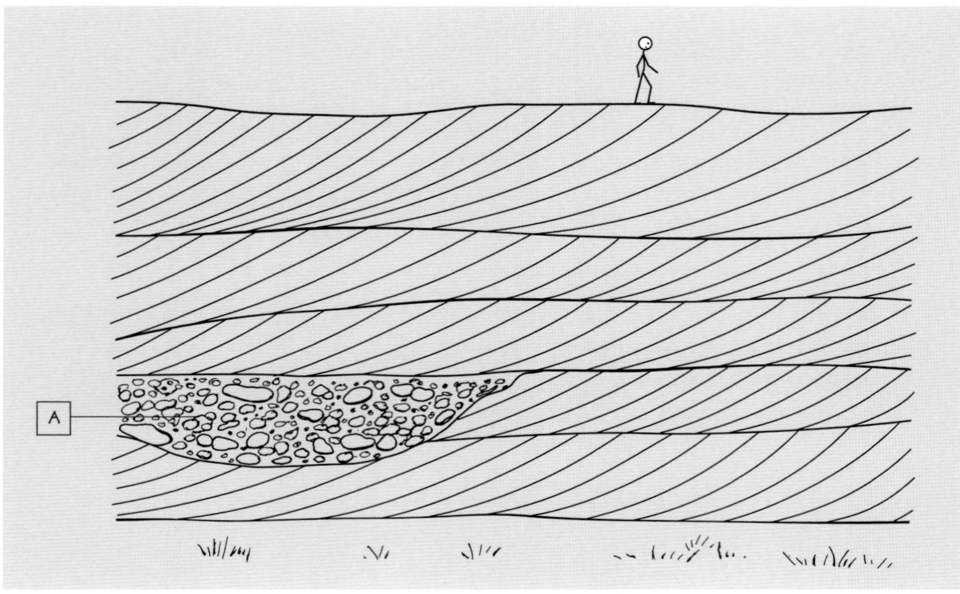

Figure 14.1 Sketch cross-section through some sedimentary rocks deposited in a desert. Label A is referred to in Section 14.2.

❑ What was the dominant wind direction and how do you know?

■ From right to left across the sketch because the cross-stratification is dipping to the left.

❑ In what way is sand transported down the steep slope of dunes?

■ The sand is transported down the steep slope by avalanching, in a process called grain flow (Section 4.2.6).

In order to preserve the dune sands, so that eventually they will become sandstones in the geological record, there has to be net accumulation of sand. So the sands have to be protected from the erosive and transportational power of winds. This can be achieved in several ways. (i) Through deposition at or near the water table (wet sands simply do not blow away), combined with subsidence, allowing more and more sand to accumulate. (ii) The growth of vegetation or the formation of surface cements, protecting the sands and preventing them from blowing away. The red iron oxide coating around the grains of RS 10 could be a form of early cement that may well have helped to stabilize these Permian sands so that they could be preserved. (iii) If the supply of sediment is exceptionally high, then the sands may accumulate because there is insufficient wind to remove them.

❑ Considering Figure 14.1 again, is it possible to tell what the three-dimensional shape of these sand dunes was? Give reasons for your answer.

■ No, it is impossible to tell from this one sketch what their bedform was like; we would need to look at many faces cutting across the bedforms in different directions.

The three-dimensional shape of every sand dune depends on *two* variables: (i) whether the wind pattern, is variable or mainly blowing in one direction (unidirectional), and (ii) whether the supply of sediment is low, high or very high.

At first sight, the three-dimensional form of desert dunes may appear very complex but the two variables described above result in four basic dune morphologies. **Barchan dunes** are crescent-shaped dunes when viewed from above. The ends of the crescent shape point downwind. They form where the wind pattern is unidirectional and the supply of sediment is low (Figure 14.2a, b, Plate 14.1). If the wind remains unidirectional but the supply of sand increases, the crests of adjacent barchan dunes will join up and grade into long, linear, asymmetric ridges called **transverse dunes** (Figure 14.2c). They are termed transverse dunes because the crest of the ridges are orientated perpendicular to the dominant wind direction. Where the wind direction is variable, very complex dune forms are found; one morphological type is the linear dune (Figure 14.2d) which forms where the sediment supply is low (similar to that for barchan dunes), and the other type is the star dune (Figure 14.2e) which forms only where the sediment supply is very high. In order to distinguish these dune morphologies in aeolian sandstones, it is necessary to collect many measurements of the direction of dip of the cross-stratification over a wide area.

When sand input is minimal, wind erosion and transport will produce large expanses of bare loose rocks. Such areas are simply referred to as stony deserts.

So, in summary, wind processes are important in shaping, transporting and depositing sand grains, together with moving all loose sediment away from an area and leaving behind bare rock and clasts that are too large to be moved. The wind is also important in a desert in helping any water that may be present to evaporate. Let us now consider the features that might form when water does enter the desert.

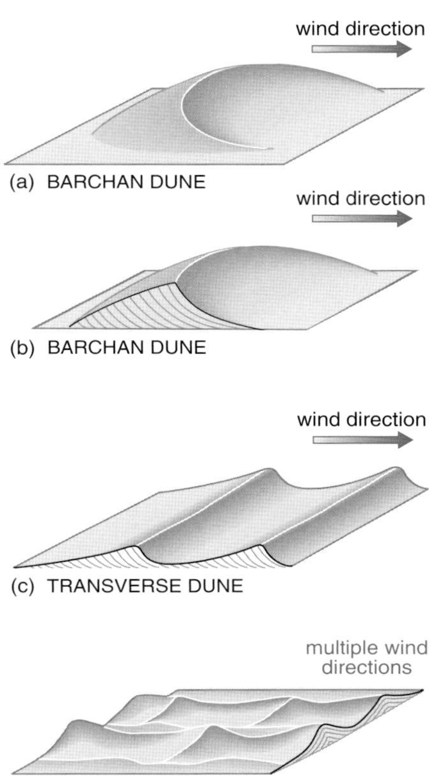

Figure 14.2 The morphology of the four basic types of desert dune in relation to the wind pattern and amount of sediment supply: (a) and (b) barchan dune; (c) transverse dunes; (d) linear dunes; (e) star dune. The typical cross-stratification pattern for barchan, transverse and linear dunes is shown. The cross-stratification pattern in star dunes would be complex and multi-directional.

14.2 WATER IN THE DESERT

Water causes some of the most spectacular and dramatic features of desert erosion because most of the sediments are not bound together by vegetation, fine-grained sediment or cement. Rainfall is sporadic, but when it occurs over a short seasonal period (e.g. Australian desert) or every few years (e.g. Sahara) it may be torrential. In desert regions within or close to mountain ranges, a rainstorm can be catastrophic. The sudden rainfall carries large volumes of accumulated sediment down from the mountains and into the desert area, and may erode a steep-sided, narrow valley called a **wadi** (Arabic for watercourse, Plate 14.2a). As the flood waters subside by infiltration and evaporation, sediment is deposited. It is usually poorly sorted and may show only crudely defined bedding, although, if conditions are favourable, planar stratification and cross-stratification like that in other fluvial deposits may be found (Plate 14.2b). For much of the time, most wadis are dry valleys that are only periodically occupied by water.

❑ Consider Figure 14.1 again. What do the rocks labelled A most likely represent and why?

■ Wadi deposits. The clasts are too large to be have been moved by wind processes and the cross-cutting nature of the contact between the cross-stratified sandstones and the wadi deposits suggests erosion of the sand dunes.

❑ On Video Band 13 *Deserts*, two other morphological features formed by the presence of water in a desert were described. What were they?

■ (i) Alluvial fans and (ii) temporary lakes or 'playa lakes'.

Alluvial fans were described in Section 13.3. Temporary lakes, often termed **playa lakes**, can form in any depressions in the desert surface as it rains and are common at the base of alluvial fans, as shown on Video Band 13. The playa lakes become infilled with fine-grained sediments that have been transported in suspension either by wind or water. If sediment has been carried by water down alluvial fans, the finest-grained sediments will be carried the furthest and will be deposited at the distal part of the fan in the flat area where the playa lake forms. The resulting sedimentary rocks are typically homogeneous or laminated mudstones and siltstones. **Desiccation cracks** are ubiquitous and form as the sediments dry out. The water may support organisms, such that both trace and body fossils may be found. If the water does not drain away, but is allowed to evaporate slowly, evaporite minerals such as gypsum and halite will be deposited and a salt pan is formed.

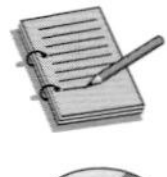

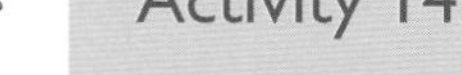

Activity 14.2

To test your understanding of Section 14 and consolidate your knowledge of deserts, you should now complete Activity 14.2.

14.3 Summary of Section 14

- Deserts form where the average rate of evaporation exceeds the average rate of precipitation.
- Deserts occur in both hot and cold climate areas of the world.
- Frosted sand grains together with moderately good sorting of the sand-sized grains and large-scale cross-stratification are diagnostic of aeolian processes.
- Sand dunes usually start to form by grains accumulating around an obstacle and producing a drift. When sufficient sand has accumulated, the dune begins to migrate, through saltation up the shallow slope and avalanching down the steep slope in a process called grain flow.
- The wind pattern and size of the sediment supply determines the morphology of the dunes that form.
- Barchan dunes are crescent-shaped with the ends of the crescent pointing down-wind. Transverse dunes are long linear asymmetric ridges with the steep side pointing downwind. In both cases, the cross-stratification is dominantly inclined in one direction.
- Sand dunes may be preserved in the desert through either deposition near the water table combined with subsidence, or through the growth of vegetation or precipitation of cement.
- Sporadic rainfall in the desert leads to the formation of wadis in which crudely bedded and often coarse-grained sediments are deposited. The presence of water also means that playa lakes can form in depressions; these become infilled with fine-grained sediments and evaporites may precipitate.
- Where mountains fringe the edge of the desert, alluvial fans and playa lakes are commonly found.

14.4 Objectives for Section 14

Now you have completed this Section, you should be able to:

14.1 Describe very broadly where deserts are found in the world today and why they have formed.

14.2 List the features that together are diagnostic of sediments deposited by aeolian processes.

14.3 Describe the processes by which sand dunes form.

14.4 State the two factors that are important in determining the morphology of the sand dunes that form.

14.5 Describe the shape and cross-stratification pattern produced by migration of barchan and transverse dunes.

14.6 Describe how sand dunes might be preserved in the desert.

14.7 Name and describe the features that may be formed by water in the desert.

Question 14.1 In the form of a Table, summarize the main morphological features of deserts and the common sedimentary facies they produce.

Question 14.2 Describe what other features you could look for in large-scale cross-stratified sandstones to help determine that they were deposited by aeolian rather than tidal processes.

15 Siliciclastic coastal and continental shelf environments

The coastline and continental shelf of both modern day and ancient continents represent a complex natural continuum of different environments; in other words, one type and shape of environment changes to another as the relative dominance of fluvial, tidal, wave, wind and biological processes changes. As we showed in Video Band 9 *Coastal processes* (Activity 1.1) together with Sections 4 and 5, these processes also govern the characteristics of the sedimentary rocks that are deposited in coastal environments. In this Section, we will consider coasts dominated by the deposition of siliciclastic rocks and in Section 16 those dominated by the deposition of carbonate rocks.

The main morphologies of siliciclastic coastlines are deltas, strandplains, barrier islands and estuaries. These are all illustrated on the right-hand side of the continent on the Sedimentary Environments Poster and Sedimentary Environments Panorama. In this Section, we will consider strandplains, barrier islands and deltas, but we do not have enough space to cover estuaries as well. As well as the coastline morphologies, we will consider the **continental shelf** which you were briefly introduced to in Block 3 Section 2.2.1. The continental shelf is the region adjacent to the edge of the continent that is covered by relatively shallow water (usually less than 200 m deep and sometimes as shallow as a few tens of metres). The continental shelf and coastal sediments are grouped together because the sediments found on the continental shelf are often intimately associated with the adjacent stretch of coastline.

15.1 Subdivision of the coastal and continental shelf environment

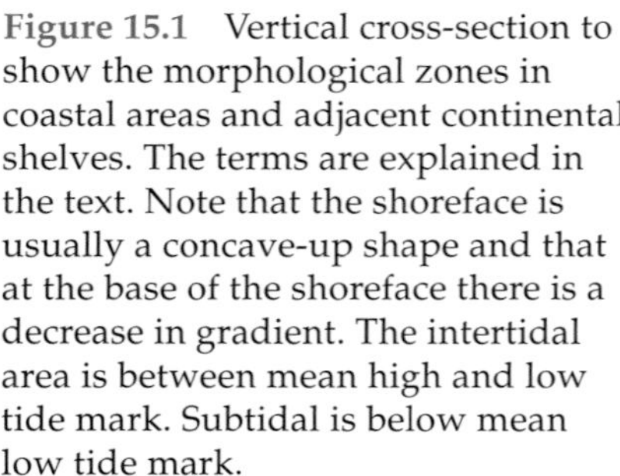

Figure 15.1 Vertical cross-section to show the morphological zones in coastal areas and adjacent continental shelves. The terms are explained in the text. Note that the shoreface is usually a concave-up shape and that at the base of the shoreface there is a decrease in gradient. The intertidal area is between mean high and low tide mark. Subtidal is below mean low tide mark.

Common to all of the coastal environments are a number of zones, defined by the position of mean high and low tide, and the depth to which wave processes affect the sediments. Figure 15.1 shows an idealized cross-section through the coast and continental shelf, showing the various zones. For completeness, we have also shown the continental slope leading down from the edge of the continental shelf into the deep ocean, even though we do not discuss these latter regions until Section 17.

The mean high tide is the average of the high spring and neap tides, and similarly, the mean low tide is the average of the low spring and neap tides. (The concept of spring and neap tides was introduced in Section 4.2.4.) The

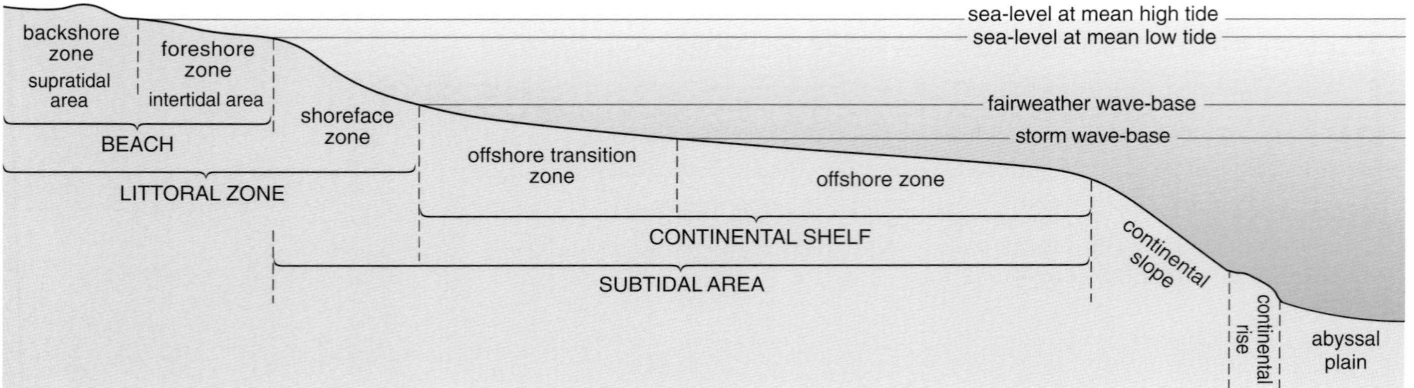

mean high and low tidal marks are important because they define the **intertidal** area of the shore, which is the area influenced directly by breaking waves (Video Band 9 *Coastal processes* (Activity 1.1)). Above the mean high tide is the **supratidal** area, i.e. the area affected only by the highest spring tides or occasional storm waves. (*Supra* is derived from the Latin word for 'above'.) At the other extreme, the **subtidal** area is the area below mean low tide, the upper part of which is totally exposed only during the lowest of the spring tides. However, you should note that the subtidal area is still influenced by tidal processes because tides produce strong currents (Section 4.2.4). We shall consider each zone in turn from the land to the deep sea. The **backshore** zone (supratidal) is the area above mean high tide; it is not covered by seawater except during very large storms. The **foreshore** zone (intertidal) is seaward of the backshore and is the area between mean high tide and mean low tide. This is the area of the coastline that is covered by seawater twice a day at high tide and uncovered twice a day at low tide. Consequently, it is a very dynamic area with the sediments and organisms constantly moving around to adjust to the changing conditions. The foreshore zone and backshore zone together comprise the **beach**. The **shoreface** is the area below mean low tide but above fairweather wave-base, and is also often very rich in organisms.

❑ Explain what fairweather wave-base represents.

■ Fairweather wave-base is the maximum depth at which there is water movement caused by waves during fairweather conditions (Section 4.2.5).

The sea-bed above fairweather wave-base is affected by waves on an almost daily basis. With increasing depth down to storm wave-base, wave-related processes operate with decreasing frequency.

❑ Explain what storm wave-base represents.

■ Storm wave-base is the maximum depth at which there is water movement caused by waves during storm conditions (Section 4.2.5).

❑ Bearing in mind that the amplitude of storm waves will be considerably greater than that of fairweather waves, because the winds are stronger during storms, why is the storm wave-base deeper than the fairweather wave-base?

■ The greater the amplitude of a wave, the larger the diameter of the orbital motion, and so the deeper the orbital motion will penetrate. The depth that waves penetrate is equivalent to about half their wavelength (Section 4.2.5).

Below the shoreface is the **offshore transition zone**; this is the area below fairweather wave-base but above storm wave-base. The area below storm wave-base to the distal edge of the continental shelf is called the **offshore zone**. The offshore and offshore transition zones both form part of the continental shelf. Beyond this is the continental slope; this is the slope down into the deep ocean basin which we will consider further in Section 17.2.

We can also describe the tidal range, i.e. the average height by which the tide rises and falls on a coastline, as either being small, where it is usually less than 2 m; or medium-sized where the range is about 2–4 m; or large, i.e. greater than 4 m.

15.2 Interaction of fluvial, tidal and wave processes

Section 4.2.4 and the video sequence *Coastal processes* on DVD 3 (Activity 1.1) introduced you to the different types of water and sediment movement that result from fluvial, tidal and wave processes. The interaction of these processes results in a variety of bedforms and sedimentary structures in coastal and shelf areas and it is the dominance of particular processes that produces the spectrum of coastline morphologies that we observe. Both tides and waves produce strong currents. To recap on these processes, try the following questions.

Question 15.1 Complete Table 15.1 by indicating which water movements are applicable to each of the processes.

Table 15.1 For use with Question 15.1.

	Wave	Tide	Fluvial
oscillatory flow			
unidirectional flow			
bidirectional flow over ~12 hours			

Question 15.2 Complete Table 15.2, indicating which process(es) may lead to the formation of the listed sedimentary structures and bedforms.

Table 15.2 For use with Question 15.2.

	Wave	Tide	Fluvial
herring-bone cross-stratification			
current-formed ripples			
wave-formed ripples			
cross-stratification			
tidal bundles			
planar stratification			
trough cross-stratification			
planar cross-stratification			
mud-draped ripples			
climbing ripples			

Let us now consider the geomorphological result of the dominance of each of fluvial, tidal and wave processes along the coastline. The resulting environments are deltas, strandplains and barrier islands. Perhaps the easiest to consider first is fluvial processes.

Examine the coastline shown on Figure 12.1 and the Sedimentary Environments Poster.

❑ Which geomorphological feature do you think will form if the dominant process at the coastline is fluvial, where there is a constant high sediment supply, and strong current flow from a river into the sea?

■ The formation of a delta would be the main result, if there is a constant supply of sediment to the coastal area by the fluvial system, and provided the wave and tide processes are not strong enough to immediately transport the sediments away.

For fluvial processes to be dominant, wave and tide processes need to be relatively weak and there also needs to be a well-developed alluvial system on land, providing a moderate to high sediment supply. This supply may result from a mountain chain that is being uplifted, or because there is a huge drainage basin with a high rainfall. The Mississippi delta which is forming today off the southern coast of North America is an example of the latter, because it drains much of the continent. In addition, the delta is also protected from waves because it lies within a semi-enclosed sea area (the Gulf of Mexico).

Along all coastlines, no matter what the geomorphology, there is always some influence from tides and waves and to a lesser extent from fluvial processes.

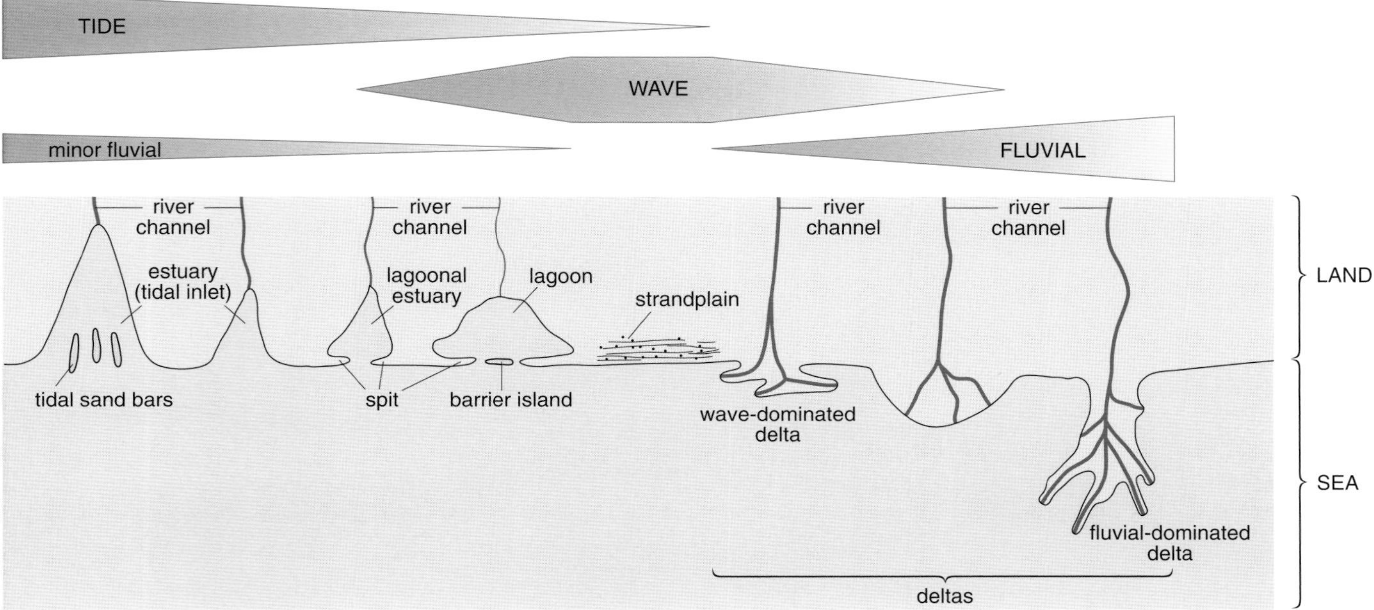

Figure 15.2 Sketch illustrating the spectrum of coastal morphological features and the dominant processes in each case. The importance of waves, tides and fluvial input is shown schematically by the boxes at the top of the Figure; tapering of the box indicates less influence.

❑ Carefully examine Figure 15.2 and identify the process that is much more important for the formation of strandplains rather than for barrier islands.

■ Waves are much more important for the formation of strandplains than barrier islands.

❑ Why do you think this might be?

■ As waves break, they move the sediment landward, thus the sediment becomes attached to the coastline rather than being separated from it.

In addition, the formation of strandplains is favoured by a high sediment supply, and a steep coastal gradient which will cause the waves to break much more abruptly.

The dominance of either wave or tidal processes along a particular piece of coastline is generally complicated since it depends on several factors. These include;

- the size and shape of the ocean adjacent to the coastline;
- the width and gradient of the continental shelf;
- the shape of the coastline.

❑ How will the size of the ocean affect the relative strength of the waves?

■ If the ocean is large and the coastline exposed to the open ocean wave processes, wave strength will be generally greater because the wind has had thousands of kilometres over which to blow (the *fetch*) and build up waves.

The coastlines of north-west Britain, Brazil and west Africa are examples of predominantly wave-dominated coastlines because they face the open Atlantic Ocean. The width and gradient of the continental shelf also affects the dominance of wave and tidal processes since narrow shelves favour wave processes, whereas wide shelves dampen wave effects and amplify the action of tidal currents. In a similar manner, the tidal range and strength of the tidal currents varies according to the shape of the ocean basin and the effect of the spin of the Earth. Thus, tidal ranges around the world differ, as shown in Figure 15.3.

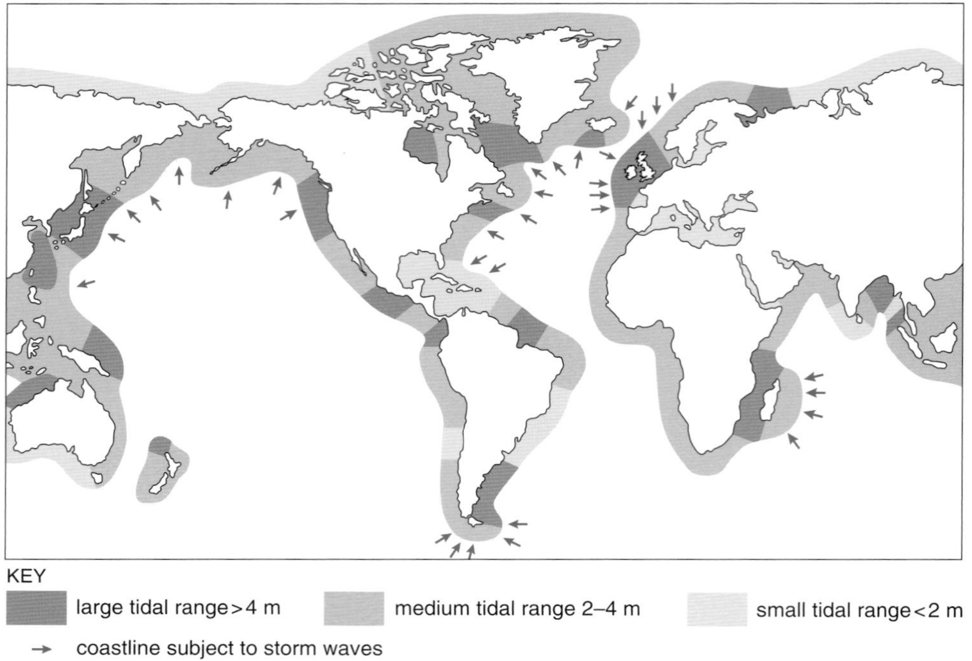

Figure 15.3 A classification of present-day coastlines around the world, based on tidal ranges.

The shape of the coastline determines whether the tidal processes are enhanced or reduced. For instance, natural ranges can be enhanced where flow is funnelled into a narrow passage. For this reason, estuaries tend to be dominated by tidal processes, though of course there is also some fluvial influence. If tidal processes are dominant along a linear strip of coastline without rivers, then extensive areas of the foreshore (tidal flats, Section 4.2.4) will be covered in interbedded thin layers of sand and clay deposited during different parts of the tidal cycle (video sequence *Coastal processes* on DVD 3 (Activity 1.1)) and offshore tidal sandbodies will form.

Waves and tides produce strong currents within the coastal and continental shelf areas. As waves strike the coastline and dissipate their energy, they create currents. Some waves strike the coast at right angles whilst others strike it obliquely, according to the prevailing wind direction and currents. As the water moves onto the foreshore, it must be balanced by that which is moving away. Thus, waves that strike obliquely generate a current that travels parallel to the coastline; this is the **longshore current** (Figure 15.4, right-hand side). Longshore currents are also created by variations in the height of the breaking waves, this in turn is governed by the weather and morphology of the coastline.

The effect of longshore currents can be seen along many stretches of coastline where groynes (breakwaters) have been erected to trap the sand on the beach. It is usual to find that at each groyne the sand is banked up high on one side while it is much lower on the other side. This arises because as the longshore current moves the sand along the beach the sand becomes trapped against the up-current side of each groyne. Direct evidence for longshore currents is difficult to find in the geological record, but nevertheless the presence of longshore currents is an important process for moving sediment along coastlines.

Waves that strike the shoreline at right angles create cells of flow so that the amount of water flowing on to the foreshore must be balanced by the amount that returns. The return flow is called a **rip current** and it operates within the first few hundred metres of the shoreline (Figure 15.4). Rip currents also form because of variations in the wave height along the coastline. Evidence for rip currents can sometimes be seen in the geological record in the form of long elongate scours called *gutter casts*.

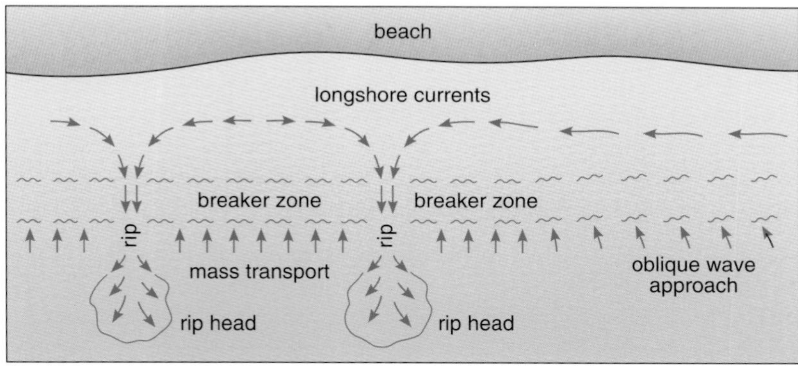

Figure 15.4 The shoreface is a zone of onshore mass transport of water by waves. This is balanced either by longshore currents (if the wave hits the shoreline obliquely), or the formation of rip currents (if it hits the shoreline normally).

Tides also create currents when they act in confined spaces; this can occur for example where the sea-floor topography changes, resulting in inequalities between the ebb and flood tide over the tidal cycle (Section 4.2.4). In such cases, sediment will be moved differentially in one direction and net transport will occur. Such tidal currents acting in subtidal areas can deposit large dunes or sand waves (Section 4.2.4). Figure 15.5 shows some large-scale cross-stratification produced by the migration of sand dunes. These structures have been preserved because they have not been subsequently reworked by wave action, so we can assume they were probably deposited below storm wave-base. Tidal currents are also responsible for scouring the continental shelf.

❑ Recalling what you read in Section 4, name another process that forms large-scale dunes.

■ Aeolian processes also form dunes.

Figure 15.5 Cretaceous large-scale cross-stratification produced by the migration of tidal dunes exposed in a quarry in southern England. Note that in the upper bed the cross-stratification indicates a strong dominant flow direction to the right because the cross-stratification is unidirectional rather than bidirectional. The long axis of the dune would have been orientated perpendicular to the tidal current.

15.3 Strandplains

If you have spent a lot of time at the British coast, you are probably familiar with strandplains. The modern beach that we visited in the video sequence *Coastal processes* in Activity 1.1 is an example of part of the strand plain environment. Strandplains are basically linear accumulations of either sand or gravel running parallel to the coastline and attached to the land.

Question 15.3 What are the dominant processes and conditions necessary for the formation of a strandplain?

Whether the strandplain is made up of sand or gravel depends on the sediment supply that is available and the wave energy. Those coastlines that frequently experience high energy storm waves tend to be made of pebbles. Storms today are concentrated in middle and high latitudes, so it is coastlines in these regions that tend to be made up of coarser-grained sediments.

Along the foreshore, the strandplain comprises one or more coast-parallel ridges or **berms**. These are separated by slight depressions or runnels (Figure 15.6). The berms and runnels are built up by the wave action. As waves approach the shoreline, the orbital motion of the parcels of water reaches as far as the sea-bed (Figure 4.18b). The frictional resistance of the sea-bed will cause the waves to increase in height and decrease in wavelength. As they break, they expend their energy by driving a thin, turbulent sheet of water (**swash**) carrying sediments up the foreshore. As the less energetic water moves back down the beach as the **backwash**, some sediment is deposited whilst the remainder is carried further seaward again. The ridges build up most when the waves are particularly large, for example during storm conditions. If the waves are smaller and break gradually as they approach the shoreline, they have very little energy left by the time they reach the foreshore so sediment is gently pushed up the beach by the swash but is not scoured and returned again during the backwash. The position of the ridges relates to the variations between storm and fairweather conditions together with spring and neap tides (Section 4.2.4).

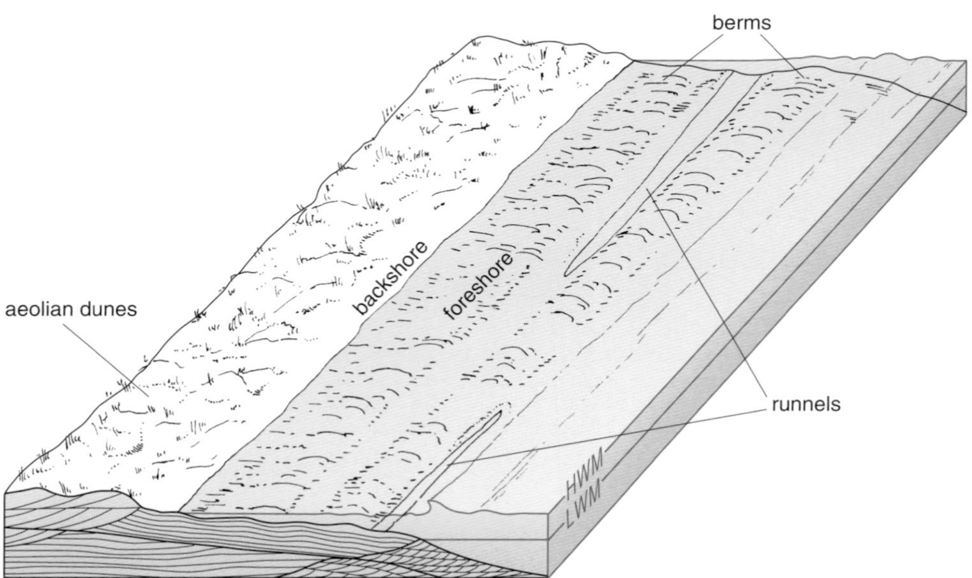

Figure 15.6 Block diagram of the foreshore showing berms and runnels.

Several factors affect the slope of the foreshore. The most important of these is sediment grain size: the coarser-grained the foreshore sediment, the steeper the profile. This is because when a wave breaks on the foreshore, the swash (water movement up the beach) is more intense than the backwash (the return flow of water down the beach). This is mainly due to the loss of water through percolation down through sediments on the swash. There is less water to flow back over the surface of the beach as backwash so it has lower energy than the swash and less material is moved offshore, resulting in net movement of sediment onshore.

- ❑ Would you expect water to percolate more readily into coarser-grained sediment or finer-grained sediment?
- ■ Into coarser-grained sediment because the pore spaces between the grains are larger.

This means that the coarser the grains, the less water is available to move sediment offshore at the surface of the beach and so the beach profile is steeper.

Wave energy also affects the foreshore profile. Higher energy waves tend to flatten the profile by removing sediment from the foreshore. What we can conclude from this is that the coarser-grained the sediment, the steeper the profile; however, for sediment of a particular grain size, the higher the wave energy, the flatter the profile. This means that a pebble beach will have a steeper profile than a sandy one, but in regions like Britain, both are likely to have flatter profiles in winter than summer because of the greater incidence of high energy waves associated with winter storms.

The constant movement of the high speed swash and backwash means that the foreshore sediments are planar stratified with current lineations indicative of the upper flow regime (Section 4.2.4). These beds are usually inclined gently seaward due to the natural slope of the foreshore (Activity 1.1, Plate 15.1). On sandy beaches, slight changes in the angle of the beach profile between summer and winter result in subtle planar erosion surfaces between groups of planar beds (see block diagram on Sedimentary Environments Poster). Wave- and current-formed ripples are also common and antidunes may be present; the latter form in small channels during the backwash and where shallow rivers and streams cross the beach.

- ❑ Explain which textural and mineralogical features you would expect in sandstones that were deposited in the foreshore area.
- ■ The foreshore area is constantly reworked, so the sandstones of the foreshore area are typically well-sorted, rounded, glassy and usually compositionally mature, though the maturity depends on the proximity of the sediment source.

The foreshore area is often colonized by a range of biota, depending on whether the surface they are living on is hard or soft. Epifaunal or infaunal bivalves and gastropods are typically common, together with various annelids (e.g. lugworms, ragworms) and arthropods (e.g. lobsters, crabs). In ancient foreshore deposits, we might expect to find the hard parts of similar animals preserved as well as their trace fossils. The high energy conditions of the foreshore will result in many of the body fossils being broken. The common trace fossils found along ancient shorelines are illustrated in Figure 9.33. The trace fossils are not exclusive to each zone and local physical, chemical and biological factors will ultimately determine what is found at each site.

Question 15.4 The holes in the intertidal zone illustrated in Figure 15.7 are the top of vertical feeding tubes made by worms, and their casts are also shown. Explain whether the vertical feeding tube or the cast made by the worm is more likely to be preserved in the geological record.

Figure 15.7 Worm tubes and casts on an intertidal zone, for use with Question 15.4. Scale bar is 16 cm long.

The backshore area of the beach, i.e. above mean high tide level, is swept by waves only occasionally, at spring high tides and during storms. Since this area is subaerial for most of the time, the sand dries out and is then reworked by aeolian processes into aeolian coastal dunes. The area is also stabilized by salt-tolerant grasses (Plate 15.3, Figure 15.6 and the video sequence *Coastal processes* (Activity 1.1)).

The shoreface zone lies below the effects of breaking waves, but above fairweather wave-base, so sediment is moved by the oscillatory motion of the water as each waveform passes. Wave-formed ripples (Section 4.2.5) are common in the shoreface resulting in small-scale cross-stratification. In addition, currents created by the waves will form both straight-crested and sinuous or curved-crested ripples and dunes.

❑ Describe what type of cross-stratification results from (a) straight-crested dunes; (b) sinuous or curved-crested dunes.

■ Straight-crested dunes result in planar cross-stratification and sinuous or curved-crested dunes result in trough cross-stratification (Section 4.2.4).

A great abundance of biota lives in the shoreface zone. This means that the primary sedimentary structures may often be destroyed by bioturbation as a consequence of the organisms burrowing and moving through the sediment. However, the frequency of reworking by wave currents means that much of this bioturbation may not eventually be preserved. Figure 9.33 shows some of the trace fossils typical of the shoreface. Biota from many phyla are found in both modern and ancient shoreface areas. They include benthic communities of arthropods (e.g. crabs, lobsters and trilobites), molluscs (bivalves, gastropods), brachiopods, cnidaria (e.g. sea-anemones, corals) as well as pelagic organisms such as molluscs (cephalopods, e.g. ammonites) and vertebrates (e.g. fish).

The area immediately below the shoreface and therefore below fairweather wave-base is the offshore transition zone. The fact that this zone is above storm wave-base means that sediment is moved around but only during storm conditions. The offshore transition zone is generally characterized by finer-grained sediments than the adjoining shoreface because it is a lower energy environment. This zone is also heavily colonized by many organisms and the sediment is usually well bioturbated because of the infrequent disruption by waves. Bedforms in the offshore transition zone include wave-formed ripples.

❑ What is the name of the almost entirely unique sedimentary structure formed by storm waves in the offshore transition zone?

■ Hummocky cross-stratification (Section 4.2.5; Figure 4.21 and Plate 4.8).

Beyond the offshore transition zone is the offshore zone; this comprises the rest of the continental shelf (Figure 15.1). The bulk of sediments deposited on the outer part of the continental shelf both today, and in the past, are silts and clays. Following deposition of the coarser-grained material close to the shoreline, most of the finer-grained sediment is carried out onto the continental shelf in suspension. Final deposition is aided by flocculation and by filter-feeding organisms ingesting the mud and redepositing it as faecal pellets. RS 27 represents a typical shelf mudstone that was deposited in an extensive continental shelf sea during the Jurassic. Nodules of mudstone cemented by calcium carbonate are common in mudstone successions deposited on the continental shelf and are thought to form during the early stages of burial of the mudstones due to the chemistry of water in the spaces between grains.

Marine mudstones are very important in the geological record because they preserve the delicate details of marine biota; their fine-grained nature means that conditions within the sediment are often anoxic and thus organic matter is more likely to be preserved. The organic matter may be the source of oil and gas. Mudstones that are rich in organic matter are dark-grey to black in colour and often they smell bituminous; an example of organic-rich mudstones is shown in Plate 15.4. Organic-rich mudrocks often contain pyrite (FeS) produced in response to the high amounts of sulfur preserved in these organic-rich sediments, due to anoxia at the time of deposition. The lateral equivalents of the beds shown in Plate 15.4 occur in the North Sea and are the source of much of Britain's oil reserves.

Siliciclastic sandbodies can be found on the outer parts of the continental shelf and they are formed by a number of different processes which include offshore transport of the sands during very strong storms and strong tidal currents.

The main facies of each of the coastal zones in the strandplain environment can be summarized in the form of a graphic log as shown in Figure 15.8.

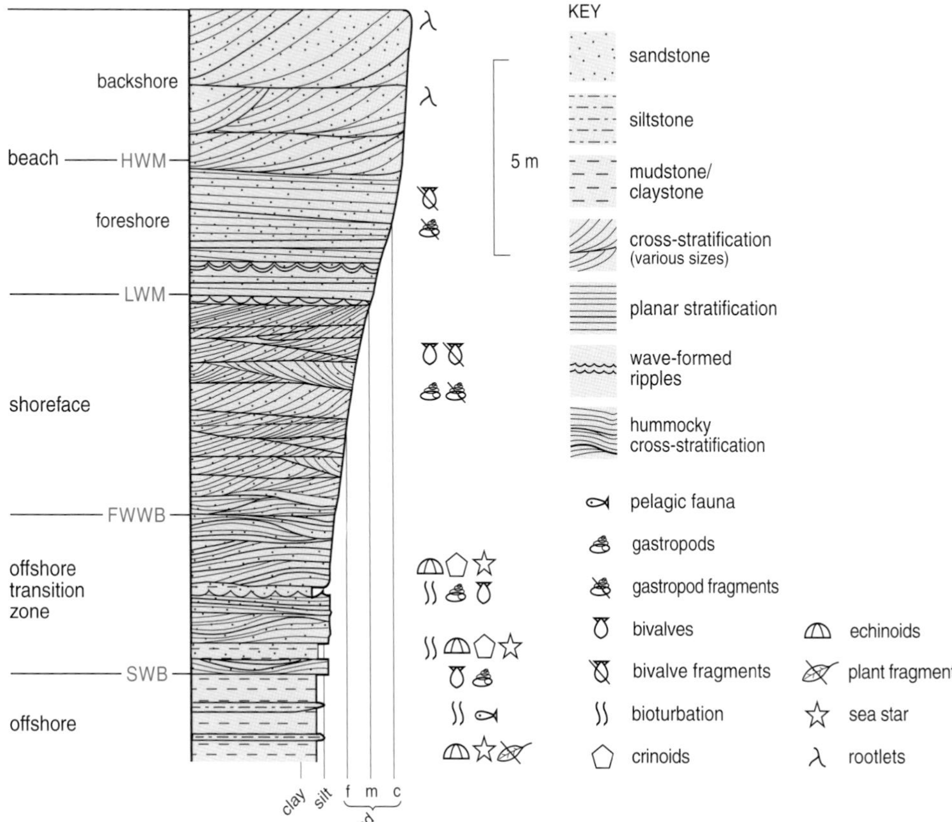

Figure 15.8 Graphic log of a prograding strandplain environment. A colour version of this Figure and a block diagram of a barrier island is shown on the Sedimentary Environments Poster and Sedimentary Environments Panorama (on DVD 3).

15.4 Barrier islands

Barrier islands are similar to strandplains in many of their features. What is different is that there is a long linear barrier offshore from the land-mass. Between the continent and the barrier there is a **lagoon**; this consists of a shallow-water area protected from major coastal currents and high amplitude waves, except during storm conditions when the barrier may be breached. The barrier can be thought of as a long, linear island situated anywhere between tens and thousands of metres offshore. Breaks in the barrier occur where rivers enter the main sea, where currents change, tidal currents flow in and out of the lagoon, or where storms have broken through the barrier (Plate 15.5). The gaps ensure that there is a passage of water and sediment between the lagoon and the open sea, but as this passage of water is restricted, and rivers from the land drain into the lagoon, it means that conditions in the lagoon are not fully marine but brackish.

❑ What features and processes favour the formation of barrier islands compared with strandplains (see Section 15.2)?

■ A lower coastal gradient, some fluvial input, longshore drift, a slightly higher tidal range, slightly lower wave power and a moderate rather than abundant supply of sand.

Seaward from the barrier, the facies are very similar to those of the strandplain system (compare Figure 15.8 with Figure 15.9). Usually, there is less wave energy than in strandplain environments, and so HCS may be entirely absent or only poorly developed in the offshore transitional zone. This is because barrier islands do not tend to form along storm-dominated coastlines, where the energy is too great to favour preservation of the barriers. The facies that are characteristic of barrier islands are those deposited within and on the shores of the lagoon, because the quiet sheltered waters of the lagoon result in the deposition of fine-grained sediments from suspension. The sheltered waters also favour animal and plant life, resulting in abundant trace and body fossils.

❑ Explain whether the body fossils will be marine forms like those found on the shoreface of the barrier.

■ Because the lagoon is not usually fully marine, species that are adapted to live in brackish water conditions will be present. They will include bivalves such as oysters, arthropods and some gastropods (Table 11.1).

The brackish but fairly quiet water conditions mean that usually there are a lot of individuals but that the species diversity is low. The topographic slope around the shorelines of the lagoon is low and the wave action weak, and so if there is also a medium to high tidal range, tidal flats (Section 4.2.4) will develop. These comprise flat, laterally extensive areas of coastline dominated by tidal processes where mainly silts and clays are deposited. The bedforms and sedimentary structures produced here will include planar stratification, small-scale cross-stratification, mud-draped ripples and clasts made from mudstone layers which have been subsequently broken up by current action.

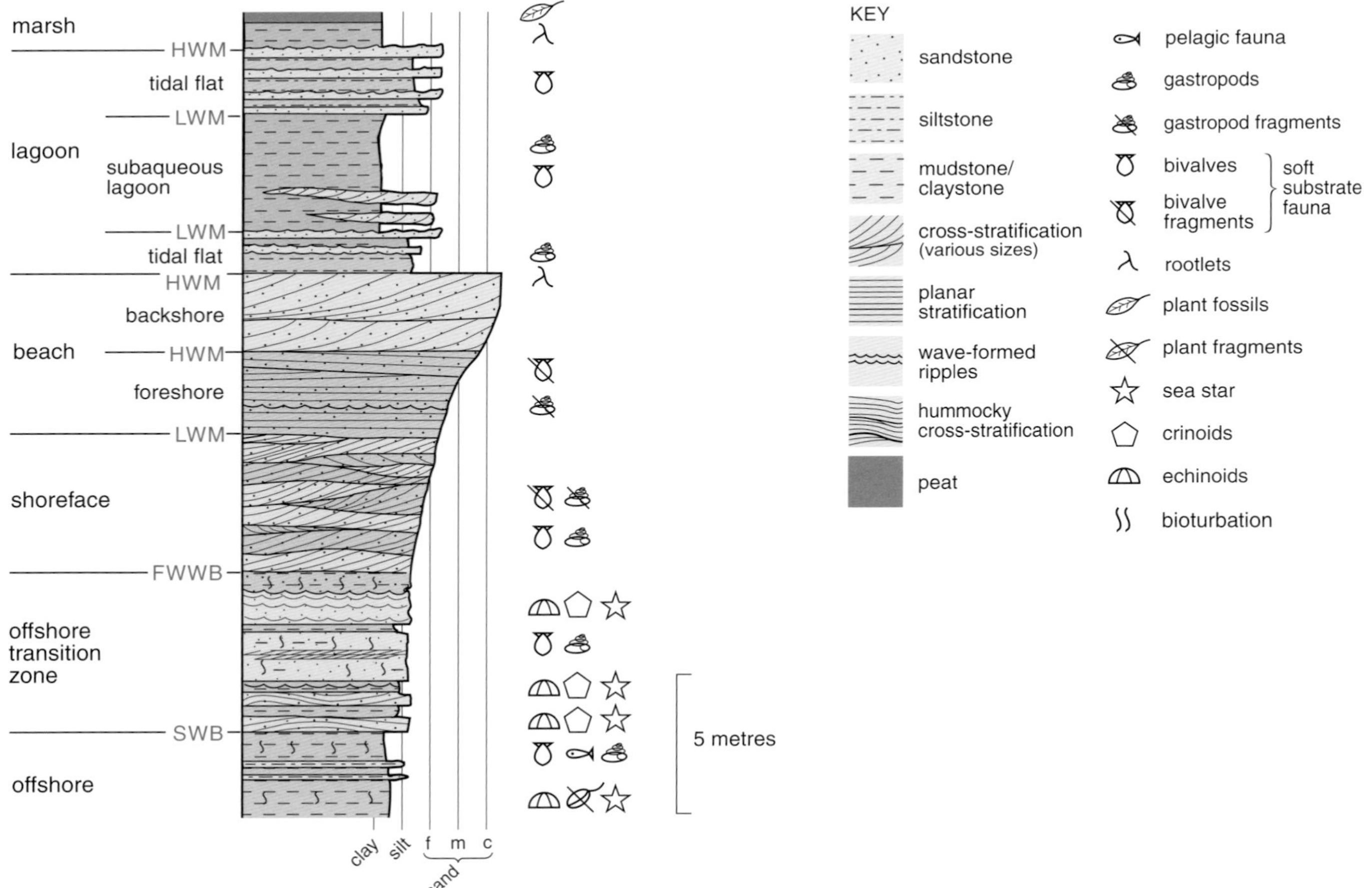

Figure 15.9 A summary log of a barrier island environment. A colour version of this Figure and a block diagram of a barrier island is shown on the Sedimentary Environments Poster and Sedimentary Environments Panorama (on DVD 3).

If the barrier is breached during storms, then sand from the barrier will spill out into the lagoon in a fan shape, a feature usually referred to as a *washover fan* (see block model and graphic log on the Sedimentary Environments Poster and Sedimentary Environments Panorama). These washover fans will then be preserved as discrete lenses of cross-stratified and planar-stratified sandstone within the lagoonal mudstone. At the back of the lagoon, the sheltered conditions and water supply is favourable to colonization by marsh plants. Accumulation of large amounts of organic material may lead to the development of peat which, with subsequent burial, may become coal. A summary graphic log showing the succession of facies typical of a barrier island environment is shown in Figure 15.9.

Activity 15.1

To clarify what you have just learned about strandplains and barrier islands, you should now complete Activity 15.1.

15.5 DELTAS

Deltas form along coastlines where alluvial processes are dominant such that the sediment supply being discharged by a river into a lake or the sea is so plentiful that it cannot be totally dispersed into the sea or a lake, by wave, current or tidal action. It was the historian Herodotus (*c.* 454 BC) who first used the Greek capital letter delta (Δ) to describe the broadly triangular area of sediments where the River Nile discharges into the Mediterranean Sea. Ancient deltaic successions are of great economic importance because they contain abundant coal oil and gas reserves. These include the Niger delta in Africa, the Mississippi delta in the USA and some deposits in the North Sea. You will have a chance to learn more about the Mississippi delta in Activity 15.2 later in this Section. During the Carboniferous, Britain was the site of much deltaic sedimentation and these sedimentary deposits now form the coal-bearing rocks that dominate parts of the Pennines, north-east England, the Central Valley of Scotland and South Wales today (units 81 to 84 on your Ten Mile Map) and are the source of almost all the British coal deposits.

❑ Why are no deltas forming around the coastline of Britain today?

■ Because the conditions in Britain today must be different from those in the Carboniferous. Today, most of the rivers entering the sea around Britain drain land areas of low- to moderate-relief and the climate is stable. The rivers therefore carry relatively small sediment loads, and the sediment that does reach the sea is easily redistributed by waves and tides.

15.5.1 CONTROLS ON DELTA BUILDING

❑ Why do rivers deposit their sediment load as they enter the sea?

■ Because the mixing of the river water and seawater aids flocculation (Section 12.1). In addition, there is a sudden increase in area over which the river water and its load can flow. As the river moves from its restricted channel to the sea (or lake), the flow expands laterally, energy is dissipated and therefore the flow decelerates so it is no longer fast enough to move the bedload and flocculated particles.

In addition to the deceleration of river flow as it enters the sea, there is another important factor that affects deposition of sediment to form a delta; this is the mixing of water masses of differing densities. River water has a lower salinity and therefore a lower density than seawater, even with dense sediment being carried in it. Thus, as a river enters the sea, the river water flows out across the top of the seawater (Figure 15.10). The result of these two effects is that the bedload of the river is deposited close to the coast, but the suspended sediment is carried well out to sea.

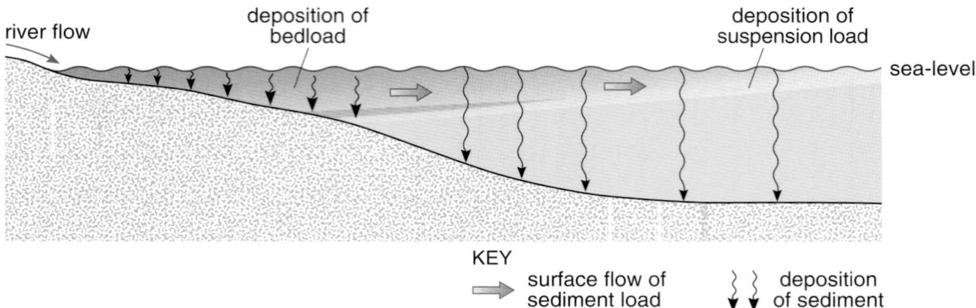

Figure 15.10 Cartoon to show what happens as sediment-laden river water enters the sea.

The bedload being carried by the river is deposited fairly rapidly because the frictional resistance caused by the river and seawater meeting will result in the current speed becoming less and mixing occurring. However, the suspension load in the river water will be carried further out to sea as part of the surface flow. The continued frictional resistance between the river water and seawater and their mixing will eventually slow the currents down and aid flocculation (Section 12.1), thus depositing the suspension load (Figure 15.10).

❑ What other factors besides density differences in the water do you think might govern the size of a delta?

■ The amount of sediment supply will influence the size of a delta, as will the power of the waves and tides. The water depth will also govern the areal extent of the delta. If the water is shallow then the delta will be able to build out over a greater areal extent than if the water is deep.

15.5.2 DELTA ARCHITECTURE AND DELTAIC SUCCESSIONS

Deltas can be divided into three topographic regions which are each characterized by a particular type of deposition (Figure 15.11). The top of the delta behind the shoreline is called the **delta plain**. The upper part of the delta plain (i.e. that nearest the main land-mass) is characterized by alluvial

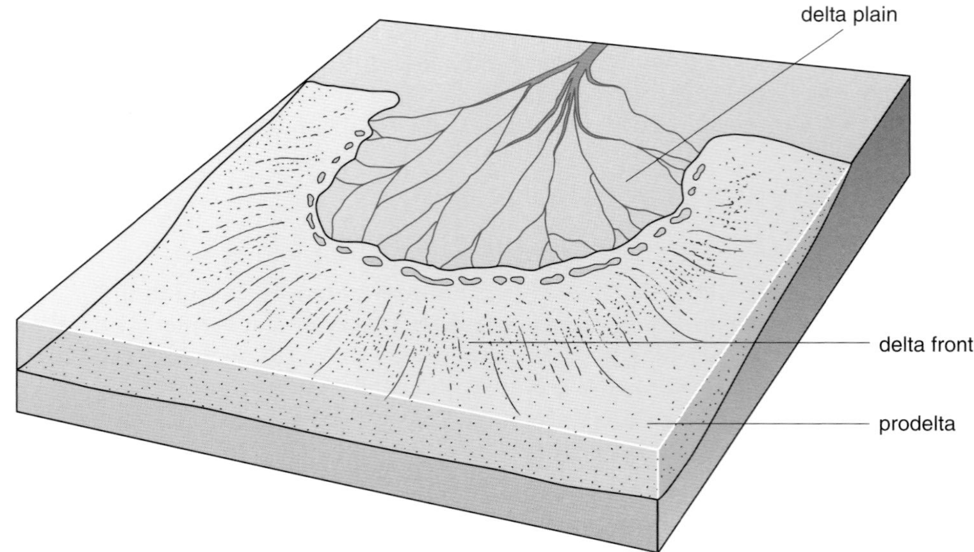

Figure 15.11 Three-dimensional block diagram to show the three main regions of deltas: delta plain, delta front and prodelta.

depositional processes and sediments similar to those described in Section 13. There is a main river channel which splits into smaller channels called **distributaries**. The position of the channels changes as levées and bars build up, and meanders get cut off. Crevasse splays are formed when the levées are breached during periods of high discharge. The lower part of the delta plain may experience some marine influence, especially tides. The distributaries divide and become more numerous on the lower part of the delta plain. Migration of river channels on the delta plain resulting in meander cut-off leads to the formation of lakes in the position of former river channels. The lakes progressively fill with mud during flooding of the river channels and thus become swamps or marshes, where (provided the climate is not arid) the favourable conditions of water and nutrients from a constant replenishment of sediment will support extensive flora and fauna. The prolific growth of new vegetation means that a vast amount of dead vegetation accumulates. This dead vegetation will be trapped beneath the new growth on the marsh or swamp, and the lack of oxygen will lead to the formation of peat. Sediment will no longer be supplied to the swamp or marsh area because the river channel has changed its course. As the sediments and the plant matter become compacted, sedimentary rocks rich in organic matter are formed and on burial the peat may become coal. Isostatic subsidence (Block 3 Sections 2.2.3 and 12.2) of the area because of the weight of accumulated sediments leads to lowering of the area and inundation by the sea. Eventually, the delta may build back out over this area and hence the whole process repeats itself.

The region where the river channels meet the sea is called the **delta front**. This is a slightly inclined area rimming the outer edge of the delta top (Figure 15.11).

❑ Recalling Section 15.5.1, describe what happens to the bedload as a river channel enters the sea.

■ The bedload will be deposited rapidly because of the slowing down of the currents caused by the frictional resistance between the freshwater and marine-water masses.

Bedload sediments are thus deposited at the mouths of the river channels, to give accumulations of sediment known as **mouth bars**. The flow of the river current gives rise to dunes and ripples on the mouth bar, which will be seen as different scales of cross-stratification in cross-section. Waves and tides may also rework the mouth bar sediments, giving rise to other types of sedimentary structures.

Slightly further offshore, on the seaward edge of the mouth bar, finer-grained sediment will be deposited (Figure 15.12). The sedimentary rocks of the distal area of the bar are usually planar-stratified siltstones and mudstones.

Plant debris carried by the river will also be deposited in these fine-grained sediments. This together with the fact that the distal part of the mouth bar remains subaqueous, encourages and supports a variety of infauna such as bivalves, gastropods, worms and crustaceans. These fine-grained sediments of the distal part of the bar are thus often extensively bioturbated.

The third outermost region of the delta is the lower part of the slope, and is called the **prodelta**. The prodelta is the site of deposition of the finest-grained sediments, chiefly clays and silts. Plant debris may also be carried here in suspension. Like the distal bar, the prodelta muds will also be inhabited by numerous marine organisms.

Through time, the delta top, delta front and prodelta will gradually build out seaward as more and more sediment is deposited. This process of building out more and more sediment in a distal direction is another example of progradation (Section 12.2.1).

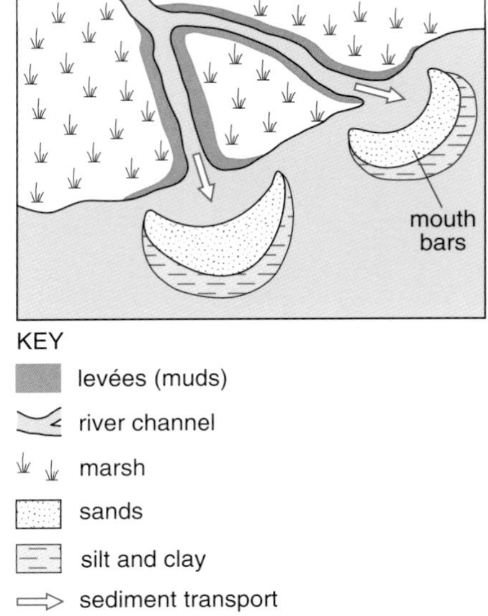

Figure 15.12 The features and sediments of the river mouth where the delta front is constructed.

The presence of a slope on the front of a delta means that the sediments are not always stable. Submarine landslides and turbidity currents (Section 4.3) may be generated and sediment will be carried across the continental shelf and down the continental slope. We shall discuss what happens to these sediments in Section 17.

Lastly in this Section, let us examine what a typical vertical succession of deltaic sediments would look like. We can do this by imagining what we would find if we drilled a hole through the delta at the point shown by the drilling rig in Figure 15.13 and assuming that Walther's Law applies (see Section 12.2.2). The first deposit that we would encounter would be peat deposited in the marsh area. The peat will be sitting on top of a soil horizon in which the plants grew.

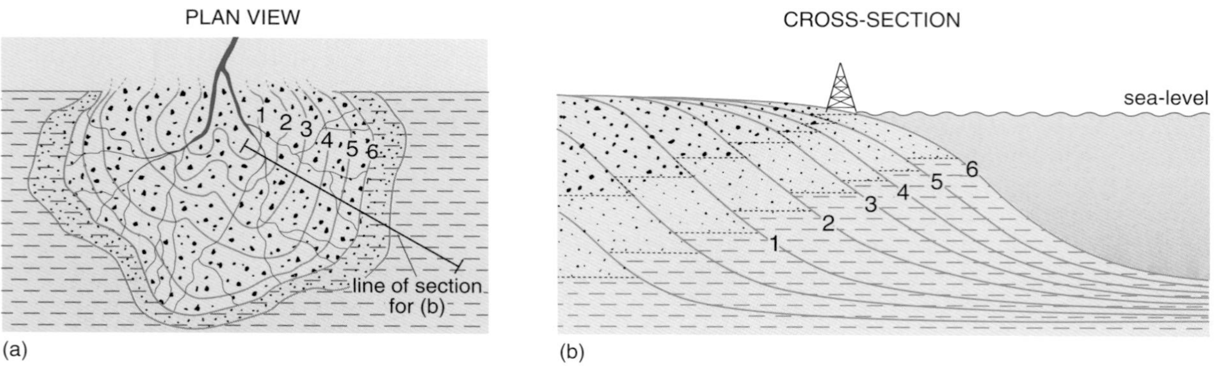

Figure 15.13 Cartoon to show how a delta builds out seaward (progrades): (a) plan view; (b) cross-section.

❑ Where did the marsh form?

■ On the site of a lake which formed in an old abandoned distributary channel.

Underneath the soil we should thus find the deposits of the old distributary channel. This will usually be a fining-upward succession of cross-stratified sandstones representing point bar deposits. The base of the channel may be marked by a lag of coarser-grained material similar to that found at the base of the channel fill in alluvial deposits (Section 13.2.1). The base of the channel is erosive and will overlie the delta front deposits of the mouth bars. This is because through time, as the delta progrades seaward (Figure 15.13b), the alluvial deposits of the delta plain will build out further and further seaward on top of the previously deposited delta front which, of course, was at a slightly lower topographical level (Figure 15.11). Continuing down through the borehole, the next deposits that we would find are the mudstones and siltstones deposited on the prodelta.

Thus, overall deltaic successions usually coarsen upwards from mudstones and siltstones of the prodelta to slightly coarser-grained sediment of the mouth bar. These will be capped erosively by the coarsest-grained sediment of all, the alluvial sediments of the delta plain. A typical deltaic succession is shown in Figure 15.14.

Activity 15.2

To test your understanding of deltas and to see some real examples, you should now complete Activity 15.2.

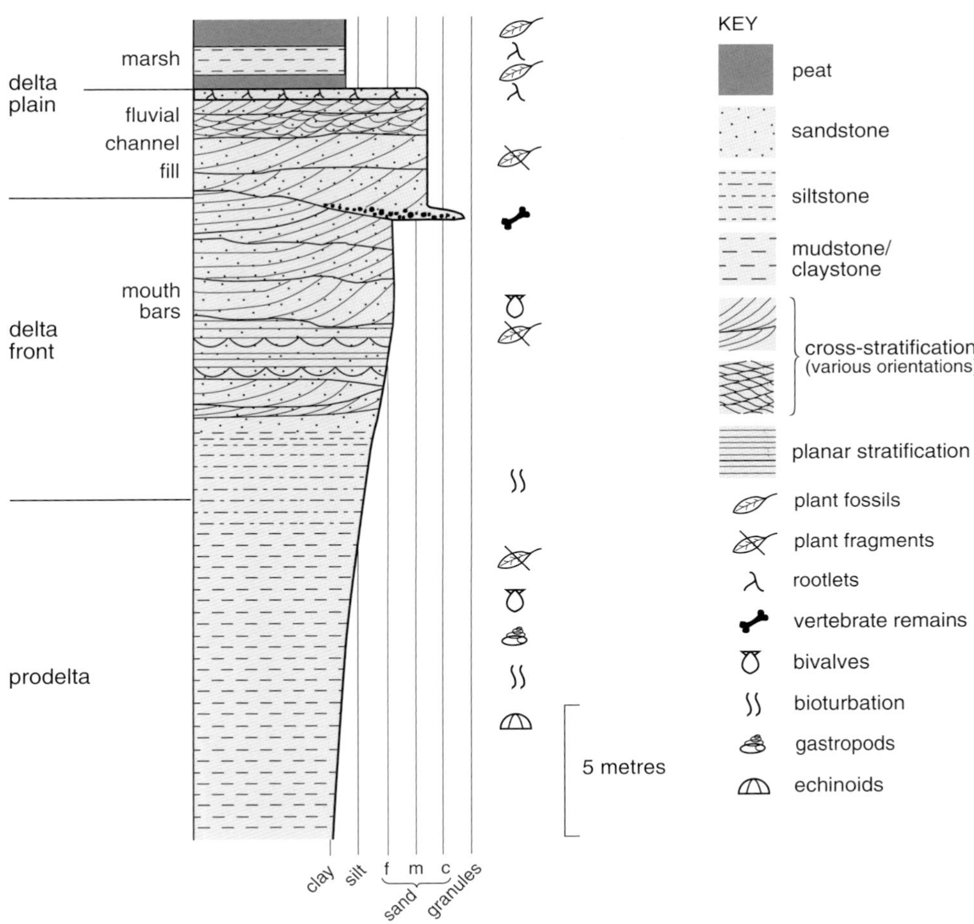

Figure 15.14 Graphic log of a typical deltaic succession. A coloured version of this Figure and a block diagram of a delta are shown on the Sedimentary Environments Poster and Sedimentary Environments Panorama (on DVD 3).

15.6 Summary of Section 15

- Physical processes along coastlines include waves, tides and currents. Currents may be derived from fluvial processes, from waves (e.g. longshore and rip currents) or from tides. Chemical processes along the coastline include flocculation and the formation of soils. Biological processes include the growth of plants stabilizing sediment and the actions of animals.
- The coastline may be split into the backshore, foreshore, shoreface, offshore transition zone, and offshore zone. These areas are determined by the position of mean high tide and low tide, fairweather wave-base and storm wave-base.
- Strong alluvial processes combined with tide and wave processes that are insufficient to transport away all the sediment away favour the formation of deltas. Strandplains tend to form in areas of strong wave action. Barrier islands form where there is moderate wave action and some tide and fluvial action.
- Strandplains comprise linear accumulations of sediment attached to the land. In sandy strandplains, aeolian dunes often develop in the backshore area. The foreshore is characterized by planar-stratified sands, the shoreface by trough and planar cross-stratified sands with some swaley cross-

stratification near the base, the offshore transition zone by hummocky cross-stratified sands and muds and the offshore zone by muds. Fauna is found throughout but will be most common in the offshore transition and offshore zones where it is not being constantly reworked by wave, tide or storm processes. The foreshore may show berms, and its profile is governed by the grain size of the sediment and storm or fair weather conditions.

- Barrier islands comprise an island separated from the mainland by a lagoon. The sediments and structures of barrier island successions are similar to those for strandlines but there is often less evidence for storms and the backshore barrier sands will be capped by muds deposited in the lagoon and on the tidal flats. During storms, the barrier may be breached, resulting in washover fans forming in the lagoon.
- Deltas have three parts: the *delta plain* which comprises a flat area dominated by alluvial deposition. The resulting vertical deposits include alluvial channel fills, overbank muds and the fine-grained sediment infill of lakes. At the distal edge of the delta plain is the *delta front*; sediments are deposited in mouth bars as the rivers emerge into the sea. The most distal part of the delta comprises the *prodelta* where the finest-grained sediments are deposited.

15.7 Objectives for Section 15

Now you have completed this Section, you should be able to:

15.1 Describe the physical, chemical and biological processes that are important in siliciclastic coastal environments.

15.2 Explain how the interaction and relative dominance of fluvial, wave and tidal processes influence the morphology of the coastline.

15.3 Describe the biota (body and trace fossils) of coastal and continental shelf environments.

15.4 Recognize and describe the main facies and features of deltas, strandplains and barrier islands.

15.5 Describe the different areas present along the coast (backshore, foreshore, offshore transition zone, offshore zone) and the processes and types of sediment typical of each.

15.6 Recognize and sketch simple graphic logs summarizing the main biological and sedimentological features of a succession that ranges from the continental shelf to backshore for (i) a strandplain and (ii) a deltaic coastal morphology.

Now try the following questions to test your understanding of Section 15:

Question 15.5 Why is the identification of HCS in a sedimentary succession important when you are trying to work out where the positions of storm wave-base and fairweather wave-base are in the field?

Question 15.6 What type of sedimentary structures in the mouth bar sedimentary deposits might tell you that they have been reworked by (a) waves; (b) tides?

Question 15.7 Explain whether the foreshore will be steeper on a sandy, storm-dominated beach or on a pebbly, fairweather-dominated beach.

16 SHALLOW-MARINE CARBONATE ENVIRONMENTS

The sedimentary environments that we have discussed so far are dominated by siliciclastic sediments. This is true of much of the Earth's surface today. However, shallow-marine carbonate sediments are an important feature of the sedimentary record and were more common in the past, chiefly during warmer climatic conditions and periods of high sea-level when shallow seas covered extensive areas of the continent, forming shallow-marine environments. By contrast, carbonate deposition is not as common today as it was, say, in parts of the Carboniferous and Cretaceous due to the current relatively low sea-level. Nevertheless, some shallow-marine carbonates are forming today, particularly in the tropics and subtropics between 30° north and south of the equator. Of particular note are the Bahamas, South Florida shelf, Arabian Gulf, and various small areas around volcanic islands in the Pacific (Figure 16.1). There are also some carbonates forming in mid-latitude regions in cool water conditions along the southern coast of Africa, Australia and New Zealand (Figure 16.1).

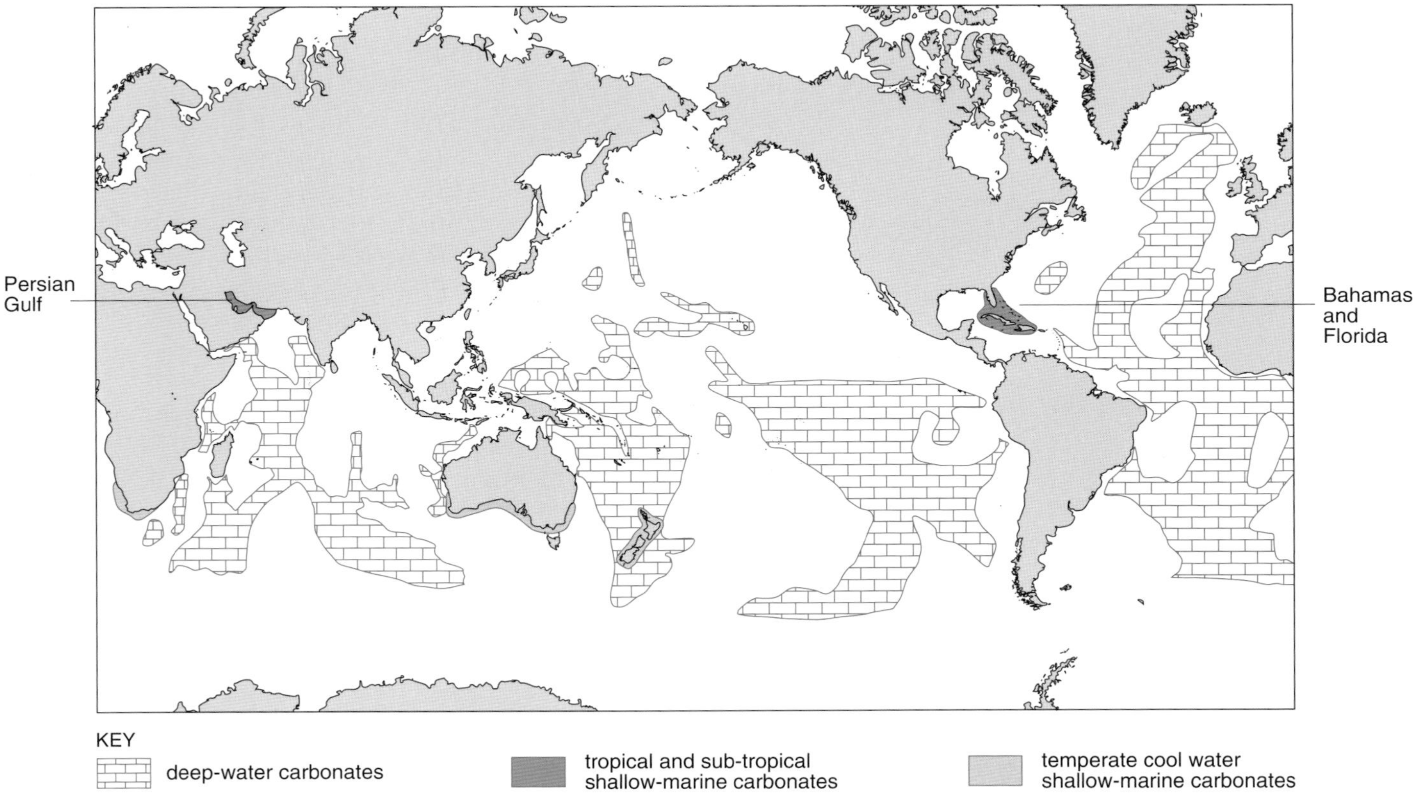

Figure 16.1 Important areas of deposition of marine carbonates around the world today. Shallow marine (this Section); deep marine (Section 17).

Carbonate and siliciclastic sediments do not always occur in isolation from each other, but in any given situation it is usual that one dominates because the presence of siliciclastic sediments inhibits carbonate production. However, mixed siliciclastic and carbonate successions are found in the geological record, where they are thought to represent changing conditions, or sediments mixed from two sources (Plate 16.1).

16.1 Controls and processes affecting carbonate deposition

In Section 12.1 the controls on sediment deposition were discussed. Tectonic and climate controls profoundly affect both siliciclastic and carbonate deposition. However, climate, together with organism biology, is particularly important for carbonates. Organism biology controls the type of sediment produced and the way in which it accumulates. It also means that unlike siliciclastics, the majority of carbonate sediment is produced *in situ* and is not transported over large distances. The main products and processes involving biota include:

- Bioclasts – the calcareous parts of the organisms preserved as clasts.
- Peloids – most peloids comprise the faecal pellets of organisms. Organisms ingest the sediment, consume the organic matter and then excrete the waste products as peloids; they are composed mainly of calcium carbonate.
- Precipitates – clasts such as ooids and micrite may be produced chemically or as a result of some biological input (see bioerosion below).
- Framebuilders – some organisms, e.g. corals, literally build strong skeletal frameworks of carbonate as they grow. This framework then supports other organisms.
- Bafflers – some organisms act as sediment baffles, trapping sediment around them; sea grasses are one example.
- Binders – algal and bacterial mats (Section 9.5) are examples of binders. Their formation as a layer over the micritic sediment holds it together.
- Bioerosion – erosion by organisms probably accounts for most of the micrite produced. Both vertebrates (e.g. parrot fish) and invertebrates (e.g. gastropods and echinoids) graze on a carbonate host (e.g. a coral reef) and in the process they break down the carbonate into the fine-grained carbonate sediment, micrite. Other animals drill holes into carbonate rocks, also producing micrite.

❑ What is the other chief way of producing micrite?

■ Direct chemical precipitation of calcium carbonate from seawater (Section 5.2).

For maximum carbonate production and accumulation, the controls have to be just right: not too deep, not too shallow, not too warm or too cold, the correct salinity and nutrient supply, not too much siliciclastic supply and, probably most importantly, the right types of organisms must be present. Table 16.1 is a summary of the different controls and subcontrols that influence carbonate deposition; it also describes why each control is important.

The dominance of biological and chemical processes in the production and accumulation of carbonates means that most of the sediment is produced *in situ*. However, like siliciclastic sediments, the carbonates may be affected by both unidirectional and oscillatory currents in the depositional environment.

❑ Would you expect some carbonate sediments to show cross-stratification, wave ripples and other current-formed sedimentary structures, and, if so, why?

■ Yes, carbonates will show all the same types of sedimentary structures as siliciclastics; some carbonates such as ooid shoals behave in almost the same way as well-sorted quartz sandstones because they are composed of small grains that can be easily moved around (Section 5.3).

Plate 16.2 shows an example of a cross-stratified oosparite from the Jurassic of the Dorset coast.

Any transport process that affects siliciclastics also affects carbonate sediments. So, carbonates, like siliciclastics, may also be affected by sediment gravity flow processes and become re-deposited as turbidites and debris flows. These types of sedimentary deposit are discussed in Section 17.

Table 16.1 Controls and subcontrols on carbonate deposition.

Control	Subcontrol	Influence and production
Organism biology	Form and function	The type and amount of carbonate clasts produced depends partly on the dominance of bafflers, binders, framebuilders and amount of bioerosion.
	Evolution	Different organisms have evolved and dominated through geological time; this controls the type of carbonate produced.
	Cool water biota	Less sediment produced by these than warm water faunas; cold water biota include coralline algae.
	Warm water biota	Major carbonate production e.g. corals, and blue-green algae.
Climate, and to a lesser extent tectonics which controls the shape and depth of the ocean basin.	Light penetration	Penetrates approximately the upper 70 m of water but light penetration is greatest in upper 10–20 m. The amount of light affects the productivity of the organisms.
	Water temperature	Affects organisms. Warm waters occur generally down to 50–100 m water depth. Temperature is also controlled by latitude and the presence of water and cold water currents. There are cold and warm water faunas; therefore water temperature controls the type of carbonate produced. See organism biology above. Ooids and micrite are precipitated only in warm waters.
	Water circulation	Influences the type of sediment produced. Good water circulation is required for some organisms to survive.
	Oxygenation	Well-oxygenated waters are essential for the growth of all skeletal invertebrates.
	Salinity	Both high and low salinity conditions reduce the biotic diversity because only some organisms are adapted to these conditions. At very high levels of salinity, most invertebrates disappear.
Tectonics and climate	Siliciclastic sediment supply	A high siliciclastic sediment supply from the land inhibits carbonate production because the organisms become swamped with sediment and cannot filter feed.
Tectonic setting		Controls the area available for carbonates to accumulate. Also controls the latitude of the area of carbonate production because tectonic plates move. Although some carbonates form at higher latitudes, the majority do so in the tropical–subtropical belt 30° N & S of the equator.

16.2 Similarities and differences between shallow-marine carbonates and siliciclastics

- ❑ In this Section, we have already discussed one similarity and one difference between siliciclastics and carbonates: what are they?
- ■ The similarity is that they are both moved around by currents and sediment gravity flows and thus contain similar sedimentary structures. The difference is that the majority of carbonates are produced *in situ* whereas siliciclastics are transported *over significant distances* prior to deposition.

There are, however, several other differences which are summarized in Table 16.2. Study this carefully and then answer the questions that follow.

Table 16.2 The differences between shallow-marine siliciclastic and carbonate sediments.

Feature	Siliciclastics	Carbonates
Sediment production	Transported from the nearby land-mass or along the coast.	Mainly produced *in situ*. Some organisms may build up structures that modify the sea-floor topography.
Grain size	Reflects the size of the particles in the source rock and the duration and type of processes that have acted on the particles during transportation.	Reflects the size of the original carbonate skeletons and chemically precipitated grains, and sometimes the type and duration of the processes of transportation and deposition.
Cementation	Sediments remain unconsolidated on the sea-floor, and are usually cemented only once they are buried.	Sediments are commonly cemented on or immediately below the sea-floor because of the amount of carbonate in solution.
Periodic subaerial exposure of sediments	Minor or no alteration.	Due to the change from seawater to freshwater in the sediments, dissolution of the carbonate and widespread cementation may result.
Mud	Indicates settling from suspension. Derived from the chemical decomposition of rocks.	Final deposition is from suspension but a large amount of carbonate mud (micrite) is produced from bioerosion. The amount of micrite also relates to the organisms that are around to produce it, either from bioerosion or from the growth of micro-organisms whose skeletons are made up of mud-sized carbonate particles.
Geological age	Erosion and deposition processes are the same at all times.	Relative tendency for calcite or aragonite to precipitate or dissolve on the sea-floor in response to different saturation levels at various times in geological history.
Latitude and climate	Sediments occur worldwide.	Most shallow marine carbonates are deposited in warm shallow water environments between 30° N and 30° S of the equator.

Question 16.1 Compare the grain size of RS 22 and RS 10. What does this tell you about the strength of the transporting current in each case?

Question 16.2 RS 18 and RS 21 have both been deposited relatively close to their source area. What features allow us to deduce this?

16.3 Carbonate platforms

The term carbonate platform is a general one used to describe a thick succession of marine, mostly shallow-water carbonates. There are five different morphological types of carbonate platform defined on the basis of their extent, general geometry (shallow or steep edges) and relationship to the continent (Figure 16.2).

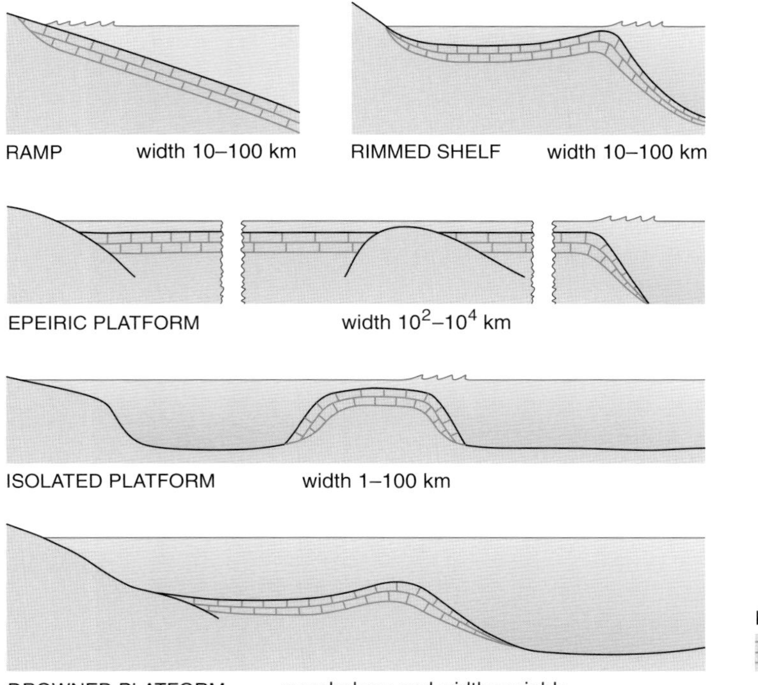

Figure 16.2 Cross-sections through the different morphological types of carbonate platform.

The morphological type of carbonate platform that develops depends on the pre-existing topography, the organisms present, and the rise and fall of sea-level. The majority of the facies will be similar on different types of carbonate platform. However, the thickness and lateral extent of the facies will change. For instance, rimmed shelves favour the growth of reefs at the edge whilst epeiric (laterally extensive shallow seas) carbonate platforms have widespread lagoonal deposits.

At the seaward edge of rimmed shelves, epeiric carbonate platforms and isolated carbonate platforms, where the depositional slope is steep, the carbonates may be re-deposited by sediment gravity flows (see Sections 4.3 and 17). Such re-deposition is much less common or absent in ramps because the depositional slope is not as steep.

16.4 Peritidal facies

We shall consider the processes and facies deposited on carbonate platforms in terms of four morphological features that may be found on all of them but are developed to varying degrees. **Peritidal** is a term used to mean 'near the influence of tides'; it includes the area above mean high water and just below mean low water. This encompasses most of a carbonate platform, except any parts of an organic growth (Section 16.5) present in this zone. We shall now consider three principal areas of extensive peritidal carbonate deposition, namely tidal flats, lagoons, and carbonate sandbodies. The presence, dominance or absence of each of these facies depends on the latitude, climate, tide influence, wave influence and the morphology of the carbonate platform.

16.4.1 TIDAL FLATS

The shoreline, or most proximal part of a carbonate platform, usually includes a tidal flat area, i.e. an area either regularly or occasionally covered by water during the tidal cycles, at which time it is dominated by weak currents and wave action. Micritic limestones dominate the deposits because of the low energy conditions but some pelmicrites and occasionally oomicrites and biomicrites (Section 6.5.3) may be preserved due to storms.

If the climatic conditions are arid, evaporite minerals, usually gypsum and anhydrite, will precipitate out of water held in the pore spaces of the sediment in the high intertidal and supratidal areas. Initially, gypsum ($CaSO_4.2H_2O$) may precipitate in the form of well-developed crystals, typically 0.1–25 cm across, similar to the desert rose variety shown in Plate 16.3. Continued high evaporation rates and the associated increase in salinity will cause the gypsum to be converted to its anhydrous equivalent anhydrite ($CaSO_4$). When this grows in carbonate sediment, it gives rise to the characteristic nodular form called chicken-wire anhydrite. This form is commonly found in many ancient sulfate deposits (Plate 16.4). Extensive areas of evaporite deposition are particularly well developed together with the shallow marine carbonates along the coast of the Arabian Gulf today (Figure 16.1).

❑ Where do you think all the mineral salts come from to form the evaporites?

■ The continued re-supply of seawater during the rising and falling tide on the tidal flats.

If climatic conditions are humid, then vegetation grows on the tidal flats and supratidal marshes; this is dominated by algal growth and (in the Recent past) mangrove trees.

The periodic exposure of tidal flat sediments gives rise to a number of other distinct sedimentary structures. These include desiccation cracks which form due to the repeated subaerial exposure, and a dried crust of sediment at the surface (Figure 16.3) which may curl up into thin flakes on drying. This may then be broken up into limestone clasts during a later flooding, and form soils.

❑ Will fossil soils be found mainly in the intertidal or supratidal area?

■ In the supratidal area, because this is subaerially exposed for longer periods of time.

The fauna are an important element of tidal flat facies. In general, the range of species is fairly restricted due to the rapid fluctuation in salinity and water depth. Typically, gastropods, bivalves worms and crustaceans adapted to the conditions may be found together with some microfossils.

❑ Will there be any trace fossils in the tidal flats, and, if so, why?

■ Yes: because of the fluctuating conditions, the worms and crustaceans will leave feeding and dwelling trace fossils.

Reworking of gastropod and bivalve shells during storms can give rise to biomicrites and biosparites. Other important elements of the biota of tidal flats are bacterial and algal mats, which when fossilized form stromatolites (Figure 5.3 and Section 9.5). These thrive in the exceptionally saline conditions, and, importantly, they bind together the fine-grained carbonate sediment. The mats with a simple planar morphology are most common, although small domes (30–50 cm in diameter) may also form. The mats and intervening sediments are often cut by desiccation cracks. Modern day mats with desiccation cracks are shown in plan view in Figure 16.4, and a cross-section through a fossilized example is shown in Figure 16.5.

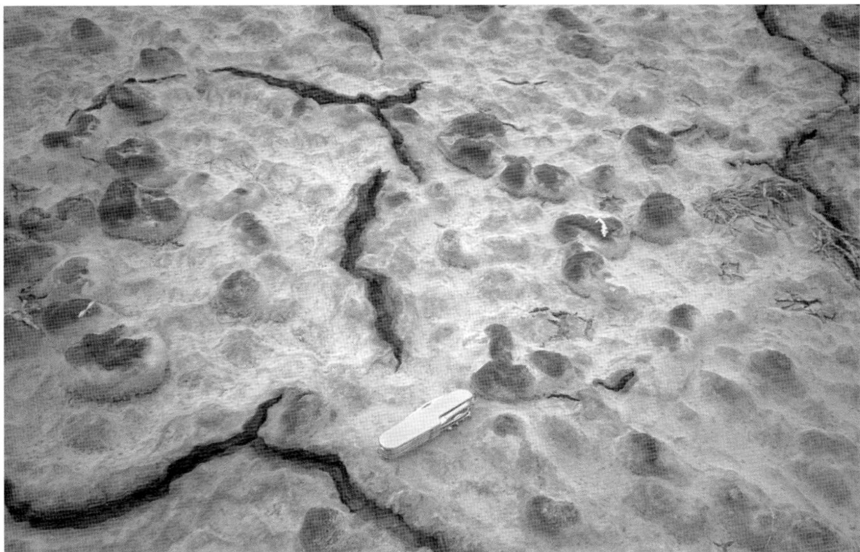

Figure 16.3 Dried crust of sediment at the surface of tidal flats. Note the desiccation cracks starting to form. Penknife shows scale.

Figure 16.4 Algal mats, Middle East. Note how the mats curl up at the edges of the polygonal desiccation cracks. Person in the distance shows scale.

Figure 16.5 Laminated planar stromatolites (dark colour) with interbedded micrite (light colour) and subvertical, tapering desiccation cracks. Glacier–Waterford National Park, USA. Finger on the left shows scale.

All the features of the tidal flat facies described above can be put together into a generalized 3D model and vertical facies succession; these are shown in Figure 16.6. In a similar fashion to siliciclastic tidal flats, as the tide rises over the carbonate tidal flats first, it fills shallow tidal channels (Figure 16.6a) in the intertidal area, and then floods across the whole broad flat intertidal area (Figures 16.7 and 16.8).

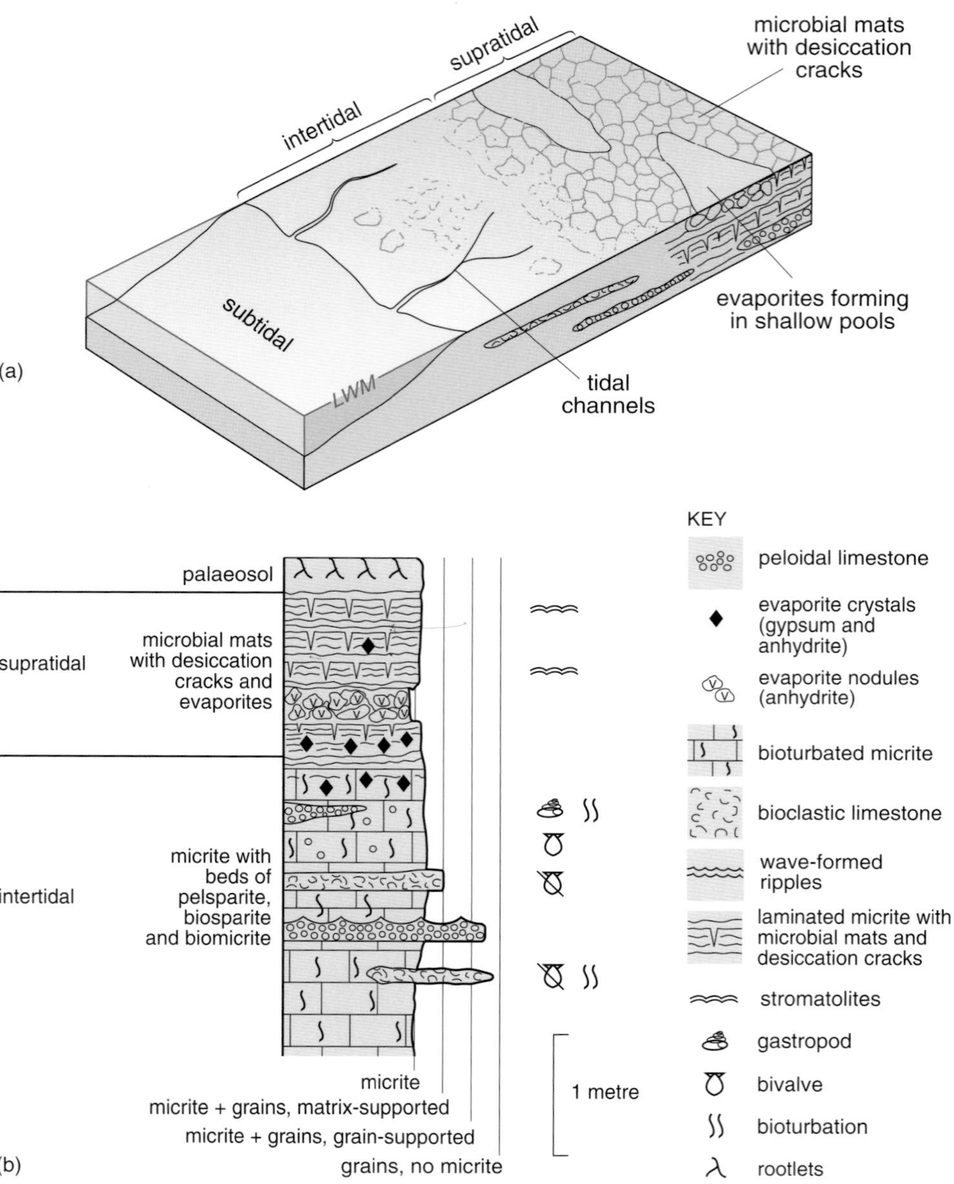

Figure 16.6 (a) Generalized block model of the facies and (b) graphic log of facies succession for an arid tidal flat.

Figure 16.7 Broad shallow tidal channel in the intertidal zone of a carbonate tidal flat, Middle East.

Figure 16.8 Flooding of a tidal flat by typically calm waters; note the raised edges of the desiccation cracks. Legs in top left show scale.

16.4.2 Lagoons

Lagoons are protected subtidal areas that lie on the landward side of a barrier. A barrier can be either some kind of organic growth, like a reef, or a stabilized bank of oolitic and/or bioclastic sediments (Figure 16.9).

Rimmed shelf carbonate platforms always have lagoons whereas ramps do not necessarily develop these features. Lagoons are typically several kilometres long and occasionally up to 100 km wide. They can be up to many hundreds of kilometres in length. They are connected to the open sea through gaps or channels across the barrier (Plate 16.5). The fact that the lagoon area is protected from the coastal currents and wave action means that it is a low energy environment (Plate 16.6). This low energy means that micrite is the typical lithology and that the rich biota may become preserved in life position. The biota will include bivalves, gastropods, crustaceans and algae. In the Bahamas and Florida today, sea grasses growing in the lagoon stabilize the micritic sediment. Peloids of faecal origin are common in the lagoonal deposits because of the abundance of biota and the lack of current reworking. During storm conditions however, waves may break through the barrier and surge through the channels creating currents in the lagoon. These currents rework the sediment into storm beds which are typically preserved as thinly bedded biomicrites.

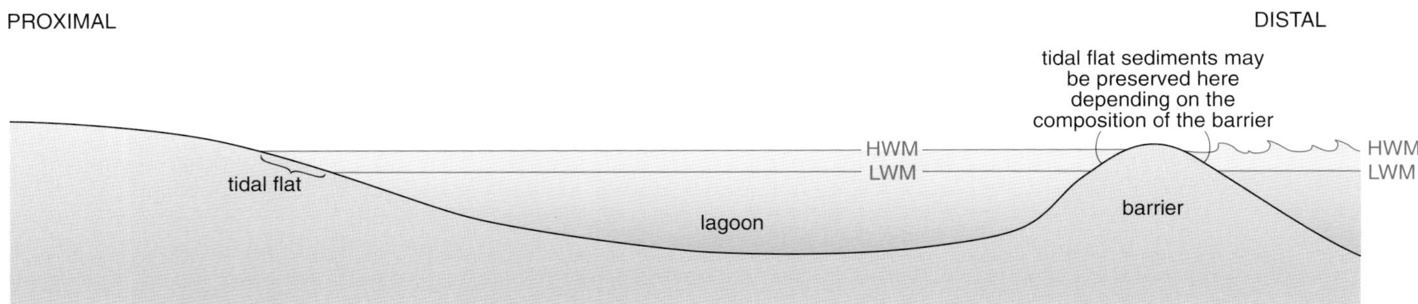

Figure 16.9 Cross-section through a lagoon showing its relationship to a barrier and tidal flat sediments.

16.4.3 CARBONATE SANDBODIES

Carbonate sandbodies are composed of ooids and/or rounded and sorted marine bioclasts (e.g. fragments of corals, bivalves, foraminiferans, algae, echinoderms, brachiopods). Carbonate sandbodies develop in both the subtidal and intertidal area of carbonate platforms subject to high tidal currents and/or wave activity.

❑ What carbonate rock types are likely to form if a carbonate sandbody is preserved, and why?

■ The likely rock types are oosparites and biosparites. This is because the constant scouring and reworking by currents is unlikely to allow the settling out of much carbonate mud (micrite) and the grains are usually either ooids or bioclasts which will become cemented by sparite during burial.

Carbonate sandbodies are very similar to siliciclastic sandbodies (as discussed in Section 15) in terms of their architecture and sedimentary structures. For instance, sandbodies may form part of the foreshore or shoreface or build up to form a barrier at the edge of the carbonate platform (Figure 16.9). Sedimentary structures are very common, being mainly cross-stratification on a variety of scales (Plate 16.2). Herring-bone cross-stratification may develop where there are tidal currents and scours and channels may form through the sandbodies; and, if they are deposited in the subtidal offshore transition zone, they may also show hummocky cross-stratification. The carbonate sandbodies may also be bioturbated.

Carbonate sandbodies may be orientated either parallel or perpendicular to the coastline. Those that form barriers or that are developed on the continental shelf tend to lie parallel to the edge of the continental shelf and shoreline and are common on the protected landward side of shelf margins where bioclastic debris may be derived from nearby reefs (Plate 16.6). Where tidal currents are strong, the sandbodies may be orientated perpendicular to the coastline and edge of the continental shelf. In the area between the sandbody ridges, muddy (micrite) carbonate sands often accumulate.

16.5 CARBONATE BUILD-UPS (INCLUDING REEFS) AND BIOSTROMES

Also important in many carbonate platforms are coherent bodies of limestone, differing from surrounding facies, and produced by sessile surface-dwelling benthos, together with variable amounts of interstitial sediment and cement. These bodies range from mounded masses that originally had positive relief on the sea-floor (**build-ups**) to tabular or lenticular beds (**biostromes**) that did not.

Of all the sedimentary deposits we have considered, build-ups and biostromes show the greatest degree of biological influence, and hence variation according to the vagaries of evolutionary history. Even where evolutionary convergence; (Section 8) appears to have cast successive organisms in similar roles at different times, they are unlikely to have been identical in all respects of their ecology.

❑ Casting your mind back to what you read about the characteristics of organisms in Section 8, why should we expect some differences between successive convergent forms?

■ Differences can be expected because of the differing inherited constraints and potentialities of different groups of organisms.

In addition, changes in climate and sea-level have also altered the environmental contexts in which organisms have grown. During times of relatively high global sea-level, for example, extensive shallow seaways spread across parts of the continents. The margins of carbonate platforms flanking these interior seaways experienced quite different regimes of wave activity and nutrient supply, to cite just two variables, from the predominantly open ocean-facing margins of today.

Hence, we can expect carbonate platform deposits to show greater variation through geological time than those formed in other depositional environments, in which more consistent physical and chemical influences tend to dominate. The uniformitarian approach to interpreting fossil examples must therefore be used with caution. On closer inspection, superficial similarities are often found to mask profound differences in detail. This in turn means that terms derived from modern examples can be misleading when applied to ancient ones, by suggesting that the latter possessed attributes which they may not in fact have had.

Perhaps the best illustration of this point is provided by the term 'reef'. You might suppose that everyone has a fair idea of what a reef is from common knowledge, so that interpreting fossil examples should not be a problem. However, the term has been a source of endless confusion – and debate. From the original nautical usage to describe any rocky submarine prominence on which a vessel might become stuck, the early geological literature effectively limited itself to those spectacular examples built in present-day shallow tropical waters by corals. For example, one of Charles Darwin's outstanding geological contributions was his study of the growth of coral reefs on sunken Pacific volcanoes.

Today, coral reefs may form prominent barriers along the margins of platforms (including both isolated oceanic atolls and those situated on continental shelves, such as the Great Barrier Reef of Australia), protecting relatively quiet lagoonal waters from the open sea beyond (Figure 16.10a). The means by which such structures achieve their rigidity, against the relentless surge of ocean waves, thus became a focus of interest among geologists studying them. Besides the initial growth of skeletal framebuilding organisms (such as various corals), and partial filling of cavities by shelly debris (in part produced by boring organisms), further consolidation of the reef framework is achieved through overgrowth by encrusting shelly organisms (such as certain kinds of algae) and cement growth (Figure 16.10b). Geologists studying reef growth have thus sought to identify 'framebuilders', 'binders' and 'sediment producers' as well as early phases of cementation.

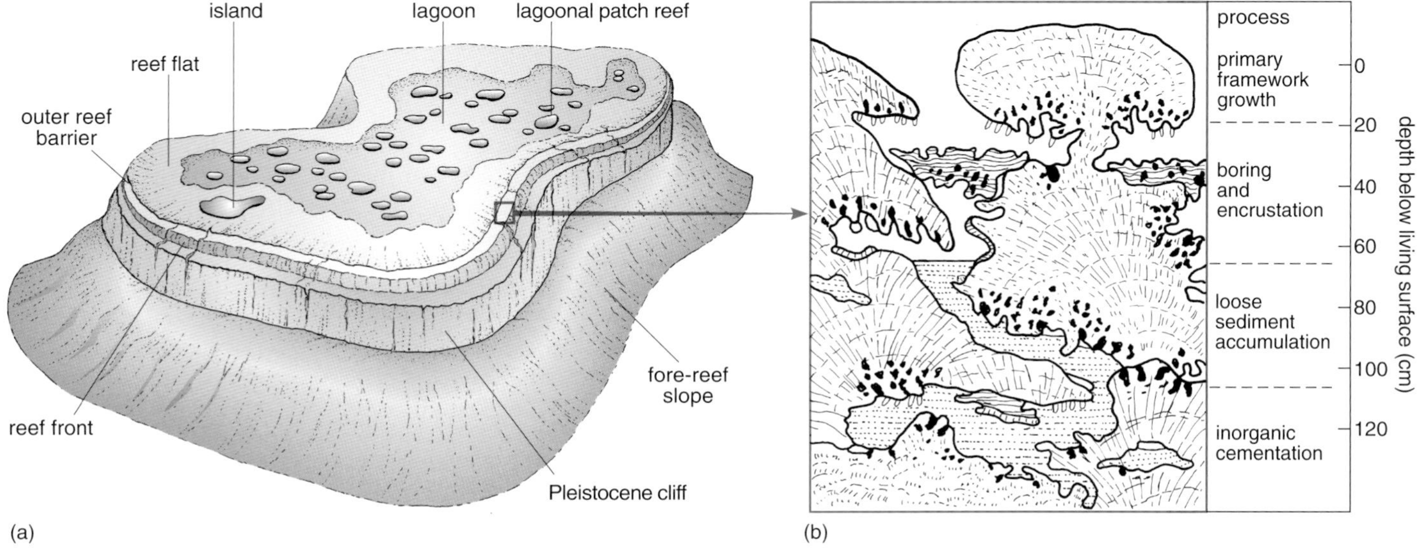

Figure 16.10 Diagram showing the typical form of a modern tropical coral reef along a platform margin: (a) overall structure; (b) reef framework and the processes involved in creating it. Pleistocene is a stratigraphic term encompassing all but the most recent of the Quaternary Period.

Geological definitions of reefs coined around the 1940s and 1950s therefore variously combined the attributes of positive relief on the sea-floor, the presence of skeletal framework, and/or wave resistance. However, such definitions subsequently had to be qualified, when detailed studies of cores drilled from reefs revealed how little (less than 30% on average) of the surface framework in modern reefs usually survives into the fossil record. Because of the destructive effects of bioerosion (by predators, grazers and borers) and storms, most of the resulting reef rock instead comprises bioclastic debris. Thus, the role of framework in modern reefs is now regarded as essentially limited to that of entrapping sediment at the surface, while cement, with or without the assistance of binders, is what stabilizes the accumulating pile beneath.

Besides this profound taphonomic alteration of reef fabrics, we now also know from drilling of rock cores and from seismic sections that much of the topography of today's reefs is inherited from the underlying rocky substrate on which the reef-builders grew (Figure 16.11). By and large, that substrate was sculpted by subaerial erosion when sea-levels were lower during the last glaciation. Indeed, many of the reefs on continental shelves turn out to be little more than veneers on such **antecedent topography**. Hence, the pronounced positive relief of modern reefs is largely an historical accident of our seeing them in an interglacial episode. With the stabilization of sea-level, lagoons could be expected to become filled in by accumulating sediment, thereby restricting reef development to a marginal fringe.

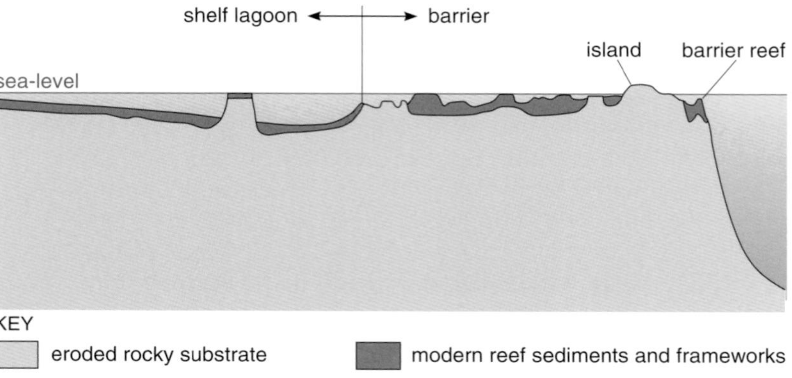

Figure 16.11 Antecedent topography of a modern reef.

More importantly, numerous problematical fossil examples fail to conform even with such qualified definitions of reefs. For example, build-ups largely composed of carbonate mudstone (**mudmounds**), with few, if any, skeletal framebuilders (Figure 16.12), appear to have grown mainly in relatively quiet waters during certain geological periods (e.g. the Carboniferous).

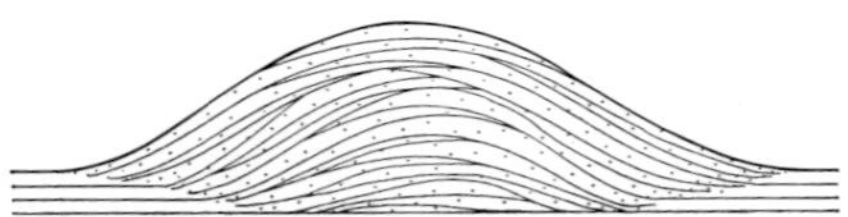

Figure 16.12 Carbonate mudmound.

Moreover, biostromes (lacking original relief) throughout the geological record show a spectrum of fabrics ranging from bound frameworks, to embedded clustered shells (which never created rigid frameworks) or even washed-out shell rubble. Hence, in recent years, geologists have increasingly tended to broaden the definition of 'reefs' to encompass such structures as well. The term has thereby become so generalized as to be virtually worthless in terms of environmental implications: there may or may not have been positive relief on the sea-floor; framework may or may not have been present; and the structure may or may not have been wave-resistant, or perhaps not even subject to much current activity in any case. We are simply left with a vague notion of any more or less *in situ* accumulation generated by crowded sessile benthic organisms, which differs from surrounding facies – and which could have arisen in a very wide range of circumstances.

Consequently, modern work on such deposits lays greater stress on synthesizing models from palaeoecological and sedimentological inferences based on fundamental constituents (which can tell us about original environmental conditions) rather than on simplistic comparisons with modern reefs (which might be quite inappropriate as analogues). By this means, important patterns of change through geological time can be recognized.

One such change relates to variation in the feeding ecology of organisms. Today, marine communities can be recognized as exploiting a range of nutrient levels in the surrounding water. Where nutrients are plentiful, abundant plankton can thrive, and benthic communities tend to be dominated by suspension and deposit feeders, along with rapidly growing fleshy algae. Because the nutrient supply (along with several other linked factors, such as temperature) commonly fluctuates, often on a seasonal basis, the constituent organisms tend to settle and grow opportunistically. They are heavily grazed and fed upon in turn by predators and scavengers with rather generalized feeding habits. Reproduction in the sessile benthos tends to be sexual, yielding large numbers of discrete, variable offspring that can disperse to favourable patches, often growing in clumps or carpets of one or a few species, as in mussel or oyster banks. Such congregations typically fail to produce upstanding frameworks, consisting instead of clusters of shells embedded in, and supported by, the surrounding sediment (Figure 16.13).

Figure 16.13 A Pleistocene fossil oyster bed in the Pleistocene of Jamaica. Exposure shown is about 1 m across.

At the other extreme, where the supply of nutrients is highly limited, plankton is permanently sparse, and the sessile benthos tends to be dominated by animals that contain symbiotic single-celled algae in their tissues, such as reef-building corals. This symbiosis ensures a tight recycling of the valuable nutrients within the hosts' bodies (Section 9.3.7), though some topping up may also be accomplished by trapping of microscopic planktonic animals. Colonial growth by asexual budding is especially common among such forms, because established colonies have a competitive advantage over newly settling larvae in spreading over available hard surfaces. The relative lack of fluctuation in environmental conditions allows the colonies to grow and prosper. The stony coral colonies also tend to be fairly robust, so withstanding the wave surge from the open oceanic waters where favourable conditions for such reef growth are most common. The framework produced by the intergrown colonies provides a complex architecture of walls and cavities, which are in turn exploited by many other species (Figure 16.10b). Paradoxically, then, today's coral reefs are among the most diverse marine communities, despite growing in 'nutrient deserts'.

Fossil examples corresponding to these rather differing communities can be detected by analysis of the palaeoecology of the constituent organisms and their sedimentological context. Fossil coral reef deposits are well represented in the Neogene. By contrast, an extinct group of highly modified sessile bivalves, called rudists, became widespread on low-latitude carbonate platforms in the Cretaceous. Despite the superficial similarity of their clustered shells to some coral frameworks, they were more similar in constitution to oyster banks. Dense congregations of rudists generated extensive biostromes of more or less *in situ* shells (Figure 16.14) implanted in shell debris from earlier generations, especially over the outer parts of the platforms. True reef frameworks rarely developed on these Cretaceous platform margins.

Figure 16.14 Detail of a biostrome of rudist bivalves, seen in vertical section.

Subdivision of all fossil examples into the two extreme states of build-ups and biostromes described above would still be grossly simplistic. A range of possibilities exists between them. For example, frameworks have evidently been generated in some cases by colonial suspension feeders in what must have been moderately or intermittently nutrient-enriched waters. Although these were sometimes quite delicate in construction (implying relatively low current energies), they nevertheless also furnished cavities that became lined by a rich diversity of other forms. Moreover, the destructive effects of shell-crushing predators as well as grazers and boring organisms appear to have intensified from Mesozoic times onwards, allowing relatively more frequent preservation of such delicate frameworks in Palaeozoic platform deposits.

The range of differing communities preserved in the fossil record shows how varied the original environments of fossil build-ups and biostromes were, and why it is therefore necessary to look at their constituents in critical detail in order to interpret them correctly. Because of the strong but varying influence of evolving life forms, a simplistic, blanket uniformitarian approach to their facies can be misleading; in this case, the present – to modify the maxim – often furnishes the wrong key to the past!

Activity 16.1

A beautifully preserved example of a fossil reef is illustrated in the video sequence *Interpreting the Capitan Reef* on DVD 3, which you should now watch as part of Activity 16.1. This explores some of the difficulties in trying to interpret fossil communities for which no close parallels exist today.

16.6 Carbonate platform facies models

In terms of the different morphological types of carbonate platform introduced in Section 16.3, the peritidal facies, carbonate sandbodies and carbonate build-ups form the basic building blocks of facies models. However, it is important to note that the development and proportion of the different facies comprising the different morphological types have varied over geological time due to changes in climate, tectonic setting and organism biology. In this Section, we shall consider two examples: a carbonate ramp from the Carboniferous Period and a modern-day rimmed shelf carbonate platform from the South Florida Shelf.

16.6.1 Carboniferous-age carbonate ramp

The Carboniferous-age carbonate ramp can be divided into four facies groups or zones, as illustrated in Figure 16.15. They comprise:

- distal micrites and biomicrites deposited mainly below fairweather wave-base;
- mudmounds (Section 16.5);
- oolitic sandbody of the shoreface and foreshore; and
- backbarrier peritidal sediments.

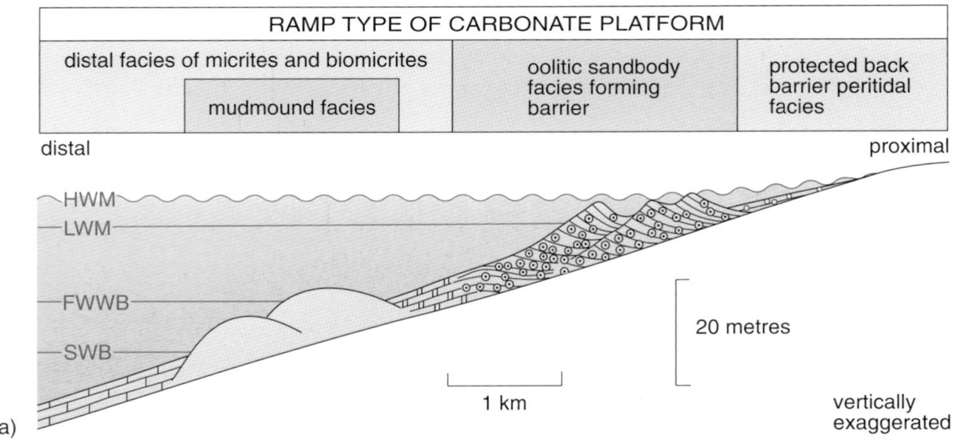

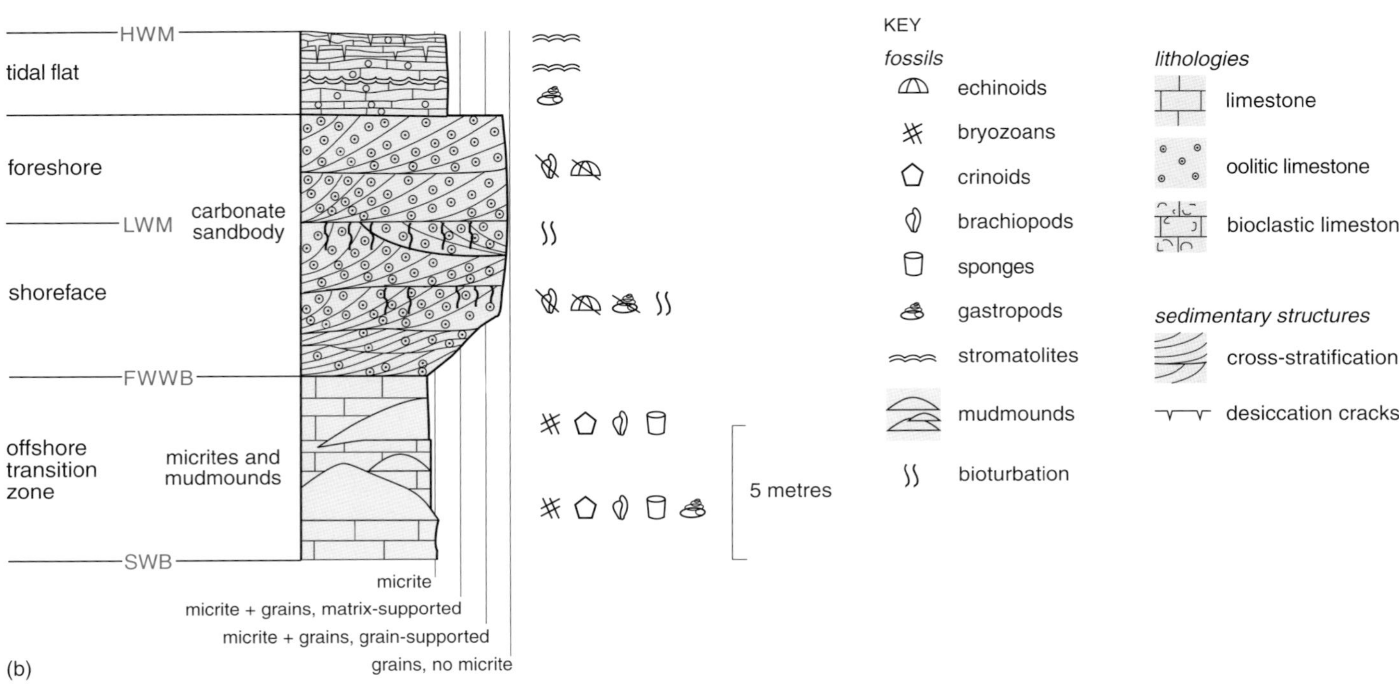

Figure 16.15 Idealized Carboniferous ramp type carbonate platform showing (a) reconstruction 2D cross-section model and (b) graphic log of a typical succession of facies assuming Walther's Law.

Question 16.3 (a) Explain whether the succession shown in the graphic log on Figure 16.15b represents a shallowing-upward or a deepening-upward succession.

(b) Based on your answer to (a), decide whether or not the ramp was prograding seawards.

Question 16.4 Using Figure 16.15, decide how the distal micritic facies and mudmound facies are related in three dimensions.

Question 16.5 What is the true dip on the ramp shown in Figure 16.15?

16.6.2 MODERN-DAY SOUTH FLORIDA RIMMED SHELF CARBONATE PLATFORM

Rimmed shelves comprise a shallow lagoon area on the shelf which is protected from the open ocean by a barrier. The barrier develops at the edge of the continental shelf and slopes away into the ocean. Figure 16.16 shows an idealized cross-section of a rimmed shelf carbonate platform based on the South Florida example. This example comprises a 5–30 km wide shelf, the distal edge of the shelf occurs at 8–18 m of water depth and this descends at a gradient of between 1 and 18° into water that is 200–400 m deep. The shelf is about 300 km long and extends south and south-west from Miami; it is covered by a discontinuous string of islands (the Florida Keys) marking the inner margin of the shelf (Figure 16.17).

The facies include:

- Peritidal and lagoonal facies which are dominated by micrites and biomicrites. Much of these have been stabilized by grasses in the lagoon and there is only poor development of intertidal–supratidal flats (Plate 16.7). Most of the shoreline is vegetated by mangroves. Some of the shoreline consists of an antecedent topography formed of Pleistocene rocks and dead reefs representing the former position of the distal edge of the shelf.
- Bioclastic sandbody facies with a barrier reef in the distal part of the lagoon.
- Patch reefs and an outer reef tract near the distal edge of the shelf.
- Carbonate-mudstones and -sandstones derived from the reef and deposited distally of the edge of the shelf (Figure 16.17b).

Question 16.6 Assuming that the sea-level were to fall, causing the carbonate rimmed shelf system shown in Figure 16.16 to prograde, complete a graphic log of the succession that you might expect to find, using the template provided in Figure 16.16b. Use the symbols provided on the key to Figure 16.16.

Activity 16.2

You should now complete Activity 16.2 to consolidate your understanding of shallow marine carbonate environments.

16.7 SUMMARY OF SECTION 16

- Shallow-marine carbonates differ from siliciclastics in the following ways: sediment is produced mainly by biological and chemical processes *in situ*; grain size is not necessarily related to the amount and/or type of transport; sediments are commonly both cemented and dissolved *in situ*; much carbonate mud is produced through bioerosion; precipitation and dissolution of different types of calcite has varied through geological time; and most shallow-water carbonates are deposited in warm shallow water near the equator.

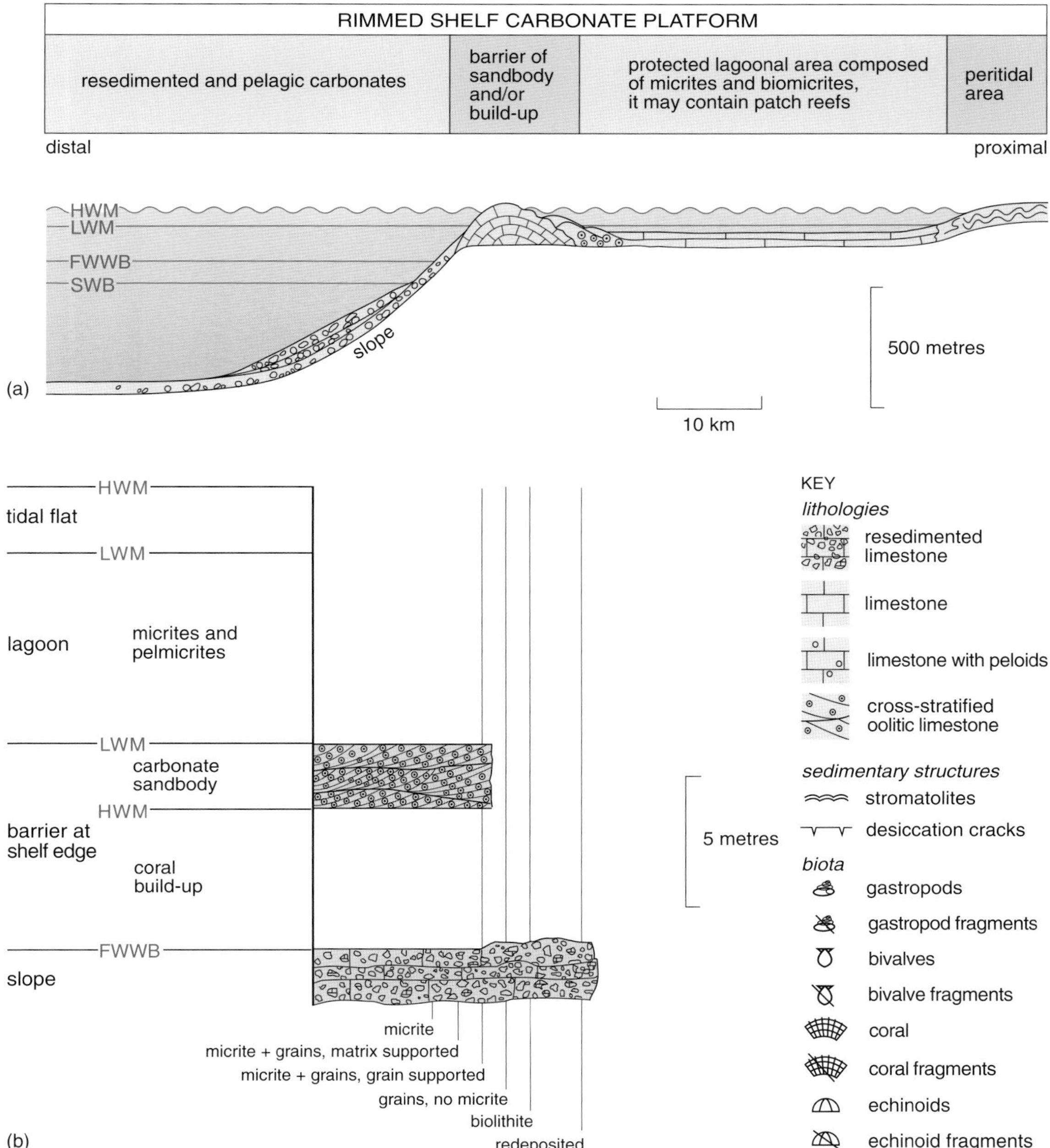

Figure 16.16 (a) An idealized cross-section through a rimmed shelf carbonate platform. (b) Partially completed graphic log for use with Question 16.6.

- Shallow marine carbonates are deposited on carbonate platforms which vary in shape and extent; these include carbonate ramps and rimmed shelf carbonate platforms.
- Carbonate deposition is controlled by the type and number of organisms; the climate which in turn influences the water temperature, circulation, oxygenation and salinity; and siliciclastic sediment supply and tectonic setting.
- The physical processes that affect shallow-marine carbonates include waves, tides and currents. The chemical processes include precipitation and dissolution. Biological processes are particularly important in shallow-marine carbonates and affect both the amount of deposition and erosion. The type(s) and number of organisms are both important.

Summary list continued on p. 201.

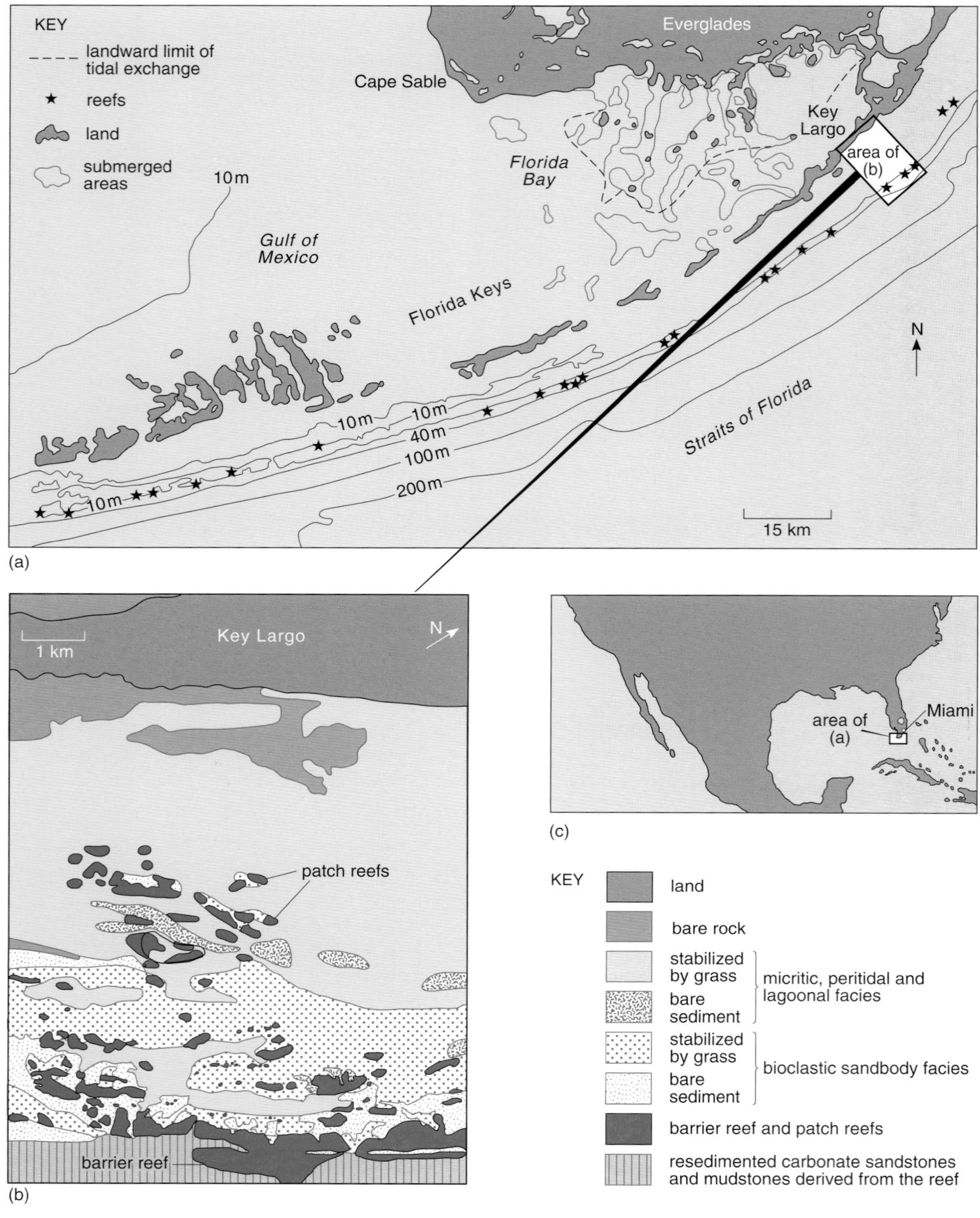

Figure 16.17 (a) Map of the Florida Keys area showing the principal topographic features. The landward limit of tidal exchange in the key refers to the proximal limit of the area under tidal influence. (b) Detailed map of the Key Largo area showing the facies. (c) Location map of the Florida Keys.

- Tidal flat deposition is dominated by the deposition of micritic limestones; these often contain stromatolites, desiccation cracks and evaporite minerals.
- Lagoonal successions are dominated by micrites, biomicrites and pelmicrites with occasional evidence of tidal processes and storms.
- The formation of carbonate sandbodies typically results in the deposition of cross-stratified oosparites and biosparites.
- Carbonate build-ups and biostromes are commonly cemented *in situ*, preserving their original features. Their form and make-up varies in time and space depending on the type and amount of organisms present and their evolutionary stage.
- Prograding carbonate ramp successions typically comprise, from the base, micrites and biomicrites deposited below fairweather wave-base, overlain by an oolitic and or bioclastic sandbody which in turn, is overlain by back-barrier peritidal sedimentary deposits.
- Prograding rimmed shelf carbonate platforms comprise, from the base, micrites and redeposited limestones, the remains of a carbonate build-up, an oolitic and or bioclastic sandbody, and lagoonal micrites with evidence of periodic subaerial exposure.

16.8 Objectives for Section 16

Now you have completed this Section, you should be able to:

16.1 Summarize the chemical, physical and biological processes that are important in shallow-marine carbonate environments.

16.2 Summarize the differences between shallow-marine carbonates and siliciclastics.

16.3 Describe the controls on carbonate deposition.

16.4 Recognize and describe the main facies, in tidal flats, carbonate sandbodies, lagoons and carbonate build-ups.

16.5 Summarize the variety of architecture and palaeoecology in build-ups and biostromes.

16.6 Give examples of graphic logs summarizing the main biological and sedimentological features of a rimmed shelf carbonate platform and a ramp type carbonate platform.

Now try the following questions to test your understanding of Section 16.

Question 16.7 Examine RS 21, RS 22 and RS 24 and explain whether they are more likely to have been deposited as part of a lagoonal succession or a carbonate sandbody.

Question 16.8 (a) Which way was the current flowing when the beds in Plate 16.2 were deposited?

(b) Explain whether the sediments in Plates 16.3 and 16.4 have been deposited by biological, physical or chemical processes.

(c) Explain whether you would find the type of minerals illustrated in Plates 16.3 and 16.4 in the carbonate environments shown in Plates 16.5, 16.6 and 16.7.

17 DEEP-SEA ENVIRONMENTS

The 'deep sea' is used to include all the area of the oceans and seas that are on the distal side of the continental shelf. At the distal edge of the very gently dipping continental shelf, there is a change in gradient to a more steeply dipping slope, which is called the **continental slope** (Figure 17.1).

The continental shelf usually dips at less than 0.1° and is typically covered in seawater about 150 m deep. By contrast, the continental slope dips at between 1–15° and has a vertical fall of about 1 km. So water depths at the bottom of the continental slope are of the order of 1000 m – quite a change! The continental slope is often deeply cut by fissures that are orientated roughly perpendicular to the continental slope; these are called **submarine canyons** and look a bit like steep-sided valleys (Figure 17.1). At the base of the continental slope is the **continental rise**. This region is slightly elevated above the rest of the ocean floor and comprises sediment that has been deposited at the base of the continental slope. Beyond the continental rise we reach the abyssal plain, which lies beyond the usual reach of most terrigenous (continental-derived) sediment.

❑ Thinking back to Block 3, what type(s) of crust would you expect to find underlying the continental shelf, continental slope, continental rise and abyssal plain?

■ There is a lateral transition between continental and oceanic crust across a continental margin (Block 3 Section 2.2.1). The continental shelf rests on thinned and stretched continental crust whereas the abyssal plain must be on true oceanic crust. The intervening slope and rise are in the transitional region. However, it is probably best to regard the continental rise as being oceanic crust buried by a fringe of sediment.

Today, and throughout much of the Phanerozoic, the oceans have covered about 75% of the Earth's surface making them, in terms of area, a very important site of sediment deposition. It is also worth noting that a significant proportion of deep-sea sediments is eventually destroyed during subduction of the oceanic lithosphere (Block 3 Sections 2.3.1 and 7.4), so they are not as common as they might be in the geological record. However, some of these sediments do become scraped off at subduction zones and stuck to the edges of the over-riding plate and then preserved in the geological record (e.g. Block 3 Figure 12.5b).

17.1 PROCESSES AND SEDIMENTATION IN THE DEEP SEA

A number of physical, chemical and biological processes affect sedimentation occurring in deep-sea areas. They can be grouped into three categories: sediment gravity flows (Section 4.3), deep-sea currents and pelagic deposition from suspension.

Sediment is moved from the continental shelf, down the continental slope (mostly via submarine canyons) into the ocean basin by sediment gravity flows. These comprise a dense fluid composed of large amounts of sediment, i.e. 20–70% mixed with seawater (Section 4.3).

❑ Will a marine sediment gravity flow be more or less dense than the seawater around it?

■ A sediment gravity flow will be more dense because it is composed of seawater and particles of *sediment* which together are more dense than seawater.

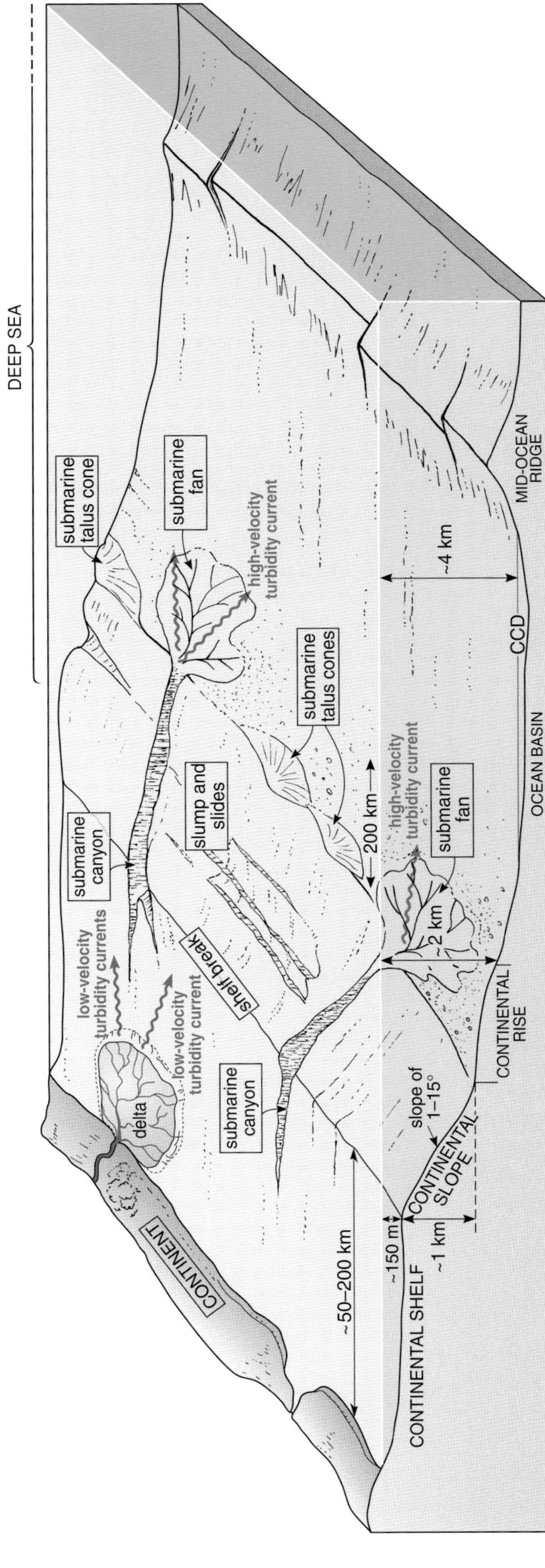

Figure 17.1 The main morphological features of the continental shelf and adjacent deep sea areas. Scale exaggerated vertically.

The greater density of the sediment gravity flow compared to the surrounding seawater means movement will be maintained through the seawater and that gravity is the driving force. Sediment gravity flows include slides and slumps, debris flows and turbidity currents. They result in deposition of large amounts of sediment at the base of the continental slope. The depositional processes will be discussed further in Section 17.2.

The second category of processes in the deep sea involves deep-sea currents. Wind blowing across the ocean surface not only produces waves but can also generate currents. Other currents are a consequence of the Earth's rotation, and slight differences in both temperature and salinity in the ocean water result in density variations in the water column and hence density-driven currents.

Deep ocean currents produced by density differences can be powerful enough to erode, transport, rework and deposit sediment in the oceans. There are also contour-hugging currents driven by winds operating in shallow-water areas, density variations and the formation of ice at the Earth's poles. These types of contour-hugging deep-sea ocean current generally flow horizontally at a particular depth within the water column rather than up or down. In other words, they follow the contours of the sea-floor rather like a footpath that goes around a hill at the same height, in contrast to say a stream which flows downhill. Such currents, irrespective of how they are generated, are called *contour currents*. In some cases, contour currents affect the continental shelf as well as the ocean basin. The sediments deposited by contour currents are called contourites but their importance in the geological record is a matter of debate. This is because evidence is scant due to the fact that only a small amount of sediment is deposited by these currents and because these deposits are then likely to become reworked by bioturbation, winnowing and down-slope debris flows and turbidites. Figure 17.2 shows the different transport directions associated with contourites compared to the sediment gravity flows (turbidites, debris flow, slumps and slides) associated with the continental slope. Contourites are not further discussed in this Course but you should bear in mind that they do play a part in transporting, reworking, and depositing sediment in the deep seas and oceans.

The third category of processes that occur in the deep sea involves predominantly biological and chemical processes. These are discussed further in Section 17.3.

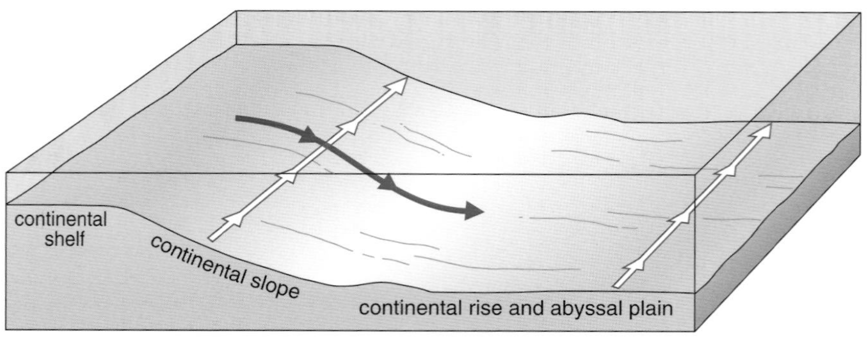

Figure 17.2 General movement direction of contour currents compared with slumps, slides, debris flows and turbidites. Note that contour currents can be present anywhere below storm wave-base.

17.2 SEDIMENT GRAVITY FLOWS IN THE DEEP SEA

17.2.1 SEDIMENT GRAVITY FLOW PROCESSES

The relatively steep gradients found on the continental slope, in submarine canyons, and along the steep distal edge to some carbonate platforms means that sediment gravity flows are the most common sediment transport processes in the deep sea. Sediment gravity flows are driven downslope by gravity, and once a sediment gravity flow is in motion it moves either as a fluid, or else it may move plastically, depending on the ratio of sediment to water. Sediment gravity flows can travel long distances: for instance, they may start by collapse of a delta front, then travel across the continental shelf and down the submarine canyon, before finally coming to rest on the abyssal plain.

As discussed in Section 4.3, there is a spectrum of sediment gravity flow types. Which type develops depends upon two factors: (i) the concentration of the sediment, and (ii) the turbulence and mechanism of the flow. The mechanism of the flow can change during its movement, so one type of flow can grade into another. However, for the purposes of this Section, we will consider only two members of the spectrum of movement processes: debris flows and turbidity currents (Sections 4.3.2 and 4.3.3). Debris flows and turbidity currents are responsible for the most common sediment gravity flow deposits both today and in the sedimentary record. It is important to realize that they can form in both siliciclastic sediments and carbonates. In both cases, the resulting deposits are the same except for their composition.

❑ Why do some carbonate platforms have a steep distal edge from which turbidites and debris flows could be initiated?

■ The build-ups found fringing carbonate platforms are almost vertical if they are built by organisms which produce a framework. This steep edge may be exaggerated by erosion during former times of lower sea-level (Figure 16.10a).

Sediment gravity flows can form anywhere that the depositional slope is greater than about 1° and where sufficient sediment is available. Indeed, both debris flows and turbidites may form in other environments besides the deep sea but it is in the latter that they are most common.

Question 17.1 Sediment gravity flows behave either as a fluid flow or plastically. Which of these do (a) turbidity currents and (b) debris flows represent?

Question 17.2 Is the sediment in (a) a turbidity current and (b) a debris flow supported by fluid turbulence or the strength and buoyancy of the matrix?

17.2.2 TURBIDITES AND THE BOUMA SEQUENCE

Turbidity currents produce a characteristic succession of beds known as **turbidites**, as they gradually slow down and deposit sediment. Each bed represents a particular part of the flow, and the size of the clasts and sedimentary structures are diagnostic of particular parts of the flow and the speed of the turbidity current. This characteristic succession of beds was first recognized and described by Arnold Bouma and is thus called a **Bouma Sequence**. Before we discuss the Bouma Sequence further, let us first consider the different sedimentary structures that form in a decreasing current speed. You may wish to refer back to Figures 4.13 and 4.14.

Question 17.3 (a) What are the three sedimentary bedforms characteristic of the lower flow regime and what sedimentary structure will each produce?

(b) If the current speed was slowing down for a given grain size, which would form first: planar deposition (upper flow regime) or current-formed ripples?

Question 17.4 If a turbidity current transports a mixture of grain sizes including pebbles, coarse-grained sand, clay, fine-grained sand and granules, and the current is slowing down, in which order would the grains be deposited?

As the current speed decreases, a fining-upward succession will form with a predictable pattern of sedimentary structures. This is exactly what we find in a Bouma Sequence. An idealized Bouma Sequence is illustrated in Figure 17.3.

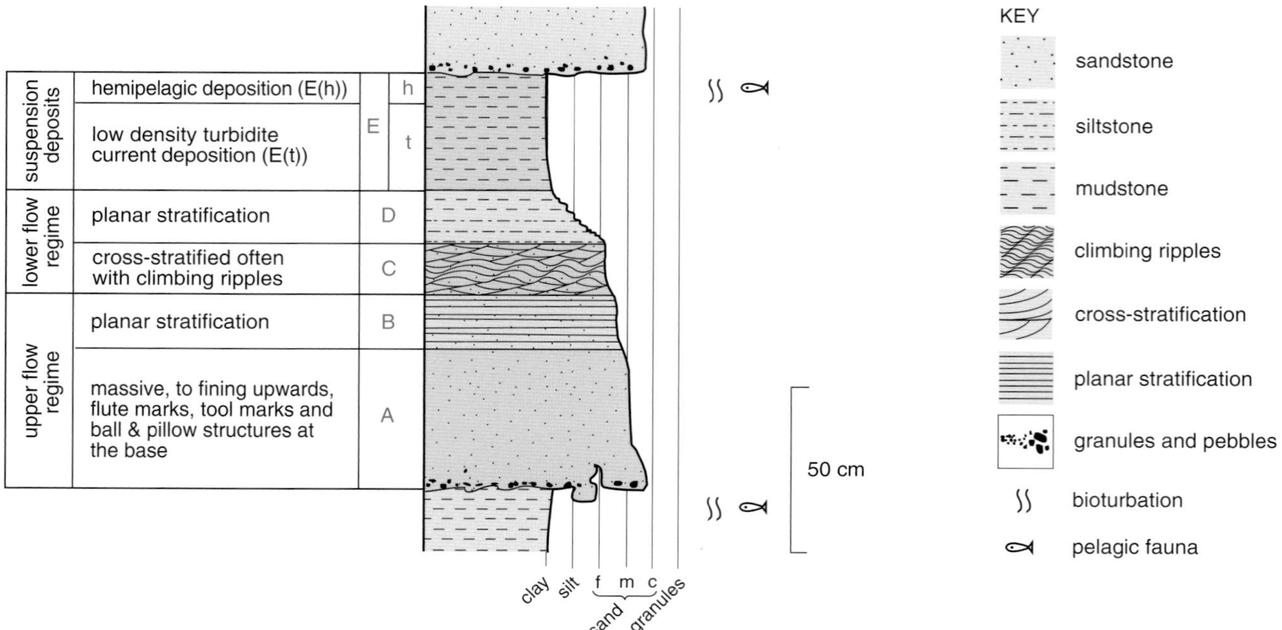

Figure 17.3 A full Bouma Sequence showing the different divisions deposited from a turbidity current, and hemipelagic deposition between intervening flows. See Figure 17.4 and text for explanation. A coloured version of this Figure is shown on the Sedimentary Environments Poster and Sedimentary Environments Panorama.

The formation of the individual divisions (A to E) in the Bouma Sequence are explained as a series of time-shot cartoons in Figure 17.4. For each time shot, you should read the text and study the cartoon. Unit E on Figures 17.3 and 17.4 is partially comprised of **hemipelagic** sediment. Hemipelagic means that it contains at least 25% of sediment derived from the land rather than directly from the sea (e.g. microfossils). These types of sediments are discussed further in Section 17.3.

If Unit E does not have time to solidify before another turbidite is deposited on top, then bedding may become deformed or convoluted because as the overlying heavier sediment loads onto the unlithified beds below, the underlying beds rapidly expel water upwards due to the increased downward pressure. This leads to the formation of ball and pillow structures like those discussed in Video Band 12 *Graphic logging* at the base of the storm-deposited sandstones.

Question 17.5 What are the sedimentary structures shown in Figure 17.5? Which part of the bed do they represent, and which way was the current flowing?

Question 17.6 In relative terms, explain how you think the period of time taken to deposit divisions A–D of a Bouma Sequence might compare to the time taken to deposit Unit E.

Question 17.7 Which division of the Bouma Sequence do you think is most likely to be bioturbated, and why?

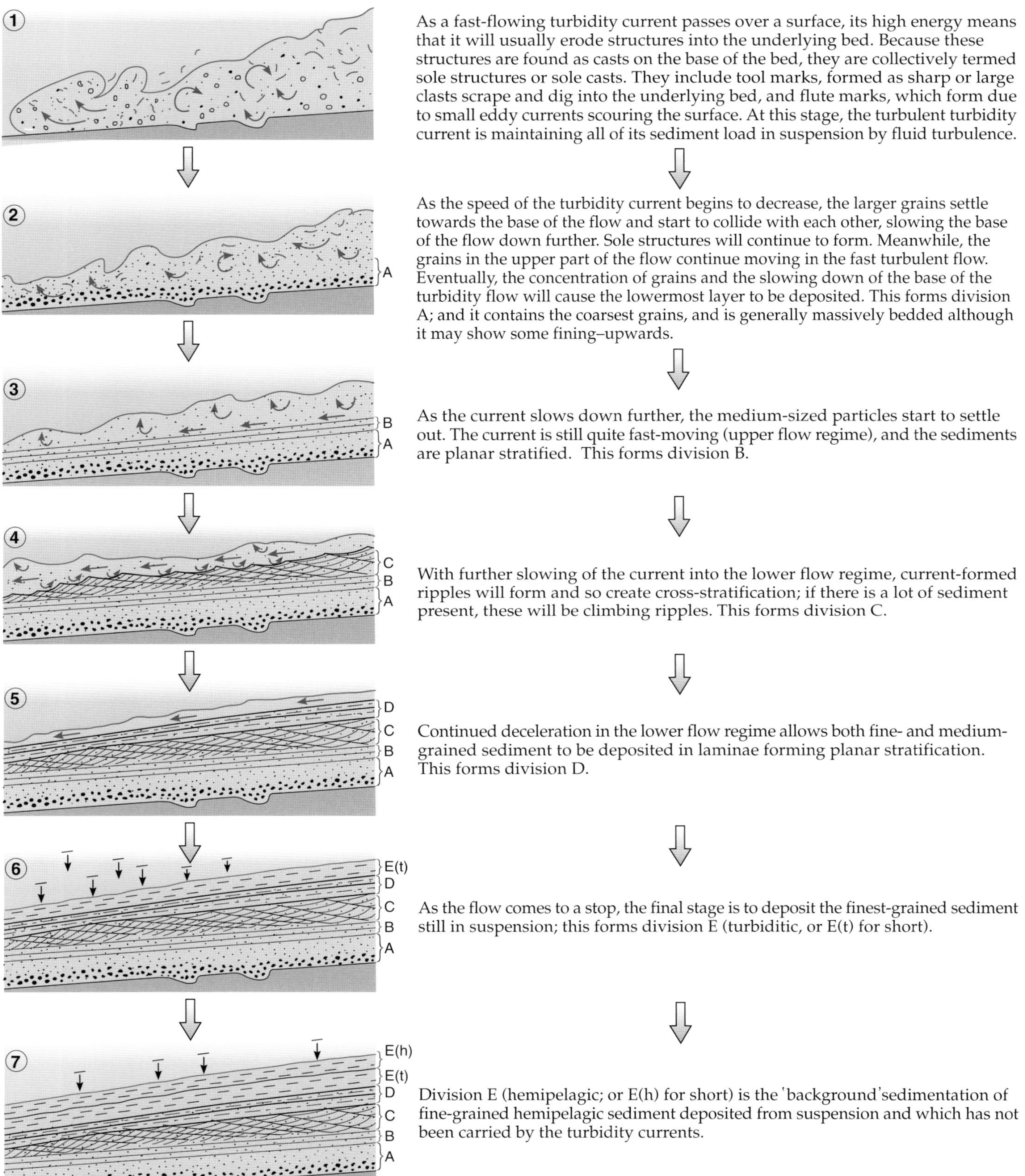

Figure 17.4 Cartoons to show the various stages in the deposition of a turbidity current resulting in a Bouma Sequence. Overall flow direction of the turbidity current is from right to left across each of the cartoons.

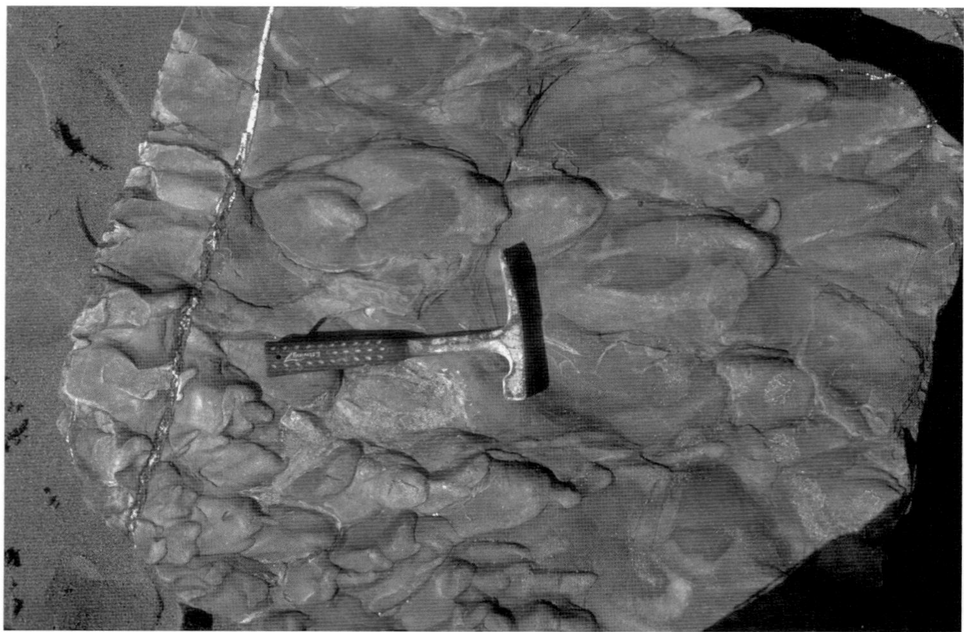

Figure 17.5 Sedimentary structures for use in conjunction with Question 17.5.

What we usually find in the sedimentary record is a whole series of Bouma Sequences stacked on top of each other, each representing a single turbidity current. An example of a succession of such turbidites is shown in Plate 17.1.

Figure 17.3 represents a complete Bouma Sequence with all of the divisions present. However, in reality, complete Bouma Sequences are not always found because of the changing conditions of the flow and the variety of grain sizes carried in different turbidity currents and incomplete sequences showing repetitions of BCDE, CDE and DE can all be found. Siliciclastic turbidites tend to be immature both texturally and compositionally.

❑ Why do you think turbidites are generally texturally immature?

■ Two factors contribute to this. One is that turbidites generally represent a whole mixture of material that has become unstable and has flowed in a turbidity current which has picked up everything in its path. The second is that the rapid deposition from suspension means that there is not enough time to sort all of the sediment completely (remember that at first *all* particles are carried in suspension by fluid turbulence).

However, if the sediment source is well rounded, then the grains in the resulting turbidites will also be well rounded.

❑ If texturally immature sediments are common in turbidites, name a rock type that might be commonly found.

■ Greywackes (Section 6.5.2 and Block 2 Section 7.1).

RS 12 is a greywacke and could well form part of a Bouma Sequence, probably division A. RS 27 could also represent the fine-grained division E of a siliciclastic Bouma Sequence. Carbonate turbidites are typically composed of clasts of various sizes of shallow-marine carbonates, mixed with carbonates that formed in deeper water.

The main factor controlling the composition of turbidites is the nature of the sediment source rather than the amount of transport. So if the sediment supply is derived from fine-grained sandstones and mudrocks, then the turbidites will be composed entirely of these. However, if there are granules and pebbles present, these will also be incorporated.

17.2.3 Debris flows and mud flows

Debris flows and mud flows are the other common type of sedimentary flow found both today and in the geological record (Plate 17.2). Debris flows and mud flows comprise large clasts ranging in size from a few centimetres to tens of metres enclosed in a fine-grained matrix of clay, silt or fine- to medium-grained sand. They can form in both siliciclastic and carbonate environments, and move plastically. The fine-grained matrix is buoyant and of sufficient strength to support a range of clast sizes as the debris flow moves. Amazingly, less than 5% matrix is apparently sufficient to provide a buoyant medium enabling the large clasts to move. Mud flows are typically matrix-supported and the clasts are usually unsorted and ungraded. Debris flows are usually grain-supported, and the clasts move by rolling and sliding within the flow; in such cases, they may then show some sorting.

17.2.4 Sediment gravity flow successions and facies models

As discussed in Section 17.2.2, Bouma Sequences are often repeated and stack up to form turbidite successions as shown in Plate 17.1. Turbidites are found in association with debris flows and mud flows at the base of submarine canyons where they form a fan-shaped feature called a **submarine fan** (Figure 17.6a). Fan sediments have been fed from the mouth of the submarine canyon which effectively forms a point source. However, turbidites, mud flows and debris flows are also deposited from linear sources (Figure 17.6b), for instance slopes that are controlled by active faulting, in which case the whole slope is unstable and therefore produces much sediment, or else at the base of a slope produced by oversteepened organic growth (e.g. a build-up), or by a morphological feature such as part of the continental slope.

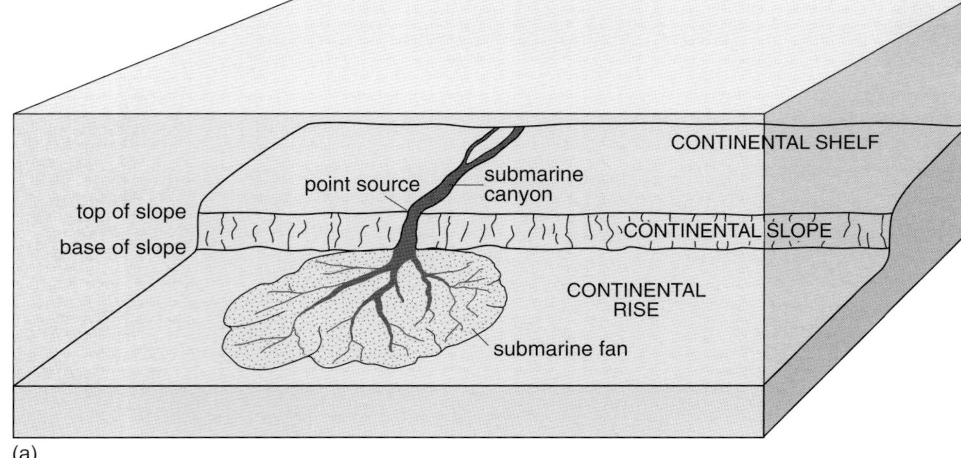

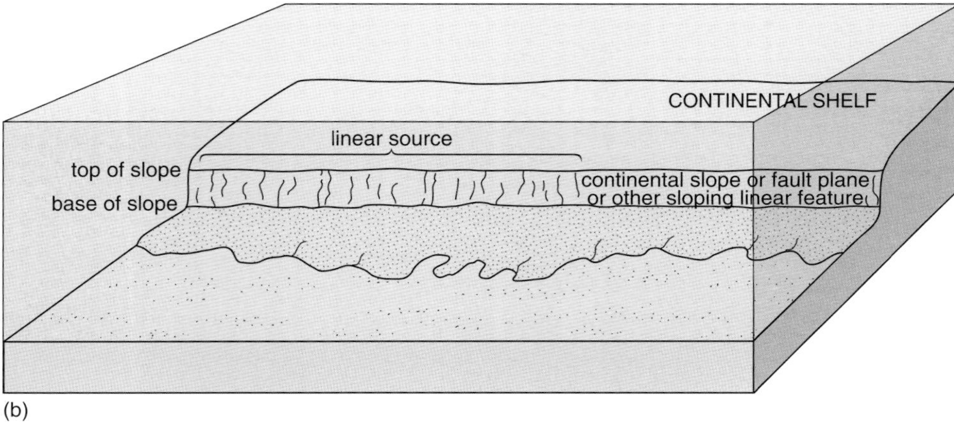

Figure 17.6 Turbidites and debris flows from (a) a point source forming a submarine fan and (b) a linear source such as a fault scarp.

- ❑ What would be the plan view shape of deposits found at the base of a submarine fault scarp or at the edge of a linear morphological feature such as a coral build-up?
- ■ For a linear morphological feature that is forming the sediment source, the deposits would be deposited in a long belt roughly parallel to the basal edge of the fault scarp or build-up, similar to that shown in Figure 17.6b.

Most carbonate turbidite, mud flow and debris flow systems are fed from linear systems because it is a linear build-up at the shelf break that provides the topographic feature from which they originate. Siliciclastic turbidite, mud flow and debris flow systems can be fed either by a point or linear source.

Two factors are important for whether or not turbidite, mud flow and debris flow successions form; these are the *slope* and the *sediment supply*. The continental slope does not constantly produce turbidites, mud flows and debris flows along its entire length. If the slope is stable, and there is no sediment supply, then there will be no deposition of this type of sediment. Submarine canyons form a conduit which transports much of the sediment from the continental shelf because their catchment area is nearer the continent, and therefore nearer the terrigenous sediment supply from subaerial erosion (see Figure 17.1). Sediment being fed to the continental shelf through deltas, strandplains and barrier islands, will be carried down the submarine canyons by currents and end up being deposited in a submarine fan. The morphology of a submarine fan produced from a point source compared to deposits from a linear source is shown in Figure 17.6a.

Submarine fans are similar to deltas in their morphology and the way in which the sediment is fed along channels to the distal parts of the fan, with the exception that submarine fans, of course, are *entirely* under water. The Sedimentary Environments Poster and Sedimentary Environments Panorama illustrate a submarine fan in the deep-sea environment. At the end of the canyon, where the gradient lessens, the sediments are dispersed from the major channel into smaller channels which serve to deposit the sediment as distal lobes. At different times, each of these channels may be active or abandoned, braided or meandering. The channels also have levées similar to alluvial channels (Figure 17.7).

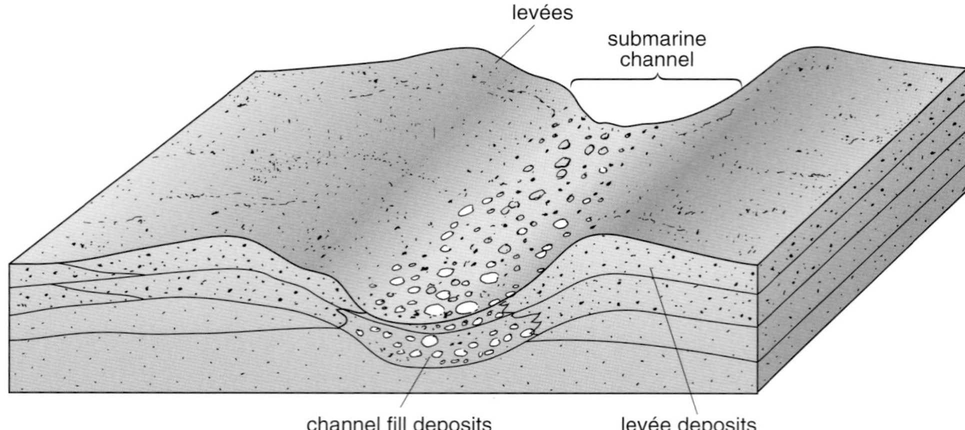

Figure 17.7 3D diagram to show the form of the channels and levées in a submarine fan.

- ❑ Which factors will influence the grain size of sediment deposited in the different areas of the submarine fan, and what will be their distribution?
- ■ The type of sediment supplied will influence the grain size distribution and this may change through time. However, assuming that there is a fairly constant supply of sediment with a constant grain size distribution, then like all sedimentary systems deposited by currents, the coarsest-grained material will be deposited nearest the sediment source because the energy is not usually sufficient to carry it very far. As the current decreases away from the mouth of the canyon, the finer-grained material is deposited. There will also be a difference in grain size distribution between the channels and their levées. The channels contain the coarser-grained material and the levées the finer-grained material.

An *idealized* graphic log assuming a constant grain size distribution in the sediment supply is shown in Figure 17.8.

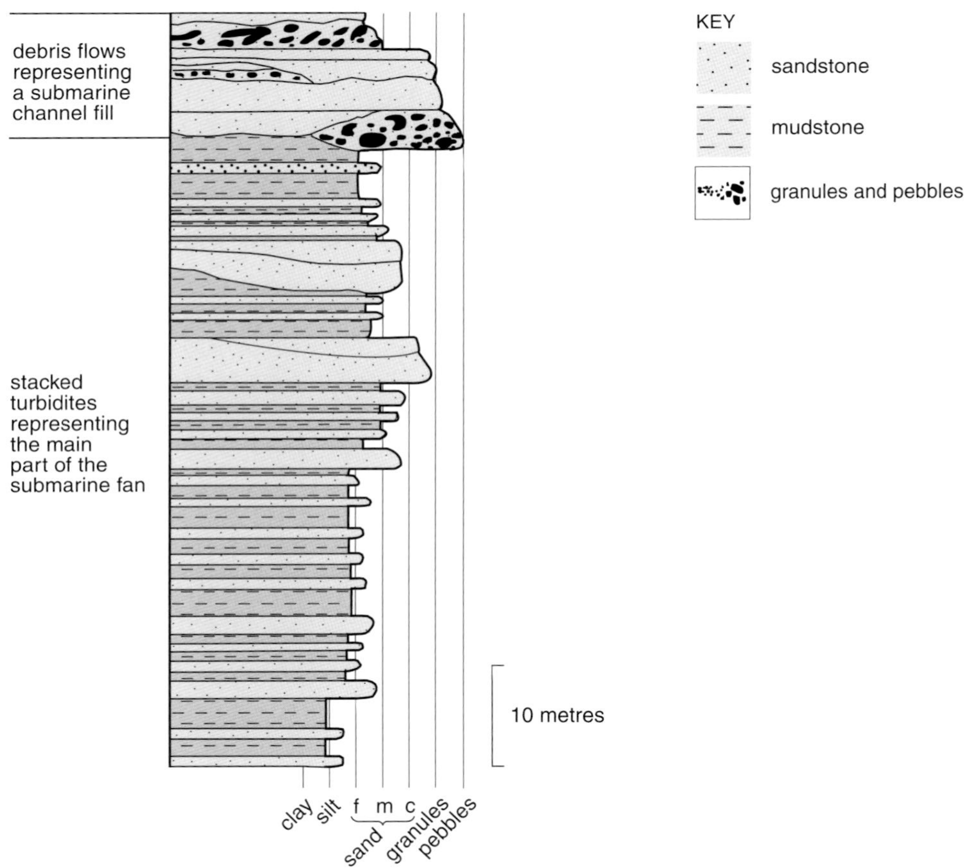

Figure 17.8 Idealized graphic log through a submarine fan assuming that there is a constant supply of sediment with the same grain size distribution. A colour version of this Figure and a block diagram showing the features of submarine fans is shown on the Sedimentary Environments Poster and Sedimentary Environments Panorama.

17.3 HEMIPELAGIC AND PELAGIC SEDIMENTS

In addition to the sediments deposited by turbidity currents and ocean currents, there is also constant deposition of fine-grained sediments some of which have been carried in suspension from the continent. These include weathered volcanic ash, wind-blown dust and clay minerals derived from the continent; these are collectively referred to as **terrigenous** sediment. Importantly, sediment is also produced biologically in the surface waters of the oceans and then falls as 'biogenic rain' to the ocean floor. Some sediment may also be produced directly by chemical precipitation. Sediments that contain less than 25% terrigenous material, and which are therefore almost entirely made up of sediments produced and deposited within the ocean and seas, are termed **pelagic**. They include chalks and some bedded cherts. Some sediments deposited on the ocean floor will contain a mixture of pelagic and terrigenous material.

❑ What is the name given to sediments containing a mixture of pelagic (up to 75%) and terrigenous material?

■ Hemipelagic (Section 17.2.2).

Pelagic and hemipelagic sediments are being deposited over more than 60% of the world's oceans today, but in general the sedimentation rate is very low, and so they often do not form particularly thick deposits.

17.3.1 HEMIPELAGIC AND PELAGIC DEPOSITIONAL AND EROSIONAL PROCESSES

Hemipelagic and pelagic sediments are deposited from suspension, so they tend to be massively bedded (i.e. the sediments show very little sign of bedding or lamination). However, slight variations in the composition of the sediment through time will give rise to lamination and bedding. The quantity of pelagic sediment produced depends on having favourable conditions to support biological activity in the surface waters of the ocean; these include light, warmth, and nutrients.

❑ In the deep sea, whether or not carbonate sediments are preserved depends on another factor. What is this and how does it affect the sediment preserved?

■ The carbonate compensation depth (CCD) is the other factor (Section 10.1.6, Figure 10.4). The CCD is the depth at which the amount of carbonate particles being supplied to the ocean is balanced by the amount dissolving, so below this depth carbonate sediments will not accumulate.

The actual depth of the CCD varies with temperature, salinity, pressure and shape of the ocean basin. It is thus variable in both time and space.

Question 17.8 Assuming that there is a source of both siliceous and carbonate-secreting organisms living and dying in the surface waters, explain which one will provide remains able to accumulate below the CCD.

17.3.2 PELAGIC SUCCESSIONS

Pelagic sedimentary deposits include the remains of calcium carbonate-secreting micro-organisms such as coccolithophores and foraminifers, and the remains of siliceous micro-organisms such as diatoms and radiolarians. The carbonate-secreting organisms can result in the deposition of chalks and other deep-water carbonates (Figure 16.1), whereas the siliceous organisms give rise to bedded cherts.

Plates 17.3 and 17.4 show two chalk successions. At first glance they might appear similar, but Plate 17.3 shows Cretaceous chalks mainly composed of coccoliths, whereas Plate 17.4 shows Neogene chalks composed mainly of foraminifers. This is because different organisms have dominated at different periods of geological time. High magnification microscope work is needed to recognize the type of microfossil of which the chalk is composed. Plates 17.3 and 17.4 also show some other features typical of pelagic chalk successions. The black nodules in Plate 17.3 are chert nodules (Block 2 Section 7.3). In this case, the silica originated from the skeletal parts of sponges, and was remobilized during diagenesis. Many chert nodules are actually the remains of large horizontal burrow systems (*Thalassinoides*). During burial, silica from sponge spicules goes into solution and as it is negatively charged it is drawn towards the burrows because they contain small amounts of organic matter which is positively charged. The silica re-precipitates in the old burrow and, once it has started to solidify, this will encourage more silica to come out of solution and hence the nodules grow.

Plate 17.4 shows obvious bedding because it is a succession of light-coloured chalks, composed of practically 100% foraminiferans, interbedded with marls which comprise about 20–30% clay minerals together with foraminiferans. In this case, the slight variation in clay content, perhaps reflecting climatic control on the rate of clay supply from the land, gives rise to the bedding. Minerals may precipitate directly out of seawater on the deep sea-floor, especially when there is very low sediment input. These include pyrite, manganese, glauconite and phosphate.

❑ What signs of life do you think may be found most abundantly in deep-sea sediments?

■ Pelagic biota are commonly found in deep-sea sediments because although they may live in the surface waters, on death they fall to the bottom where general lack of reworking by currents means there is good preservation potential.

❑ What factor will limit the preservation of calcitic remains in the deep sea?

■ The position of the CCD (Section 17.3.1). If the calcitic fossil falls below the CCD, it will not be preserved.

Activity 17.1

Now complete Activity 17.1 to consolidate your understanding of deep-sea environments.

17.4 Summary of section 17

- Physical processes in the deep sea include sediment gravity flows and deep-sea currents. Biological processes include the life and death of micro-organisms in the surface waters. Direct chemical precipitation also occurs in the deep sea.
- Decelerating turbidity currents deposit sediments with particular features. This idealized succession of units is called a Bouma Sequence; these fine upwards and can be divided into six units deposited during distinct stages of the waning turbidity current flow. The basal unit (A) is the coarsest-grained and is massive. The base of Unit (A) often shows evidence for erosion (sharp contact, flute marks, tool marks) and rapid deposition (ball and pillow structures). Unit B is planar-stratified, and Unit C is cross-stratified following the formation and migration of ripples. Unit D is planar-stratified, while Unit E(t) comprises the finest-grained sediments carried by the turbidity current and is deposited in the final stage of the

turbidity flow. Unit E(h) comprises hemipelagic, fine-grained sediment deposited after the turbidity current has ceased. Units A to D represent bedload deposition, Unit E suspension deposition. Units A and B represent deposition from the upper flow regime whereas Units C and D are from the lower flow regime.

- Sediment gravity flows are driven by gravity so they require a slope to start them in motion. A slope dipping by about a degree is sufficient to set sediment gravity flows in motion.
- Sediment gravity flows may come from a point source, for instance, in the deep sea down a submarine canyon. The sediments from a point source are deposited in a fan shape (submarine fan). The fan itself will be dissected by channels along which the sediment is transported.
- Sediment gravity flows may be fed from a linear source. The resulting deposits form a linear mound at the base of the slope.
- Large areas of the deep sea are the site of deposition of both hemipelagic and pelagic sediments. Hemipelagic sediments contain greater than 25% terrigenous material (e.g. marls, unit E(h) of turbidites) whereas pelagic sediments contain less than 25% (e.g. chalk). The pelagic material is either produced biologically in the surface waters (i.e. the remains of micro-organisms such as diatoms and coccolithophores) or by direct chemical precipitation.

17.5 OBJECTIVES FOR SECTION 17

Now you have completed this Section, you should be able to:

17.1 Describe the physical, chemical and biological processes that are important in deep-sea environments.

17.2 Describe and draw a summary graphic log showing all the divisions of a Bouma Sequence and explain how each division forms.

17.3 State the features of, and describe the difference between, point source and linear source sediment gravity flow successions deposited in marine environments.

17.4 Explain what is meant by hemipelagic and pelagic sediments and give examples of each.

Question 17.9 In the form of a Table, summarize the four different depositional processes important in the deep sea and their sedimentary deposits.

Question 17.10 Explain whether submarine fans are more likely to form when sea-level rises or when it falls.

18 A MATTER OF TIME

At the start of this Course, in Block 1 and the video sequence *Geological time* on DVD 1, we reviewed how geologists came to comprehend the immensity of geological time. By now, you will be used to thinking in terms of millions, tens of millions or even hundreds of millions of years when considering the span of history that can be read from rocks and their fossils. With this in mind, students often ask a fundamental question when standing in front of a quarry or cliff face exposing sedimentary rocks. This question is 'how long did that bed take to form?', rather than 'how long ago did it form?'.

The 'how long did it take?' question usually cannot be answered by quoting a precise time measured in hours or years, but by applying the understanding of geological processes that you have gained in this Block, you should be able to suggest a 'ball-park' estimate of the time it took for a given sedimentary feature to form.

Look at the two graphic logs shown in Figure 18.1. How quickly would the sandstone intervals have been deposited? In hours, days, months, years, decades, centuries?

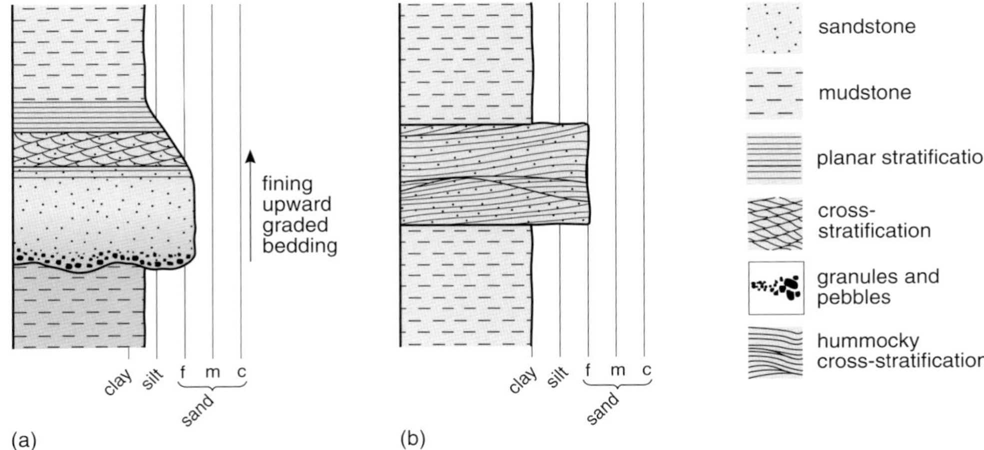

Figure 18.1 Graphic logs of sandstone units occurring between beds of mudstone. (a) *c*. 1 cm thick. (b) *c*. 50 cm thick.

The turbidite succession in Figure 18.1a was deposited by a density current – a mixture of sediment and water. Once the turbulent part of the current had passed, most of the sand held in suspension would have been deposited in a few hours to form the Bouma Unit A (Figure 17.3) – remember that turbidity currents may reach speeds of 50 km hr^{-1} (Section 4.3.3). The top part of the sand unit showing planar stratification and cross-stratification (the Bouma B and C units; Figure 17.3) would probably have taken longer to be deposited, as the lower speed tail of the turbidity current may have continued to flow for a day or more. But the mud left in suspension after the flow had long gone may have taken weeks or even months to fall to the sea-floor to form the Bouma E(t) unit. Moreover, the mud deposited from suspension as normal deep-sea background sedimentation would be deposited even more slowly, possibly over as much as hundreds or thousands of years (the E(h) unit).

❑ Consider the entire succession shown in the log in Figure 18.1b: which parts of it represent the longest periods of time?

■ By far the greatest amount of time – 99% or more – is recorded by the mudstones. This is because they were deposited very slowly from suspension over hundreds or thousands of years, whereas the sandstone with HCS is a storm bed, resulting from a single event that lasted for hours or at most a few days.

We also need to consider the time significance of the contacts between beds. If the base of a sandstone bed is erosional, as is the case with the turbidite unit shown in Figure 18.1a, then the contact between the sandstone and underlying mudstone may represent a long period of time, during which fine-grained sediment was deposited, and later removed as the turbulent head of the turbidity current passed by. Likewise, the contact of the HCS sandstone and underlying mudstone in Figure 18.1b will represent a long period of time if storm-generated currents eroded muddy sediments before the deposition of the sandstone.

Figure 18.2 shows the time duration of some of the sedimentary features described in this Block, ranging from individual laminae to successions of sediment deposited in depositional environments such as river systems, deltas and submarine fans.

❑ Do successions of sediment deposited over long periods in major sedimentary environments (such as those on the Sedimentary Environments Poster) contain a continuous record of the past?

■ They do not: there are many gaps in the record due to non-deposition or erosion.

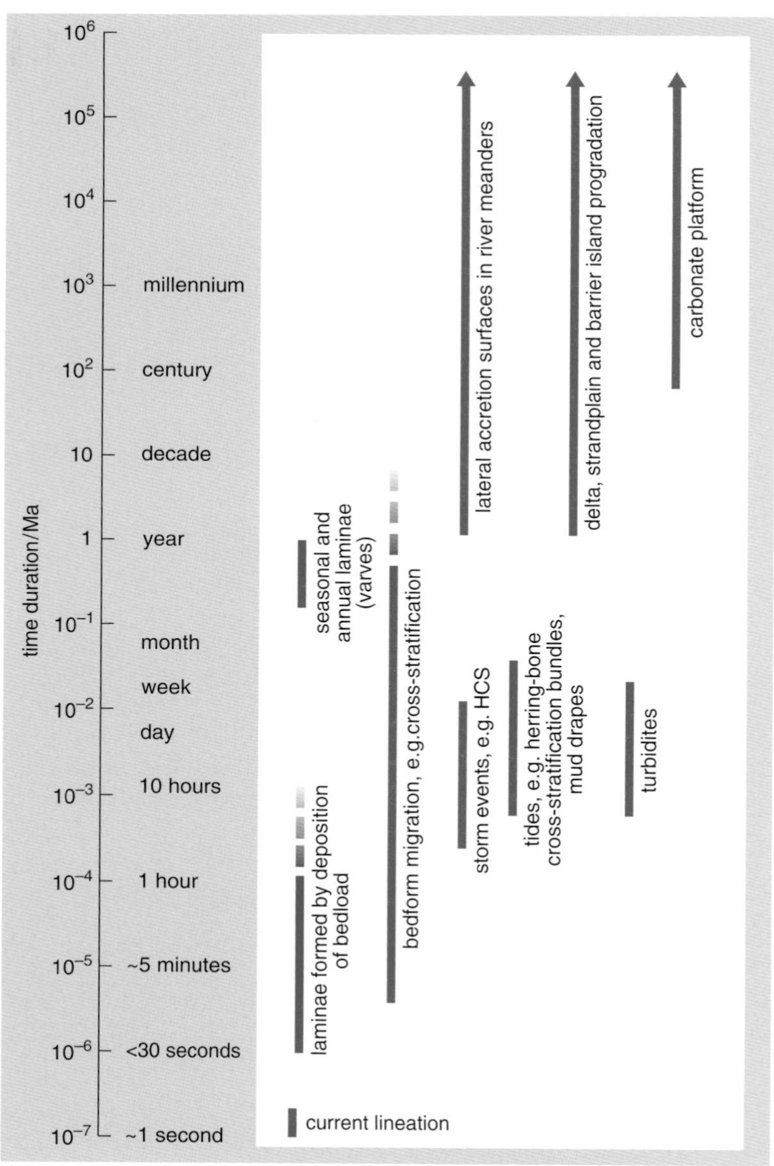

Figure 18.2 The length of time taken to form a variety of sedimentary structures and successions discussed in this Block. The time-scale is in years plotted with successive divisions representing a ten-fold increase in time.

So virtually all sedimentary successions represent the net accumulation resulting from periods of deposition, non-deposition and erosion. An environment in which continuous deposition leaves behind a complete record is most unusual. One only has to think of familiar environments such as rivers and coasts. The former are subjected to seasonal changes in discharge and occasional floods, and the latter are affected by tidal currents that change daily, plus varying wave action ranging from zero to storm force. Even in the depths of lakes and oceans, where a continuous rain of suspended sediment might be expected, the resultant sedimentary records can be interrupted by deep-water currents such as turbidity flows.

As well as considering how long beds took to form, it is worthwhile considering rates of deposition. The rate can be exceedingly high for the formation of, for example, cross-stratification, or storm beds. It can be as much as 100 000 metres per thousand years, but this enormous rate only lasts for very short periods of time, so the net rate of sediment accumulation over long periods of time ($>10^4$–10^5 years) is likely to be only a few centimetres per year or even less. In fact, the longer the time period considered, the slower the net rate of accumulation becomes. This relationship is shown in Figure 18.3, which is a compilation of rates of deposition measured over different periods of time.

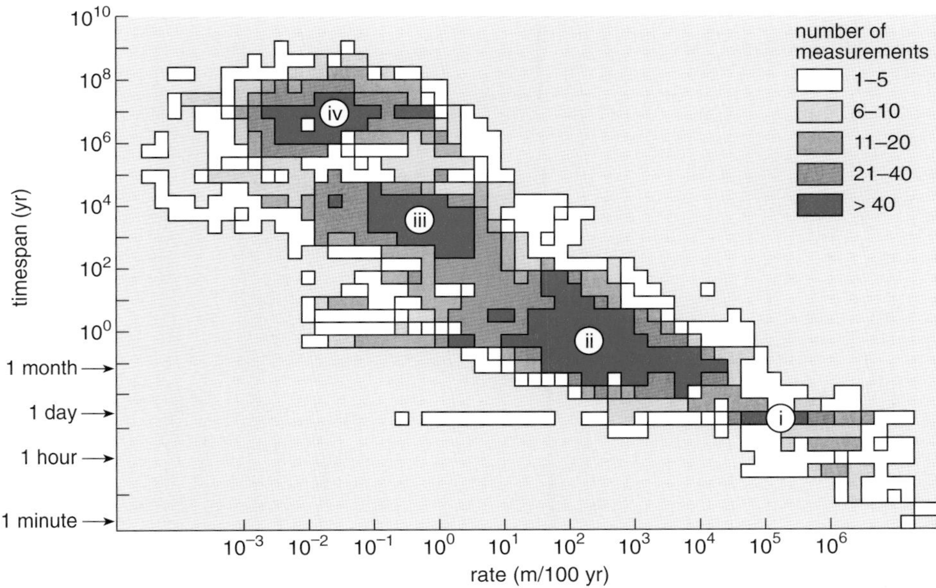

Figure 18.3 A plot of 25 000 measured net rates of accumulation in various environments against the time-span over which they were determined. The measurements cluster around four methods of observation: (i) continuous observation in modern environments; (ii) periodic measurements at measuring stations in modern environments; (iii) successions of rocks up to 50 000 years old dated by radiocarbon dating; (iv) older successions dated using fossils and radiometric dating.

Now that you have reached the end of this Course, you have had some training to enable you to interpret the rock record using the observation → process → environment approach. There is plenty of evidence to work on, but it is important to remember that it is usually incomplete, so you need to keep in mind the consilience approach implicit in the following quotation:

> 'The past is as biased as propaganda, as misleading as a fairy tale. The geological record contains a highly distorted picture of what has happened. Of shallow marine sedimentation we can learn endlessly, but of what took place in the high

mountains, on the continents, of upland forests, we can find only the faintest shadows. And some things which we may suspect have happened in the past, we may find no trace of at all.

As scientific optimists, we always tend to concentrate on what we can hope to know. Our paradigm is the successful detective, painstakingly searching for clues, to solve the crime. Try focusing on what we cannot know. Put yourself in the position of the criminal evading detection. How much of geological history is an irretrievable journey, a perfect crime? …

The bias of history is, of course, not just a preserve of geology. We know far more about how the Romans built their drains than how they constructed their roofs. We know far more about prehistoric cemeteries than prehistoric nurseries. Far more of treasure hoarders than of treasure spenders, far more of the literate than illiterate. It is the job of geohistorians to try to make amends for these distortions; to understand, for example, how and where catastrophic events can freeze and preserve the most ephemeral, the most fragile the most vulnerable: like the buildings, possessions, even the fleeing inhabitants of Herculaneum, preserved by the eruption of Vesuvius.'

Extract from 'Editorial', *Terra Nova*, vol. 8, pt. 5, p. 398, 1996.

Answers to questions

Some points in these answers are illustrated on the Sedimentary Environments Panorama and Digital Kit. Where we think you might find this helpful, we have printed the DVD icon. You only need to refer to the DVD if you need further clarification.

Question 1.1

(a) *Observations*

Property		Observations for RS 10
Composition	Grains	Quartz, rare white feldspar
	Matrix	Very little fine-grained intergranular material
	Cement	Red soft material = iron oxide coating the grains Quartz (note, however, the rock is poorly cemented since it is easy to dislodge individual grains from the rock as you handle it)
Texture	Crystalline or fragmentary	Fragmentary
	Grain size	750 μm is the most abundant grain size (coarse-grained sand)
	Grain sorting	Very well-sorted
	Grain morphology:	
	shape or form	Equidimensional
	roundness	Rounded to subrounded
	sphericity	Spherical
	Grain surface texture	Frosted
	Grain fabric	No preferred orientation; grains closely packed; grain-supported
Fossils		None seen in specimen
Sedimentary structures		None seen in specimen

(b) *Processes*

(i) The sand-sized grains indicate a moderately high energy deposition.

(ii) (1) is correct: RS 10 is well-sorted and the grains are rounded and spherical, with very little finer matrix material, indicating that either the processes responsible for its deposition sorted out the grains leaving larger and smaller particles elsewhere or that the rock represents sediment grains derived from a previous rock in which the grains were of equal size and well-rounded. It is impossible to tell from this one specimen which of these scenarios may apply.

(iii) The grains were most likely to have been deposited by wind because the surface texture is frosted.

(iv) The lack of any fossils tends to support the fact that it was deposited by wind.

(*Comment:* In general, fossils tend to be more abundant and better preserved in rocks which have been laid down under water. However, the specimen that you have is very small so the chances of finding a fossil are limited thus we cannot entirely exclude the fact that there may be fossils in these rocks, and we would need to examine the exposure to be sure.)

(v) Sedimentary structures are absent. This tells us that either the size of the sedimentary structures is greater than the specimen or that there are none at all.

(*Comment:* We would need to see a larger specimen or the exposure from which the specimen came to be sure. This demonstrates that it is important to make observations at a range of scales. The lack of sedimentary structures in RS 10 means that we cannot deduce anything about processes in this case.)

(c) *Environment*

RS 10 has been transported and deposited by wind processes. The most likely environments of deposition are a desert or possibly the sand dunes at the back of a beach.

Question 3.1

Physical weathering, which produces rock and mineral fragments, and *chemical weathering*, which leads to the production of new minerals and products in solution (Block 2 Section 7).

Question 3.2

Olivines: isolated tetrahedra; *pyroxenes:* chain silicates; *amphiboles:* double-chain silicates; *micas:* sheet silicates; *feldspars* and *quartz:* framework silicates (Block 2 Sections 4.3.1, 4.4.1, 4.4.2, 4.5.1, 4.6.2 and 4.6.1 respectively).

Question 3.3

The basalt and the granite, because they contain mafic minerals (although they are likely to form only a small proportion of the granite). Chemical weathering of these minerals releases Fe^{2+} ions which are immediately oxidized to Fe^{3+} and deposited as ferric oxide. The quartzite would be composed almost entirely of quartz and so contain no mafic minerals.

Question 3.4

Olivine and Ca-feldspar → pyroxene → amphibole → K-, Na-feldspars → mica → quartz.

Question 3.5

As K-feldspar is a lot less resistant to chemical weathering than quartz, but is found in the sandstone in a pristine, unweathered condition, not much chemical weathering can have taken place. (*Comment:* however, weathering must have been sufficient to decompose any other minerals that happened to be present in the source rock.)

Question 3.6

You would expect the gabbro to chemically weather faster than the granite. This is because gabbros contain Ca-rich plagioclase, pyroxene and maybe some olivine, the minerals which are least resistant to chemical weathering. Granite, on the other hand, contains quartz, alkali feldspars and either mica or amphibole, the minerals which are most resistant to chemical weathering.

Question 3.7

(a) Physical weathering is by far the dominant process in Plate 2.1 which shows a mountain slope thickly covered by very large, shattered rock fragments, with bare rock exposures in the background. There is no sign of any fine-grained sediment, soil or vegetation. The evidence in Plate 4.4 suggests that chemical weathering is the dominant weathering process. The photograph records a well-vegetated, verdant landscape. There must be good soil cover, sufficient to support arable farming (note the fields in the right background) as well as small trees and shrubs (left background). There are no signs of rock exposures and the low hill profiles are rounded, not jagged.

(b) The rock exposures in the background of Plate 2.1 are well jointed, allowing good access for percolating water, aiding frost shattering. The soil waters of the well-vegetated soils in Plate 4.4 are likely to be more acidic than rainwater because of the release of organic acids, which increase the rate of chemical weathering of rock fragments and mineral grains in the soil (as well as the bedrock beneath the soil).

Question 3.8

See the completed Table 3.2 below.

Table 3.2 The end products of the chemical weathering of the main groups of igneous silicate minerals (completed).

Mineral group	Products in solution	New materials	Residual minerals
olivines	metallic ions, silica	ferric oxide	none
pyroxenes	metallic ions, silica	ferric oxide	none
amphiboles	metallic ions, silica	ferric oxide, clays	none
biotite mica	metallic ions, silica	ferric oxide, clays	none
muscovite mica	metallic ions, silica	clays	none
feldspar	metallic ions, silica	clays	none
quartz	a little silica	none	quartz

Don't forget, however, that weathering doesn't always proceed to completion, so micas (especially muscovite) and K-feldspar may be found in some sedimentary rocks.

Question 3.9

(a) The gabbro (RS 19) contains a little olivine, but mainly pyroxene and Ca-rich plagioclase feldspar. The quartzite (RS 5) is composed entirely of quartz.

(b) The olivine and pyroxene in RS 19 should chemically weather to produce metallic ions (including Fe^{2+}) and silica in solution. The Fe^{2+} would be oxidized immediately to form insoluble ferric oxide (new material). Clay minerals (new material) would also form from the decomposition of the plagioclase feldspar.

Chemical weathering of the quartzite (RS 5) can only lead to the release of quartz grains, by solution along the grain boundaries, and the production of a little silica in solution. The quartz grains would not decompose and so would form an insoluble residue.

Question 4.1

They would be smaller. This is because air is both less dense and less viscous than water.

Question 4.2

(a) About 0.5 m s^{-1} (see Figure A4.1).
(b) In suspension (see Figure A4.1).

Question 4.3

Air has a much lower viscosity and density than water. It requires a higher wind speed to move grains of a given size in air (i.e. by wind) than in water.

Question 4.4

The sandstone with a history of transport only by water would be likely to be texturally less mature. This is because in general water is less effective than wind at sorting sedimentary material (Section 4.1.2). It also cushions grains during collisions and so slows down their rate of abrasion, which means they are likely to be less well-rounded (Section 4.1.3).

Question 4.5

The current flowed from the top left towards the bottom right of the photograph. Current ripples are asymmetrical, with the steep slope facing down current. In this photo the steep slopes of the ripples face towards the bottom right of the photograph.

Question 4.6

(a) About 0.9 mm (coarse-grained sand) (see Figure A4.2).
(b) About 0.16 to 0.17 mm (fine-grained sand) (see Figure A4.2).

Question 4.7

(a) The clay and silt will be deposited from suspension.
(b) Unless the ebb current is very strong, it is unlikely to be eroded because of its cohesive properties (see Figure 4.7).

Question 4.8

You should be able to distinguish about four neap periods. The thicker sand laminae with poorly defined, darker clay drapes represent the spring tides, and the thinner sand laminae with well-defined clay drapes represent the neap tides.

Question 4.9

Saltation of fine- to medium-grained sands and suspension of silts are the dominant processes during aeolian transport, whereas rolling and sliding, as well as saltation and suspension transport are important processes during water transport. As sand grains saltate to greater heights in air than in water, and strike the grains on the surface beneath them at higher speeds, they cause coarser grains to move forwards by surface creep. This process does not occur in water.

Question 4.10

(a) As the steep slope of the dune in Figure 4.23a is facing left, it follows that the dune is migrating from right to left. Conversely, the steep face of the arrowed dune in Figure 4.23b is facing right, so the dune is migrating from left to right up the slope of the much larger sand mountain.
(b) Sand is accumulating to the right of the clumps of seaweed, so the dominant wind direction is from left to right.

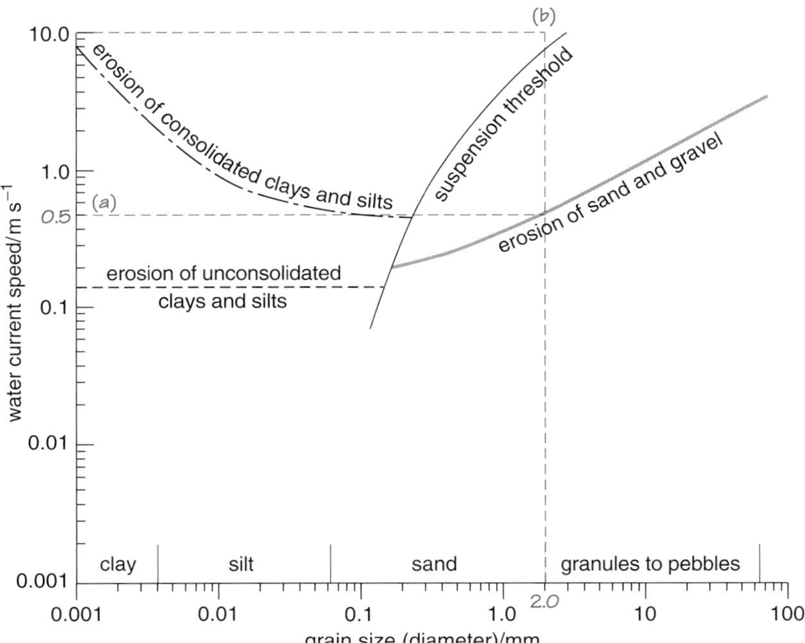

Figure A4.1 Answer to Question 4.1.

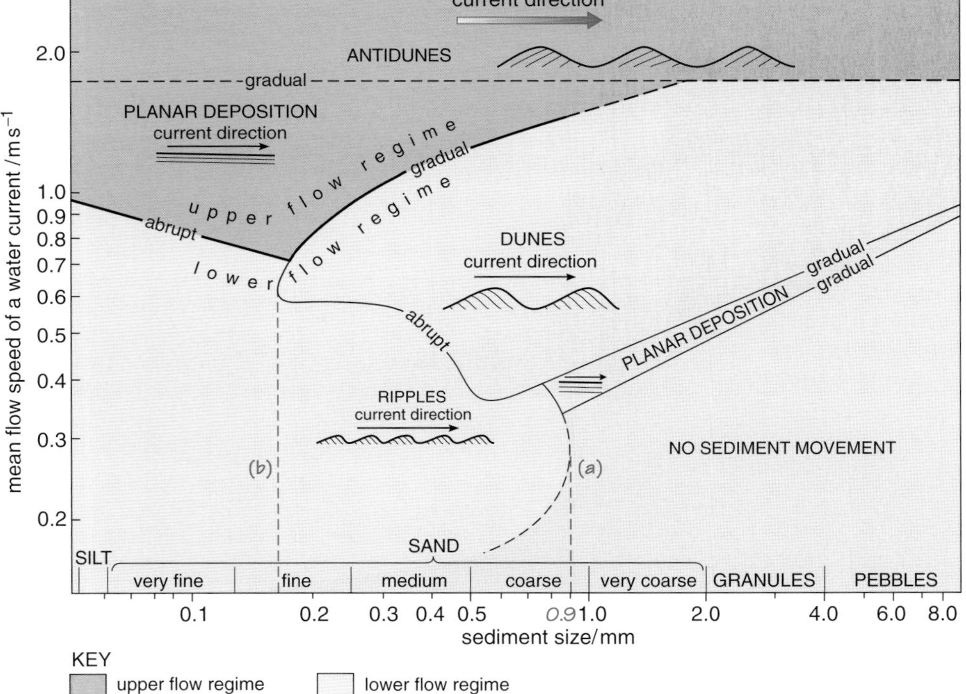

Figure A4.2 Answer to Question 4.6.

Question 4.11

The most important evidence would be frosted surfaces to the sand grains (Section 4.1.3). The sandstones would also be texturally very mature, but bear in mind that sands which have been subjected to long periods of wave transport are also likely to be texturally mature, so by itself textural maturity would not be unequivocal evidence. Even if you found frosted grains, you would have to satisfy yourself that this texture was not inherited from the source rock for the sediment. A lack of marine fossils would also be a useful clue, but again would not provide enough proof by themselves.

Question 4.12

They would be texturally immature. The high viscosity of the flow would prevent differential settling of larger fragments, so the deposits would be poorly sorted. It would also protect particles to a certain extent from abrasion during collisions with each other or rock surfaces beneath the flow. So, unless the particles were already rounded as the result of a previous phase of transport, they would remain fairly angular.

Question 4.13

The flute casts appear to become less pronounced towards the top left-hand corner of the photograph, and some can be seen to become wider. Therefore, the current must have flowed from the bottom right-hand corner to the top left-hand corner.

Question 4.14

See Figure A4.3.

(a) Between about 0.2 and 0.3 m s^{-1}.

(b) As bedload.

(c) In suspension.

(d) To about 0.06 to 0.07 m s^{-1}.

(e) From the bedload.

Question 4.15

According to Figure 4.14, there would be the gradual development of planar stratification, giving way gradually to subaqueous dunes, and then giving way gradually to antidunes (see Figure A4.4).

Question 4.16

(a) At the extremes of high and low tide there is no current flow, so the current speed is 0 m s^{-1}. This condition is reached when the curve cuts the horizontal time axis. (i) High tides occur at approximately 4 hours and 17 hours, before the current begins to accelerate on the ebb tide. (ii) Low tides occur at the start of the tidal cycle (0 hours) and at 12 hours, before the current begins to accelerate on the flood tide. The next low tide is at about 25 hours, just into the beginning of the next 24-hour period.

(b) At around 0, 4, 12, 17, 25 and 30 hours. These are the times when there is no current movement, and so fine-grained sediments can be deposited from suspension.

(c) Herring-bone cross-bedding is produced by the preservation of bedforms developed by currents which reverse direction at regular intervals, therefore the potential exists for it to develop in the Chesapeake Bay sediments. However, the ebb tide is significantly stronger than the flood tide (e.g. the maximum ebb current speed is about 0.75 m s^{-1} around 8 hours into the cycle, but the maximum current speed during the next flood current is only about 0.25 m s^{-1} around 14 hours into the cycle). This means that any subaqueous dunes produced by currents during a flood tide are likely to be eroded by the currents during the next ebb tide, and so herring-bone cross-bedding would not develop.

Question 4.17

(a) As cross-stratification produced in the lower flow regime is inclined in the direction of current flow, the current must have flowed from right to left.

(b) The current speed must have varied because the cross-stratification at the base of the exposure (representing deposition on subaqueous dunes) is separated by planar stratification which gives way to cross-stratification again. Subaqueous dunes and planar deposition occur at different flow speeds (see Figure 4.14). (Note that it isn't clear from the photograph whether the flat bed represents planar deposition in the upper or the lower flow regime.)

Question 5.1

Atmospheric CO_2 in solution (see Equation 3.1).

Question 5.2

(a) (i) Fragmental texture means that a rock consists of grains, or clasts, of pre-existing minerals or rocks which have been bound together in some way by cement or matrix (Block 2 Section 5).

(ii) No, because they form by crystallization from solution which means that they will show an interlocking crystalline texture, even though they are sedimentary in origin.

(b) For water to evaporate to the extent required for evaporites to crystallize implies that the climate must have been both hot and arid.

Question 5.3

The very fine grain size of micrite implies that the limestone must have been deposited in low energy conditions. You should recall that sediments with a grain size as fine as micrite must have been deposited from suspension (Section 4.2). They would have been winnowed away had there been current or wave action.

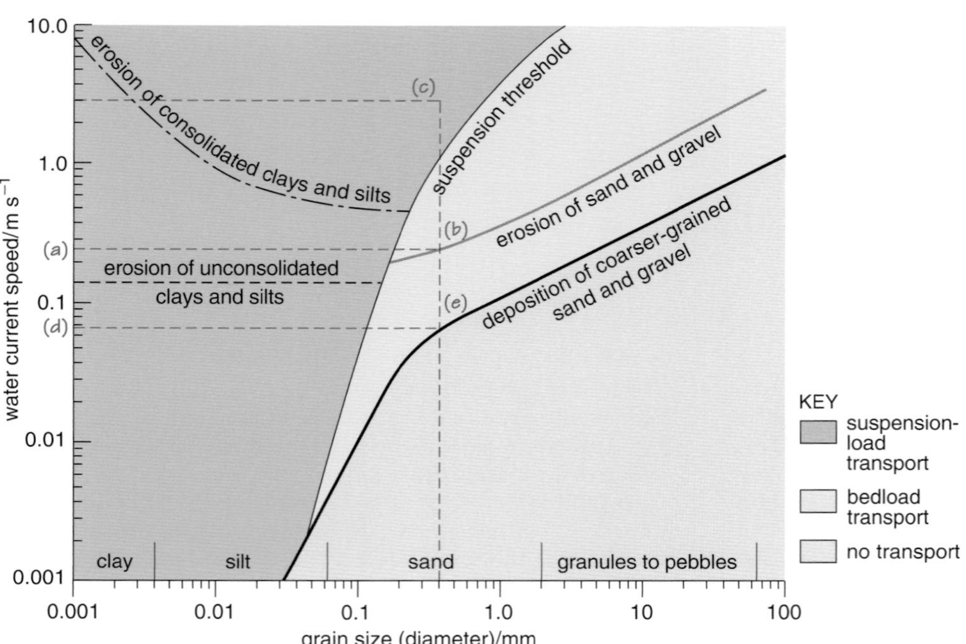

Figure A4.3 Answer to Question 4.14.

Question 5.4

(a) The internal structure of the ooids (Figure A5.1) consists of a series of concentric layers. Within each layer are fine lines radiating outwards. By contrast, the peloids are more or less uniform in appearance, devoid of any internal structures.

(b) See Figure A5.1. The different material at the centre of each ooid represents the nucleus round which $CaCO_3$ has accreted, and this progressive accretion is reflected in the layering.

(c) The rounded shape of ooids results from the way in which they are rolled around on the sea-bed.

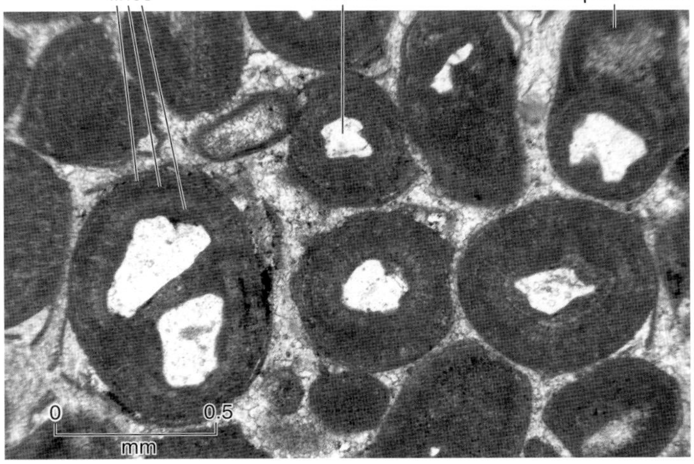

concentric growth rings separated by fine radial lines

quartz grain acting as nucleus

two ooids amalgamated to form a compound ooid

0 0.5 mm

Figure A5.1 Ancient ooids viewed in thin section (plane-polarized light) showing the internal structure of an ooid, and a quartz grain acting as the nucleus for an ooid.

Question 5.5

They are probably wave-formed because they are linear and bifurcate. The lighting on the ripples makes it difficult to see whether they are symmetrical.

Question 5.6

During at least part of the Triassic, Cheshire must have been within some kind of marine basin which had restricted access to the open sea, so that evaporation of the water could take place. In order for this to happen the climate must have been very hot and arid.

Question 5.7

Ooids and some aragonitic sediments are formed when $CaCO_3$ is precipitated from seawater. From what you have read in Section 5, you might have suggested that, as $CaCO_3$ is less soluble in warm water than cold water, its precipitation would be favoured in the warmer waters of lower latitudes. Also, less CO_2 is likely to dissolve in warm water than in cold water, and this is likely to drive Equation 5.1 to the left so that more $CaCO_3$ forms.

Question 5.8

You might infer that the sea was warm, as $CaCO_3$ is less soluble in warm water than cold water, and so its precipitation is enhanced (see answer to Question 5.7). So you might also infer that the climate in Britain was significantly warmer than at the present day. The presence of ooids, by themselves, does not necessarily imply high energy conditions for deposition as they can form biochemically in quiet waters. However, the presence of the cross-stratification confirms that currents were active, and in water sufficiently shallow for bedforms to be produced at the sea-bed.

Question 6.1

The silica released by pressure dissolution during burial.

Question 6.2

Sutured or concavo-convex grain contacts.

Question 6.3

From the chemical decomposition of mafic silicate minerals (Section 3.3). It is precipitated under oxidizing conditions.

Question 6.4

It could have been deposited in a high energy environment, as micrite was either unable to settle, or has been winnowed out.

Question 6.5

Muddy sandstone: clay and quartz, with quartz forming the highest proportion.

Sandy limestone: quartz and calcite, with calcite forming the highest proportion.

Calcareous mudstone: calcite and clays with clays forming the highest proportion.

Figure 7.7 in Block 2 shows that the mineral occurring in the highest proportion determines the name of the rock: quartz – sandstone, calcite – limestone, and clays – mudstone.

Question 6.6

Grains: bioclasts, faecal pellets, and ooids.

Intergranular material: micrite and sparite.

Question 6.7

The arkose must have undergone much more compaction before cementation than the sandstone. Most of the grains show concavo-convex contacts, and a few show sutured contacts; the proportion of pore space is greatly reduced. The sandstone displays only a small proportion of concavo-convex contacts and has quite a lot of pore space, filled with silica cement (see Activity 5.1a). (*Comment*: Did you deduce that the fine-grained material filling the intergranular spaces in the arkose is probably clay, derived from the post-depositional breakdown of feldspar?)

Question 6.8

(a) *Claystone/mudstone:* the major constituents are clay minerals, therefore it is a claystone/mudstone; because of the bioclasts, it could be called a shelly claystone/shelly mudstone or a bioclastic claystone/bioclastic mudstone.

(b) *Muddy limestone:* the major constituents are calcitic, therefore it is a limestone, but it also contains a significant proportion of clay. You could also classify it as a muddy biomicrite as it contains a high proportion of bioclasts (shell fragments), together with a micritic and clay matrix.

(c) *Conglomerate:* the dominant components are rounded, gravel-sized clasts.

(d) *(Medium-grained) feldspathic sandstone:* it is not an arkose because it contains less than 25% feldspar.

(e) *Pelmicrite.* (See Table 6.2.)

Question 7.1

The intact nature of the fish carcass shows it was not attacked by any scavenging animals as it lay on the bottom, while the lack of disturbance also of the sedimentary lamination by burrows or surface trails indicates an absence of bottom-dwelling animals in general. Thus, we may infer that there were no animals living in, on or even near the sea-floor. Yet the fish had evidently grown to a healthy size somewhere. Although it might be argued that it could have been washed in from elsewhere, the presence of all the delicate structure of the fossils and the presence also of faecal strands here indicates that some fish, at least, were living in the overlying waters.

Question 7.2 

(a) The horizontal burrows and the borings into them both record the living activities of (different) organisms, and so are trace fossils. The oyster shells are the remains of organisms and are thus body fossils (in this case, macrofossils, because they are readily studied by the naked eye).

(b) The burrows would originally have been excavated, and backfilled, beneath the sediment surface. Subsequently, precipitation of a cementing mineral within and around these filled-in burrows turned them into nodules. Later erosional activity must have partially exposed them, however, to have allowed encrustation by oysters, prior to reburial when clay deposition resumed. Hence deposition was interrupted by some erosional removal of sediment. The restriction of the oysters and the borings to the former burrow-fills shows that the latter were already hard when exposed, in contrast to the surrounding clay, indicating that cementation of the nodules had already taken place. Exactly such an instance is illustrated in the Sedimentary Environments Panorama, with the trace fossil *Thalassinoides*, which you should have a look at when convenient.

Question 8.1

The two groups inherited a gas-filled shell, conferring buoyancy (c) and hence the ability to swim (a), from a common ancestor, so these features should be considered homologous. However, coiling of the shell (b), which ensured the maintenance of stability (d), evolved independently, and so should be considered analogous.

Question 9.1 

You should have assigned the specimens as follows:

coral	FS E	bivalve	FS B
ammonoid	FS D	trilobite	FS G
gastropod	FS A	echinoid	FS C
brachiopod	FS F	crinoid	FS H

If you had problems with any of these identifications, run through the key again using the labelled images of the replicas in the 'Digital Kit' on DVD 1. It is also the Digital Kit that is referred to in Questions 9.2, 9.4, 9.13 and 9.18.

Question 9.2 

You should be able to see the pallial line about 8 mm within the ventral margin of the shell. Notice that the pallial line is parallel to much of the ventral margin of the shell, but that at the posterior end it has a sharp inward kink.

Question 9.3

Circomphalus lived within the sediment and fed on food particles suspended in seawater that was drawn into the inhalent siphon. It was thus an infaunal suspension feeder. Though stationary most of the time, it was capable of movement, so it could be called intermittently vagrant. Now enter *Circomphalus* in the correct box of Table 9.2.

Question 9.4 

The structure of the skeleton can be divided into two types of radial zone on the basis of the ornamentation: (1) wide zones, each with two prominent rows of large as well as many smaller knobs, alternating with (2) narrower and more subdued zones bearing smaller knobs, flanked on each side by paired pores, looking like double pin-pricks.

Question 9.5

Pseudodiadema was epifaunal, moving across the sea-floor using its spines and tube feet. It was essentially a grazer and probably fed on algae or sessile epifauna. Now enter *Pseudodiadema* in the correct box of Table 9.2 (vagrant epifaunal grazer, though possibly also predatory).

Question 9.6

As it passed through the indentation, exhalent water (plus faeces) would have been directed obliquely back to the left and thus away from the head, instead of forward, directly over it, as might have happened were no indentation present.

Question 9.7

The probable currents are indicated in Figure A9.1. Both specimens illustrate the importance of maintaining a clear separation of the clean inhalent water from the oxygen-deficient exhalent water, bearing faeces.

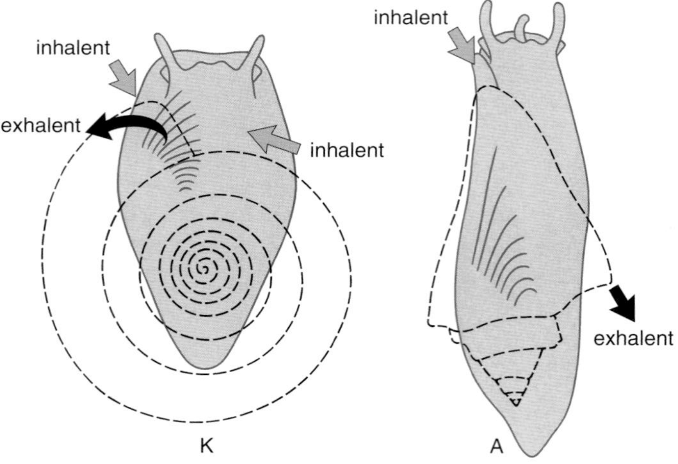

Figure A9.1 Inferred respiratory currents in FS K and A.

Question 9.8

FS K probably moved around over the sea-floor. With two unelaborated inhalent current sites, it is unlikely that it could have located food accurately by smell and so it is improbable that it hunted vagrant animals. It is most likely to have fed on algae or detritus on the surface. FS A, in contrast, had a well-developed extensible inhalent siphon, so there is a good possibility that it might have been a predator or scavenger. Since it could have burrowed, it may have preyed on other burrowing organisms. (In fact, its modern relatives do burrow into sandy sediments and feed on infaunal prey such as bivalves, which they smother with the foot.) You could thus classify FS K as a vagrant epifaunal grazer (or possibly collector), and FS A as probably a vagrant infaunal predator.

Question 9.9

You should have entered both FS F and L as sessile epifaunal suspension feeders, since both lived on the sea-floor and strained off food particles from the water.

Question 9.10

If the arms were held out radially in an umbrella shape, then their repeated branching would have filled in the entire area to create a comprehensive filter; their construction was thus well adapted for feeding in this manner. (If you want to check this idea, notice that the distance between successive branching points on the arms is similar to the radius of the top of the cup, such that the distance between neighbouring branches would have remained fairly constant as they fanned outwards.)

Question 9.11

You should have identified FS H as a sessile epifaunal suspension feeder.

Question 9.12

You should have entered both FS E and M as sessile epifaunal predators on microscopic prey, or even suspension feeders; though, as you have just seen, supplementation of their diet by symbiotic algae remains a speculative possibility.

Question 9.13 

Lenses face out in all directions around the front, outer sides and back of the eyes. The slight inclination of each bank of eyes also suggests that the lenses could have detected light changes above the animal. Thus, the trilobite seems to have had a field of view that went all round and over it. The only zone not directly visible would have been beneath the animal.

Question 9.14

You should have identified FS G as a vagrant epifaunal deposit feeder (collector).

Question 9.15

(i) Five-fold symmetry, (ii) possession of a water vascular system, and (iii) an endoskeleton of many plates of calcite with sponge-like microstructure.

Question 9.16

In the Home Kit, the phylum Mollusca is represented by the classes Gastropoda and Bivalvia, as well as the Ammonoidea, which is a subclass (subdivision of class) belonging to the Class Cephalopoda. In gastropods and bivalves, the muscular foot is used for locomotion, though in ammonoids it is modified into tentacles used to grasp food and, together with the funnel, used in jet propulsion. Both gastropods and ammonoids have a radula, but this has been lost during the course of evolution in bivalves.

Question 9.17

The slight bump on the slab represents the cast of a shallow scour hollow in the underlying mud. The scour can only just have reached the level of the network of galleries where their relief is greatest, since here the full thickness of the tubes was sand-filled. The abrupt margin of this part of the trace fossil indicates the limit of where the scour gained access to the network. The unexposed galleries beyond that limit remained unfilled by sand, and so were not preserved on the sole of the turbidite. At the other end of the trace, by contrast, the scouring action removed most of the network, such that the relief of the sand fills appears to fade gradually into the cast of the scoured surface.

Question 9.18 

The trilobite first made the resting trace (*Rusophycus*) and then later left it, making a furrowing trace (*Cruziana*) across the sea-floor. The sequence could not be the other way round because of the direction of the V-shaped scratch marks of the *Cruziana*.

Question 10.1

Those with multiple skeletal elements (echinoderms and arthropods) would have been particularly prone to break-up, as the soft tissues holding them together decayed, unless they were rapidly buried (see Section 9.3.6). The paired valves of bivalves would likewise have been liable to separation, particularly since the ligament tends to open the shell on death as the muscles decay (Section 9.3.1). Brachiopods would have been less likely to do so. In fact many have a ball-and-socket arrangement of the teeth and sockets, so that the valves can be separated only if the teeth are fractured.

Question 10.2

Infaunal animals such as the bivalve (FS B) and the irregular echinoid (FS J) would have had high intact fossilization potential, provided they died in their burrows; several centimetres of sediment would have had to be eroded to exhume them. Occasional shallow burrowers such as the trilobite (FS G) might sometimes have been preserved intact, though the chances of exhumation and break-up would have been greater than with the fully infaunal specimens. The epifaunal forms would have tended to behave differently according to skeletal construction: the gastropod (FS K) has only a single skeletal element, which would have had a high probability of intact preservation, but the crinoid (FS H) would have had a low potential because of its multicomponent skeleton.

Question 10.3

The internal mould of *Circomphalus* would show, in opposite (positive) relief to the shell interior, features such as the adductor and anterior retractor muscle scars, and the pallial line. You can verify this for yourself by gently pressing some softened modelling clay into the inside of the specimen and then withdrawing the mould thus made. (It helps to wet the shell first with detergent.)

Question 10.4

(a) Although the test is largely intact, the spines are all missing, as are also the small plates immediately around the anus (within the genital plates). The small plates around the mouth, as well as the teeth, appear to be missing, too – allowing the test to be filled with sediment – though the oral surface, including the sedimentary fill, is partially obscured by encrusting worm tubes. These points were mentioned in Section 9.3.2.

(b) The most likely hypothesis is (ii). The few weeks' delay before initial burial could account for the loss of spines, teeth etc. (Section 10.1.2), while the encrustation by worm tubes of the oral surface, including the sedimentary fill, points to subsequent re-exposure – a frequent occurrence in shallow marine environments (Section 10.1.7). In (i), we could have expected to see the intact skeleton, with its spines etc. In (iii), by contrast, the plates of the test would also have fallen apart, while in (iv), the test would have sustained some damage, though spines could still have been attached.

Question 11.1

The absence of any benthic fossils and indeed of any bioturbation in the shale (shown by the preservation of the laminations) indicates that conditions were hostile at least to macroscopic organisms at the bottom. The streamlined (nektonic) ammonoids would have lived in the overlying water (Section 9.3.4), and would indicate, in contrast, that conditions there were generally suitable for life. The most likely explanation for the difference in conditions for life would be that oxygen was seriously depleted or absent at the sea-floor but available in the seawater above (hopefully, you recognized this situation as being very similar to that of the Permian deposits with the fish fossils discussed in Section 7). Further support for depletion of oxygen at the sea-floor is provided by the organic and pyrite contents of the sediment, which imply reducing, and thus anoxic, conditions (Section 10.1.2).

Question 11.2

With a thickness of 500 m deposited over 5 Ma, the net rate of deposition was:

$500\ \text{m}/(10 \times 10^6\ \text{yr}) = 5 \times 10^{-5}\ \text{m yr}^{-1} = 5 \times 10^{-2}$ m per 1000 years = 0.05 m per 1000 years = 5 cm per thousand years. For a thickness of 600 m, the rate is 6 cm per thousand years. Note that this is the *net* rate; however, the presence of numerous gaps in the record means that any 5 cm interval taken at random may represent a shorter time (if no gaps are present in it) or a longer time (if they are).

Question 11.3

The valves are gaping open. Had the brachiopod been buried alive, the shell would be tightly closed. The shell of the dead animal thus either lay freely at the surface or became re-exposed. Incidentally, the encrusting tubes themselves do not provide evidence of post-mortem exposure; they could equally well have formed during the life of the brachiopod – which is actually what we are trying to establish.

Question 11.4

(a) The absence of sedimentary lamination in the darker layers, in contrast to the sandy layers, implies that it has been destroyed, and the mottling, consisting of sectioned tube-fillings of paler, sandy material, suggests that intense bioturbation was responsible.

(b) While the bases of the sandy layers are sharp and simple, their tops seem to merge irregularly into the bioturbated layers. This pattern suggests that the sandy layers were deposited directly onto eroded surfaces, but that silty mud deposited on top became thoroughly mixed in with the upper parts of the sand layers as a result of bioturbation.

(c) From the latter finding, we can infer that there was plenty of burrowing infauna, so the bottom waters cannot have been anoxic.

(d) Hence the lack of burrowing in the sandy layers must reflect rapid deposition, presumably from the waning storm currents that generated the HCS structure, while the bioturbated layers reflect much longer periods of relative quiet, when burrowing obliterated any current-generated sedimentary structures. These sediments were therefore probably deposited above storm wave-base (explaining the sandy layers with HCS), but below fairweather wave-base (explaining the predominance of bioturbation in the intervening layers). These environmental limits were discussed in Section 4.2.5.

Question 12.1

They are both limestones but their lithology and fossil content is distinctly different, therefore they represent different facies. RS 21 is a biomicrite with gastropods and bivalves and RS 24 is a biosparite full of echinoid fragments.

Question 12.2

(a) The bold black line X–X′ represents a time plane where facies co-exist together; it is a moment in time when the top of the beach (i.e. all four facies), moved from their old position to their new position.

(b) The thin black lines labelled A–A′, B–B′, C–C′ and D–D′ represent individual facies boundaries.

(c) Lines W–W′, Y–Y′ and Z–Z′ also represent former positions of the top of the beach; like X–X′, they are also time planes, Z–Z′ being the oldest and W–W′ the youngest.

Question 12.3

For RS 10 and RS 20 your specimen may have a slightly different grain size than mine but this is what I found: RS 10 has a grain size of about 750 μm and because it falls in the middle of the grain size range for coarse-grained sand shown on the grain-size scale it would plot exactly on the coarse-grained sand graduation of the graphic log. RS 20 has a grain size of about 250 μm; it would therefore plot half way between the fine- and medium-grained graduations. RS 27 has no visible grains in it, and is below the resolution of the grain sizes shown on the small card. It is in fact a mudstone and would therefore plot just to the right of the clay graduation on the graphic log.

Question 12.4

RS 21 is a biomicrite since it contains bioclasts in a micritic matrix. Most of the bioclasts are not in contact with each other so it would plot on the micrite with grains, matrix-supported, graduation. RS 22 is an oosparite so it has no matrix (just sparite cement) so it would plot on the 'grains, no micrite' graduation.

Question 12.5

If an area of land is subsiding and sea-level is rising, it is likely to produce less sediment as the rivers have less material to erode due to the decrease in elevation of the river profile and the lesser amount of land area.

Question 12.6

If the climate becomes cooler and wetter; (i) the area of evaporite deposition will become less and eventually diminish because the evaporites require warm dry conditions for deposition, (ii) a siliciclastic coastline adjacent to a poorly vegetated land-mass will receive more sediment because the increased run off of water from the land will transport more sediment down to the coastline.

Question 12.7

(a) A completed graphic log is shown in Figure A12.1.

(b) Beds A and B both contain marine faunas. This suggests that they were deposited in a marine environment. Bed C

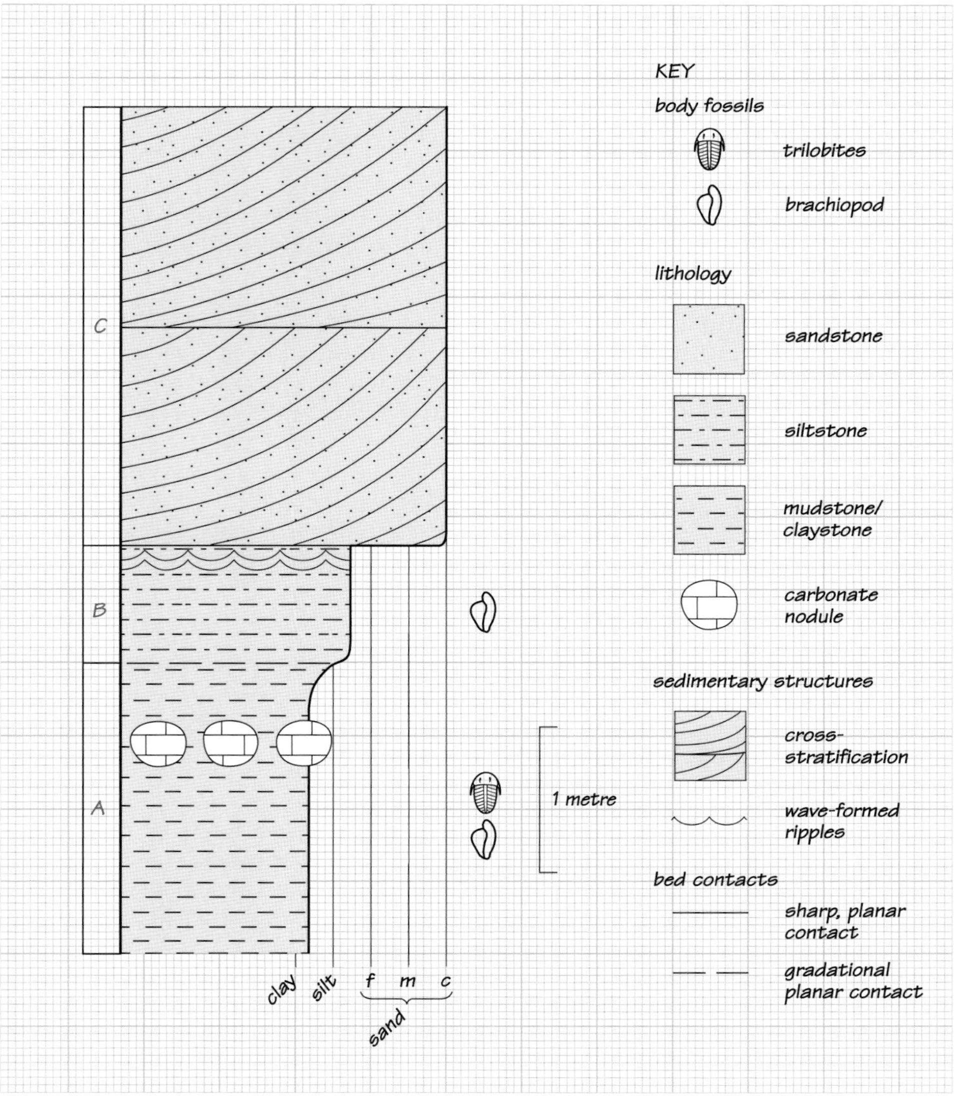

Figure A12.1 A completed graphic log for use with Question 12.7.

contains frosted sand grains, which suggests that it has been deposited by wind, and large-scale cross-stratification that is also common in wind-blown sediments. Bed C therefore most likely represents a non-marine environment.

(c) The contact between Beds B and C is sharp and could be termed an unconformity (see Block 1 Section 2.4.2), or an erosion surface. The sharp contrast between marine and non-marine beds suggests that there are strata missing.

Question 13.1

(a) No. If precipitation shows seasonal peaks, then the level of the water table the surface will show seasonal variations too. When there are marked seasonal variations then streams may flow only intermittently, or else start to flow higher up a valley in the wet season than in the dry season, leaving the upper part of the valley a 'dry' valley in the dry season.

(b) No. Rocks with a large amount of interconnected pores will allow more infiltration than rocks with a small amount of poorly connected pores. So the water table will lie closer to the surface when the drainage basin is underlain by, say, granite (which has very few pores) or clay (which has poorly connected pores) than when the underlying lithology has a large amount of interconnected pores, such as coarse-grained and poorly cemented sandstone.

Question 13.2

(a) Thick flood plain deposits are not very common, because at times of high discharge further braiding or switching of channel direction will occur rather than flooding, and whatever flood deposits are formed are liable to be eroded away later.

(b) Organic debris will be rare, too. Because sand bars and channels change location frequently, it does not allow much chance for vegetation to grow. Also, many of the geographical localities that favour braided stream development (e.g., semi-arid regions and glaciated highlands) are climatically unsuited to extensive vegetation.

Question 13.3

(a) As meanders migrate gradually down river, then by tracing the successive positions of a meander, such as the one shown at the top of Figure 13.15, we can work out that channel D is the oldest (1765), channel B is 1820–30, channel C is 1881–1893 and channel A is 1930–1932.

(b) Feature X is an ox-bow lake.

(*Comment:* The shift in meander position between 1830 and 1881 (51 years) is greater than during any of the other time periods – 1765 to 1820 (55 years), 1893 to 1930 (37 years). This suggests that channel C must have cut across the neck of the meander.)

(c) Point Y will be a point bar. The sedimentary structures will include lateral accretion surfaces, trough cross-stratification and small-scale cross-stratification formed by migrating ripples.

Question 14.1

Morphological feature	Sedimentary facies
sand dunes	cross-stratified sandstones
playa lake	mudstones with desiccation cracks, evaporites
wadis	coarsely stratified conglomerates, breccias and coarse-grained sandstones
alluvial fans	coarsely stratified conglomerates, breccias and coarse-grained sandstones

Question 14.2

The other features to look for in large-scale cross-stratified sandstones to determine if they are deposited by aeolian rather than tidal processes are: frosted grains; very high angle to the cross-stratification; lack of marine fossils; inverse grading within individual beds; lack of other features diagnostic of tides, such as mud-draped ripples and tidal bundles.

Question 15.1

	Wave	Tide	Fluvial
oscillatory flow	✓		
unidirectional flow		✓	✓
bidirectional flow over 12 hours		✓	

Question 15.2

	Wave	Tide	Fluvial
herring-bone cross-stratification		✓	
current-formed ripples	✓	✓	✓
wave-formed ripples	✓		
cross-stratification	✓	✓	✓
tidal bundles		✓	
planar stratification	✓	✓	✓
trough cross-stratification	✓	✓	✓
planar cross-stratification	✓	✓	✓
mud-draped ripples		✓	
climbing ripples	✓	✓	✓

Question 15.3

To form a strandplain, it is necessary to have an abundant supply of sand or gravel, some tidal action and strong waves.

Question 15.4

Only the tube which descends below the top of the sediment is likely to be preserved. The worm cast will probably be washed away by the next wave that washes over the beach.

Question 15.5

HCS is diagnostic of the offshore transition zone and this is defined as the zone below fairweather wave-base and above storm wave-base. It is rare that HCS forms anywhere other than in the offshore transition zone.

Question 15.6

(a) If the mouth bar sediments are reworked by waves, diagnostic features will include wave-formed ripples, and cross-stratification indicative of deposition of sediment on both sides of the ripple.

(b) If the mouth bar sediments are reworked by tides, the sedimentary structures may include tidal bundles, herring-bone cross-stratification (although this is not diagnostic) and other evidence for periodically changing current strength, including ripples with mud drapes.

Question 15.7

The foreshore will be steeper on a pebbly fairweather-dominated beach because storm currents tend to flatten the foreshore and coarser-grained sediment forms steeper-gradient beaches than fine-grained sediment.

Question 16.1

RS 22 and RS 10 are both coarse-grained, but in RS 22 (an oolite) the grains have been precipitated chemically and are therefore not related to the strength of the current although their precipitation must have involved precipitation in agitated water. The coarse-grained nature of RS 10 together with the good sorting and rounding suggests fairly strong wind currents.

Question 16.2

RS 18 contains feldspar indicating that it has not moved far from its source area because the feldspars quickly break down. RS 21 contains delicately preserved gastropods in micrite, which suggests that this material, too, has not moved far from its source.

Question 16.3

(a) The succession represents a shallowing-upward succession because shallower-water facies are shown near the top of the graphic log.

(b) The fact that successively more proximal facies lie on top of more distal ones means that the carbonate ramp system must have prograded seaward.

Question 16.4

The mudmound facies forms isolated 'islands' within the distal micrite facies. See Figure A16.1.

Question 16.5

Using the scales on the Figure, the dip on the ramp is 20 m in 3.25 km which is equivalent to a dip of *c.* 0.35°. It is very shallow indeed.

Question 16.6

A graphic log typical of progradation of the rimmed shelf carbonate platform and derived from the cross-section shown in Figure 16.5a is shown in Figure A16.2. We do not expect yours to be identical to ours, but it should be broadly similar.

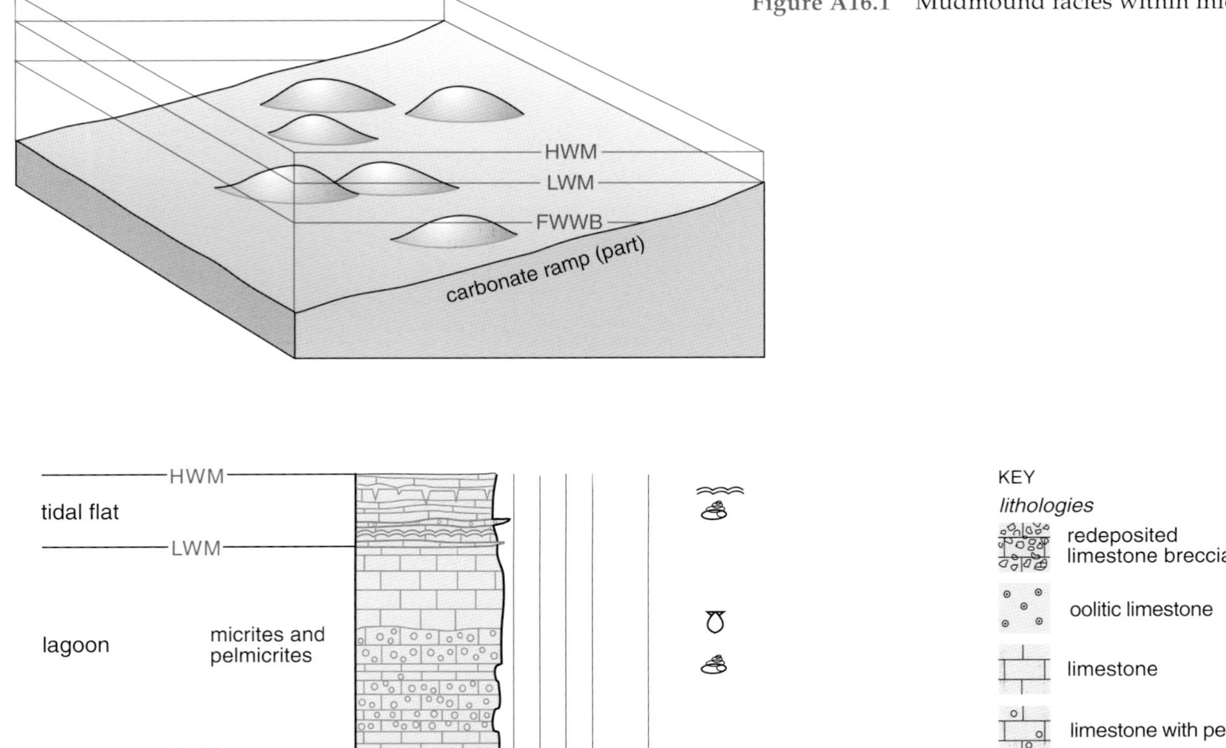

Figure A16.1 Mudmound facies within micritic facies.

Figure A16.2 Completed graphic log of a typical prograding rimmed shelf carbonate platform.

Question 16.7

RS 21 is most likely to have been deposited in a lagoon since it contains only one species and has a micritic matrix indicating conditions were suitable for sediment to be deposited from suspension. The rock shows no signs of strong wave or current activity because the bioclasts are intact and not broken up. RS 22 and RS 24 are most likely to have been deposited as part of carbonate sandbodies since they both contain grains in a sparite cement. In RS 24, the original bioclasts have been broken up and brought in, suggesting current and or wave activity since they are all disarticulated and crinoids would not live in such high energy conditions. In addition, crinoids could not tolerate waters with abnormal salinity like those that might be present in the lagoon (Table 11.1).

Question 16.8

(a) The current was flowing from right to left, because the cross-stratification, representing the steep slope of the migrating bedform, is dipping towards the left.

(b) Plates 16.3 and 16.4 are evaporites in different crystal forms and they have both been deposited by chemical processes.

(c) No, you would not find evaporites in the environments illustrated in Plates 16.5, 16.6 and 16.7 because all of these contain vegetation indicating a humid climate whereas evaporite deposition is favoured in an arid climate.

Question 17.1

(a) Turbidity currents behave like fluid flows because all of the sediment is kept in suspension by turbulence (Section 4.3.3).

(b) Debris flows behave plastically (because of the greater amount of sediment) (Section 4.3.2).

Question 17.2

(a) The sediment in a turbidity current is supported by fluid turbulence (Section 4.3.3).

(b) The sediment in a debris flow is supported by the strength and buoyancy of the matrix (Section 4.3.2).

Question 17.3

(a) The lower flow regime is characterized by planar deposition, ripples and dunes; these will give rise to planar stratification and two scales of cross-stratification respectively.

(b) As the current speed slows down, planar deposition (upper flow regime) would form first and then ripples.

Question 17.4

As the current slows, the coarsest-grained sediments will be deposited first because they are heavier. The succession from the base would be pebbles, granules, coarse-grained sand, fine-grained sand with clay at the top.

Question 17.5

The sedimentary structures in Figure 17.5 are flute casts (Section 4.3.3). As such, they represent the underside of the base of a Bouma Sequence (probably division A but it could be B or C if the Bouma Sequence is incomplete). The current was flowing from right to left because the deepest and narrowest part of the flute is carved out by the eddy first.

Question 17.6

Divisions A–D represent a much shorter deposition time than division E. Divisions A–D are deposited rapidly from a turbidity current whereas Unit E settles out slowly, the subdivision E(t) being composed of fine-grained sediments in suspension in the turbidity current, with E(h) being deposited from hemipelagic material over time and therefore representing the slowest deposited unit of all.

Question 17.7

Division E is the most likely division to be bioturbated because: (i) it is the only part that is deposited slowly; any fauna around when the turbidity current was actually passing would have been swept up with it; (ii) division E is the part that is most likely to contain organic matter, which is generally fine-grained, so if there are any trace fossils they would be confined to this division; (iii) as discussed in the answer to Question 17.5. Division E represents a relative pause or slowing down in deposition; it therefore might allow time for the fauna to recolonize the sea-floor.

Question 17.8

Only siliceous sediments will accumulate below the CCD because the carbonate particles will dissolve.

Question 17.9

Depositional process	Sedimentary deposits
sediment gravity flow	turbidites, debris flows, mudflows
contour currents	contorites
pelagic deposition	fine-grained biogenic sedimentary deposits
direct chemical precipitation from seawater	minerals, e.g. pyrite, manganese

Question 17.10

Submarine fans are more likely to form during sea-level fall, because the rivers on the continent and deltas will have to re-adjust their profile by eroding more sediment (Section 12.1). This will result in an increase in sediment supply to the continental shelf and hence to the base of the continental slope to be deposited as submarine fans.

ACKNOWLEDGEMENTS

Evelyn Brown, Angela Coe, Dave Rothery, Peter Skelton, Bob Spicer, Ros Todhunter and Chris Wilson (Open University) are thanked for providing most of the photographic Figures and Plates in this Block. Jim Ogg (Purdue University, USA) is thanked for allowing us to reproduce some particularly spectacular photographs of aeolian and carbonate features. John Taylor (Open University) is thanked for much patience and help in drafting the graphic logs and block diagrams. Finally, we wish to acknowledge the contribution made by Simon Conway Morris to the *Fossils* Block of the previous *Geology* course, S236, which has formed the basis of parts of Sections 7–11.

Grateful acknowledgement is also made to the following for permission to reproduce material in this Block:

Cover image copyright © Derek Hall; Frank Lane Picture Agency/Corbis; *Figure 4.2c* from *SAND* by Siever © 1988 W. H. Freeman and Co., used with permission; *Figure 4.3* Prothero, D. R. and Schwab, F. *(1996) Sedimentary Geology* © 1996 W. H. Freeman and Co., used with permission; *Figures 4.5 and 4.22* Fritz, W. J. and Moore, J. N. (1988) *Basics of Physical Stratigraphy and Sedimentology*, John Wiley & Sons, Inc.; *Figures 4.7 and 4.8* Sawkins, F. *et al.* (1978) *The Evolving Earth: A Text in Physical Geology*, 2nd edn, Crowell Collier and Macmillan Ltd; *Figure 4.14* adapted from Ashley, G. M. (1990) 'Classification of large-scale subaqueous bedforms: A new look at an old problem', *Journal of Sedimentary Petrology*, **60**(1), Society for Sedimentary Geology; *Figure 4.15* adapted from Tucker, M. E. (1991) *Sedimentary Petrology: An Introduction to the Origin of Sedimentary Rocks*, Blackwell Science Ltd; *Figure 4.16* adapted from Dyer, K. (1986) *Coastal and Estuarine Sediment Dynamics*, copyright © 1986 John Wiley & Sons Ltd; *Figure 4.17* Stride, A. H. *et al.* 'Offshore tidal deposits: sand sheets and sand bank facies', in Stride, A. H. (1982) *Offshore Tidal Sands*, Chapman & Hall; *Figure 4.21b* Walker, R. G. 'Foothills and plains in the area between', in Field, B. C. and Drumheller, A. (1982) *Eleventh International Congress on Sedimentary Field Excursion Guide Book*, International Association of Sedimentologists; *Figure 4.23a* courtesy of Spectrum Colour Library; *Figure 4.25* adapted from Lowe, D. R. (1982) *Journal of Sedimentary Petrology*, **52**, Society for Sedimentary Geology, reprinted by permission of SEPM; *Figure 4.26* J. Best, University of Hull; *Figure 4.27b* D. E. B. Bates, University of Wales; *Figure 4.27c* courtesy of J. E. Saunders; *Figures 5.1a and 5.2a* Roger Till; *Figure 5.3* courtesy of R. N. Ginsburg; *Figure 9.3* from Raup, D. M. and Stanley, S. M. (1978) *Principles of Palaeontology*, copyright © 1978, 1971 W. H. Freeman and Co., used with permission; *Figure 9.6* Wilbur, K. (1964) 'Shell formation and regeneration', in *Physiology of Mollusca*, **1**, pp. 243–282, Academic Press; *Figure 9.8a* Trueman, E. R. (1975) *The Locomotion of Soft-Bodied Animals*, Edward Arnold; *Figure 9.8b* from Symposium of the Zoological Society of London, No. 22; *Figure 9.9* Smith, A. B. (1980) 'Structure, function and evolution of tube feet and ambulacral pores in irregular echinoids', *Palaeontology*, No. 23, The Palaeontological Association; *Figures 9.10 & 9.19* Clarkson, E. N. K. (1980) *Invertebrate Palaeontology and Evolution*, Allen and Unwin, reprinted by permission of Kluwer Academic Publishers B. V.; *Figure 9.18* Kennedy, W. J. and Cobban, W. A. (1976) *Special Papers in Palaeontology*, **17**, The Palaeontological Association; *Figure 9.21* Brower, J. C. (1973) *Palaeontographica Americana*, Paleontological Research Institution, Ithaca, NY; *Figure 9.27* Raup, D. M. and Seilacher, A. (1969) 'Fossil foraging behaviour: computer simulations', *Science*, **166**, 21 Nov., pp. 994–5; *Figure 9.28b* Kent Chamberlain, C. (1971) 'Morphology and ethology of trace fossils', *Journal of Paleontology*, **45**, No. 2, The Society of Economic Paleontologists and Mineralogists and The Paleontological Society; *Figure 9.31b* Basan, P. B. and Scott, R. W. (1979) 'Morphology of *Rhizocorallium* and associated traces …', in *Palaeogeography, Palaeoclimatology, Palaeoecology*, Elsevier Scientific Publishing; *Figure 9.33* Frey, R. W. and Seilacher, A. (1980) 'Uniformity in marine invertebrate ichnology', *Lethaia*, **13**, Scandinavian University Press; *Figure 10.1 Iowa Geological Survey Report of*

Investigations, No. 5, Iowa Department of Natural Resources, Geological Survey Bureau; *Figure 10.4* Tucker, M. E. (1991) *Sedimentary Petrology: An Introduction to the Origin of Sedimentary Rocks*, Blackwell Science Ltd; *Figure 10.6* Fortey, R. (1991) *Fossils: The Key to the Past*, Natural History Museum; *Figure 10.7* Dave Martill, University of Portsmouth; *Figure 11.1* McKerrow, W. S. (1978) *The Ecology of Fossils*, Duckworths, reproduced by permission of the author; *Figure 11.3* Ager, D. V., *Principles of Paleoecology*, McGraw-Hill; *Figure 11.5* Fursich, F. T. and Palmer, T. (1984) 'Commissural asymmetry in brachiopods', *Lethaia*, **17**, Scandinavian University Press; *Figure 13.14* Emmons, W. H. (1960) *Geology: Principles and Processes*, copyright © 1960 McGraw-Hill Book Co. Inc.; *Figure 15.4* Walker, R. G. and James, N. P. (eds) (1992) *Facies Models: Response to Sea Level Change*, Geological Association of Canada; *Figure 16.1* adapted from Jenkyns, H.C. (1986) 'Pelagic environments', in H. G. Reading (ed.) *Sedimentary Environments and Facies*, Blackwells; *Figures 16.2, 16.17 & 17.8* Tucker, M. E. and Wright, V. P. (1990) *Carbonate Sedimentology*, Blackwell Scientific Publications; *Figure 16.10b* Briggs, D. E. G. and Crowther, P. R. (eds) (1990) *Palaeobiology: A Synthesis*, Blackwell Scientific Publications; *Figure 16.11* Purdy, E. G., *American Association of Petroleum Geologists Bulletin*, Vol. 58, American Association of Petroleum Geologists; *Figure 16.12* Lees, A. (1964) *Philosophical Transactions of the Royal Society of London*, **247**, No. 740, July, The Royal Society of London; *Figure 18.3* Sadler, P. M. © 1981 University of Chicago Press.

INDEX

Note: page numbers in **bold** are for terms that appear in the *Glossary* while page numbers in *italics* are for terms contained within Figures. *Pl.* indicates Plate numbers.

abnormal growth 134
aboral surface **86**
abrasion 24, 123
abyssal plain 202, 205
adaptation **69**, 70, 90
adductor muscle scars **83**
adductor muscles **82**, 83, **97**
aeolian **18**, 24, 25, 28
  bedforms 42–45, *43*, *44*
  cross-stratification 44
  deposition 27
  dunes 43, 44
  ripple 42
  transport 20, 21
aerobic
  bacteria 120
  decay **120**
algae 78, 79, 100, 105, 122, 195
algal and bacterial mats 141, 184, 189
alluvial **139**, 150
  environments 150–161
  fans **160**, 162, 164, *pl. 13.3*
  processes 177
aluminosilicate minerals 59
amber 127
ambulacra **87**
ambulacrum **87**, 88
ammonites 95, 105
ammonoids 71, *71*, **73**, 93–95, *93*, *95*, 133
amphibians 105
amphibole 12, 13
anaerobic
  bacteria 120, 127
  decay **120**
  respiration 121
analogy **70**, 93, 130
anemones 104
anhydrite 54
annelids 173
anoxic **13**, 54, 57, 68, 120, 127, 174
antecedent topography **194**, 198
anterior **81**
Anthozoa **104**
antidunes **33**, *33*, 34, 173
anus 81, 87, 90, 94, 97
appendages 101
aragonite 54, 81, 85, 124, 126
arkoses 63
arthropod moults 118
Arthropoda **104**
arthropods 113, 133, 173, 174
articulate brachiopods **96**
asexual reproduction 100
assemblages of trace fossils 114
Asteroidea **111**
atolls 193

backshore *166*, **167**, 173, *pl. 15.3*
backwash **172**, 173
bacteria 79, 120–121, 127
bacterial and algal mats 188
bafflers 184
Baltic Sea 134
barchan dunes **163**, *pl. 14.1*
barrier island **139**, 175–177, *176*, *pl. 15.5*
bars **158**, 179
baryte 126
beach **167**, *166*
bedded cherts 53
bedforms **30**–45, *33*, 56
bedload **20**–23, *20*, 22, 26, 30, 38, 151, 178
  deposits **26**, 27, 29
  deposition 29
  transport 20, 26
beds 29
behaviour 107
belemnites 123
benthic 195
benthos **78**, 192
berms **172**
bilateral symmetry **74**, 81, 85, 90
binders 184, 193
bioclastic debris 194
bioclasts 52, 54, 55, 64, 184, 192
bioerosion 184
biogenic 56
  limestones 57
  sediments **52**
biological
  attack 122
  weathering 10
biomicrites **64**, 188
biosparites **64**, 188, 192
biostromes **192**, 194, 196
biota **5**, 184
biotite 13
bioturbated/bioturbation **107**, 134, 174, 179
birds 105
bivalves **74**, 77, 81–85, 96, 97, 103, 111, 118, 125, 126, 130, 133, 173, 188, 196
body fossils **66**, 72–106
bodyplans **70**, 90, 99, 102, 103, 110
bone bed 124
boring 85, 92
  bivalves 133
  organisms 122, 196
borings 66, 68, 106, 132
boulders 62
Bouma Sequence **205**–208, *206*, *207*
brachiopods **74**, 78, 96–97, *96*, 104, 107, 119, 126–7, 130, 133, 136, 174
braided **139**, 153
  rivers 157–159, *158*, *pl. 13.2*
braiding **153**
branch 101
branching arms 97
breakdown of shells 122
budding 100
build-ups **192**–196
burial 99, 105, 118–120, 124
burrowers 78
burrowing 90, 92
  bivalves 84, 132–3
  crustaceans 110
  cycle 84
  suspension feeders 85
burrows 66, 68, 106, 110, 113

Ca-feldspar 13
calcite 60, 85, 98, 124, 126
  cements 60
calcium
  carbonate 124
  hydroxyapatite 124
Cambrian 102
carbonaceous fossils 105
carbonate 152, 183
  build-ups 192–196
  classification 63–64
  cement 60
  compensation depth (CCD) **124**, *124*
  deep marine 212–213
  platforms **139**, 187, 192, 197–200, 205, *pl. 16.5–16.6*
  ramp 197
  sandbodies 192
  sediments 53, 54
  shallow marine 183–201
  siliciclastics 186, *pl. 16.1*
Carboniferous Period 177, 197
cassiterite 126
casts *125*, **126**
catastrophic burial **119**, 135
cellulose 121
cement 59, 192
cementation 59–61, 193
cemented bivalves 132
cephalopods **93**, 103, 123, 133
chalk 89, 106, 126, 212, *pl. 17.3–17.4*
channels 152–153, 192
charcoal 105

checklist for sedimentary rocks *7*
chemical
  attack 124
  weathering 11–14
chemofossil **66**
cherts 57, 61, 212, 213
Chondrites *106*
Chordata 102, 105
cilia **79**, 83, 91, 96
Circomphalus 81–85, 125
class(es) **102**
classification
  of fossils 70, 72–117
    by habitats 78
  of sedimentary rocks 62–64
  table for marine invertebrates 76
clay 23, 28, 59, 62
clay minerals 12
climate 140–141, 152–153, 192
climbing ripples **31–32**, *32*
Cnidaria **104**, 174
coal 54, 57, 121, 177, 179
coarsening upwards 27
cobbles 62
coccolithophores 212
coccoliths 212
cockle 134
cohesive 26
cohesiveness **23**
collagen **86**, 98
collectors **79**, 85
colonial **73**
  growth 105, 195
colony **100**
communities 131
compaction **58**–60, 111, 128
compositional
  immaturity 63
  maturity **13**, 14, 28
compound eyes 101, 104
concretions 61
consilience **9**, 25, 45, 68, 69, 115
consolidation 23
consumers 131
continental
  rise **202**
  shelf 34, 46, **166**, 167, 169, 174, 175, 192
  slope 46, **202**, 205
contour currents 204
controls
  on deposition 140–142, 184–185
convergent evolution 64, 192
coral reefs 193
corallites **73**, 100
corals **73**, 99–100, *99*, 104, 133, 141
crabs 77, 100, 104
crayfish 101
Cretaceous 89, 196
crevasse splays **155**, 157, 179
Crinoidea 103
crinoids **75**, 97–99, *98*, 119, 133
  feeding posture 98
cross-stratification **30**, 31, 33–36, *35*, 39, 40, 56, 156, *162*, 164, 174, 177, 179, 192, *pl. 4.1–4.3, 4.5, 4.6–4.9, 16.2*
cross-stratified 156
crustaceans 101, 102, 104, 121
Cruziana 111–115, *112*
crystalline texture 7, 14
current lineations **33**, 173
current-formed ripples **30**–31, *31*, 40, 43, 156
current-oriented belemnites 123
currents 151
  deep-sea 202–205, *204*
  longshore 170–171, *171*
  tidal 171
cuttlefish 93

death 118–120
debris flows *45*, **46**, 204, 209, *pl. 17.1–17.2*
decay 120–121
deep sea **139**, 202–214, *203*
deep-freezing 127
defence posture 102
delta **139**, 177–181, *178*, *180*, *181*
  front **179**
  plain **178**
density 18, 46
  currents *45*, **46**
deposit feeders **79**, 85, 89, 107, 108, 110, 113, 115, 127, 131
deposition **20**, 25–51
deserts **139**, 162–165
desiccation 127
desiccation cracks **164**, 188, 189
detritus **79**, 111
  collectors 94
development 89
Devonian 105
diagenesis **58**, 60, 61, 118, 125–128, 213
diatoms 212
diductor muscles **97**
Diplocraterion 113–114, *114*, 134
discharge **152**, 158, 179, 217
disintegration 120–121
dissolution *124*
distal **140**
distributaries **179**
dolomite 60, 61, 127
dorsal **81**, 90
  valve 96
drainage basin 150, 152
drift 44, *44*
drill hole 119
dunes **32**–34, *43*, 44, 162–163, *163*, 171, 173, *pl. 4.4*

ebb tide 36
Echinodermata 103
echinoderms **75**, 86, 97, 110, 119, 133, 135
echinoids **75**, 77, 80, 85–90, *86*, 121, 123, 131, 133, 134
  anatomy 86
  endoskeleton 86
  spine 87
Echinoidea *103*
ecology *77–80*, 130
enamel 105
encrusting bivalves 133
endoskeleton **86**
energy of environment **27**, 29, 62
environment 139–140
environmental models 148
epifauna **77**
erosion 17, 22, 120, 140, 162
evaporation 162
evaporites 54, 152, 164, 188, *pl. 16.3–16.4*
evolution 67, 69, 72, 89, 105
evolutionary convergence 69, 192
exceptional preservation of soft tissues 127
exhalent
  currents 97
  siphon 83
exoskeleton **75**, 81, 86, 100, 104
external mould **125**
eyes 101
  compound 101, 104

fabrics 7, 27–29
facies 142–149, 197–200
  succession 190
faecal
  pellets **53**, 57, 131, 174, 184
  strands 67, *pl. 7.1*
  string 108
faeces 91
fairweather wave-base (FWWB) **41**, 42, 134, 167, 174, 197
family **102**
fauna 179
feeding
  classification 79
  methods 78–80, *79*
  posture of a crinoid 98
feldspars 12, 13, 59, 60, 63
feldspathic sandstone 63
filtration fan 98
final burial 124
fining upwards 27
fish 78, 94, 105, 127
  scales 124, *pl. 7.1*
five-fold symmetry **75**, 85, 87, 110
flints 61
flocculation **141**, 174, 177, 178
flood
  plain **150**, 155
  tide 36
flora 179
Florida Keys 198, 200
flotation 121
fluids **18**, 19

flute marks **47**–48, *48*
fluvial **28**, 150–161, *169*, 167
  erosion 150
foot **83**, 90
footprints 106
foraminifers **105**, 106, 212, 213
foreshore **167**, *166*, 172, 173, 197
fossils 7, **66**, 66–138
  assemblages 132
  fish 67
  trace 106–117, *114*
fossilization 118–128
  exceptional preservation 127
  potential **121**, 130
  spirit level 125
fragmental texture 14, 54
framebuilders 184, 193
frost shattering 10, 11, *11*, 14
frosting 24
FS A 73, 90–92, 115
FS A–H 75
FS A–M 77, 80
FS B 74, 115, 122
FS C 74, 75, 85–90, 116, 129, 136
FS D 73, 93–95, 115, 125
FS E 73, 99–100, 116
FS F 74, 78, 96–97, 116
FS G 74, 75, 100–102, 111, 112, 116, 117
FS H 74, 75, 97–99, 116, 122
FS I 81–85, 97, 103, 122, 125
FS J 85–90
FS K 90–92, 115, 122
FS L 78, 96–97
FS M 99–100
FS N 109, 110, 117
FS O 110, 117
FS P 111, 112
FS Q 136, 137
fungi 121, 122, 127
funnel **94**

gas 177
gastropods **73**, *77*, 90–92, *90*, 126, 130, 133, 184, 188
  anatomy 90
  predator 119, 92
genus **102**
geological time 215
gills **83**, 90, 91, 94, 101, 103
  in gastropods 91
glacial
  sediments *pl. 2.4*
  transport 49
graded bedding **27**
grain
  flows *43*, 44, **45**–46, 163
  morphology 7, *pl. 2.6*
  size 18, 145, 147
  sorting 7
  surface texture 7, *pl. 2.6*
grain-size 62, *62*, 63, 145–147
grain-supported 27, 28, *28*, 46

granules 62
graphic logs **144**–148, *146*, *157*, *159*, *175*, *176*, *181*, *190*, *197*, *206*, *211*, *215*, *227*, *229*
graptolites **105**
graptoloids 105
grazers **79**, 100, 123, 196
grazing 135
Great Barrier Reef 193
greywackes 63
growth lines **82**, 126
gutter casts 171
gypsum 54, 164

haematite 126
halite 54, 164
head 81, **90**, 94
head-shield **75**, 112
helical flow 154
hemipelagic **206**, 212–213
herring-bone cross-stratification **36**, 192, *pl. 4.5*
holdfast **97**
homologies **70**, 80, 89, 93, 100, 103
homologous structures 75
hummocks 41, 42
hummocky cross-stratification **41**, *41*, 174, 192, *pl. 4.8*
hydrogen sulfide 121

ice sheet 49
impact ripples *42*, 42–**43**
in situ **17**, 53, 55, **132**
inarticulate brachiopods **96**, 133
indented growth lines 91
infauna **77**, 107
  animals 107
  crustaceans 131
inhalent
  current 97
  siphon 83
insects 101, 104
interambulacra **87**
interambulacrum **87**
internal mould **73**, 125
intertidal 105, **167**, *166*
iron oxide cement 60
irregular echinoids **89**, 120
isostatic 179

jaws 94
jellyfish 77, 78, 104
jet propulsion 94

K-feldspar 13
keel **93**
Key Largo 200
key system 72
Kimmeridge Clay Formation 135
kingdoms **102**
Kosmoceras 95

ladder-back ripples 41
lag deposits **23**, 124, 135, 153
lagoons **175**, 191–192, 198
lakes 179
lamina **29**
laminae **29**, 30, 67
laminar flow **19**, *19*, 29
laminated structure 29
land plants 105
larvae 78
lateral
  (anatomy) **81**
  accretion surfaces **156**, *156*
levées **155**, 156, 179
life position **130**, 132
ligament **82**, 83, 97
  groove **82**
lignin 121
limestones 52–57
  classification 63–64
limpets 90
linear dunes 163
lips 83
lithification **58**–61
lithified 60
lobsters 100
locally displaced fossil 132
Lockeia 111, 114
longshore current **170**–171
lower flow regime **33**, 34, *34*
lunar tidal cycle 37

macrofossils **66**, 72, 105
mafic minerals 12, 60
mammals 105
mangroves 198
mantle (anatomy) **82**, 90, 103
  cavity *83*, **90**, **96**, 103
  tissue 96
mass
  mortality **119**
  movement 152
matrix 27, 28, 59, 63
matrix-supported 28, *28*, 46, 209
meandering rivers **139**, 153–157, *153*, *pl. 13.1*
meanders **153**, *154*, 155, 179
meshwork skeleton 86
Mesozoic 93, 105
micas 12, 13
Micraster 85–90
micrite **54**, 57, 60, 61, 64, 184, 191
microfossils **66**, 105, 188
migration of burrow systems 113
Mollusca **103**
molluscs 108, 133, 174
morphology **7**, 82
  in fossils **77**
  of coastal zones *166*
  of coastline *169*
  of deltas 178–180
moulds 125
moulting **100**

mouth
  anatomy 81, 87, 90, 92, 94, 96, 97
  bars **179**, *179*
mucus 79, 83
mud
  drapes **37**, *37*, 177, *pl. 4.6–4.7*
  flows *45*, **46**
mudmounds **194**, 197
mudstones 57–58, 61
muscovite 13
mussels 85, 130, 134

natural selection 69
nautilus 71, *71*, 93
nekton **78**
Neogene 81
neomorphism **126**
Nereites 108–111, *109*, 114
nodules 61, 174, 213
normal marine salinity 97, 133
nutrient levels 195
nutrients 195

observation → process → environment concept 5
octopuses 93, 103
offshore transition zone *166*, **167**, 174, 176, 192
offshore zone *166*, **167**
oil 177
olivines 12, 13
ooids 55–57, *55*, 64, 192
oolitic 55
oomicrites **64**, 188
oosparites **64**, 192, *pl. 16.2*
opal 126
oral surface **86**
orbital speeds 38, 41
order **102**
Ordovician 105
organic-rich sediments 174, *pl. 15.4*
organism biology 184
oscillatory currents 184
overgrowths **59**, 61
ox-bow lake **154**–156, *155*
oxidizing environment 25
oxygenated conditions 121
oyster bed 195
oysters 74, 85

paired limbs 104
palaeobiology **67**
palaeocurrent **35**
palaeoecology **130**
palaeoenvironments **8**
palaeontology **67**
Palaeozoic 93, 96, 97, 99, 100
Paleodictyon 109–111, 114
pallial
  line **82**
  sinus **82**
parasites **79**
patch reefs 198
peat 179
pebbles 62, *pl. 2.2*
pedicle 97
pelagic **78**, **212**, 212–213
pelmicrites **64,** 188
peloids **53**, *53*, 55, 64, 184, 191
pelsparite **64**
perforated plate **87**, 98
peritidal **187**–192, 197
Permian 67
Phanerozoic 105
photic zone **108**
photosynthesis 131
photosynthetic bacteria 105
phyla 102, 103
phylum **102**
physical weathering 10
phytoplankton 53, 57
planar
  cross-stratification **35**, *35*
  deposition 33, 34, 40
  stratification **33**, 34, 164, 176, *pl. 15.1–15.2*
  stratified 173
plane of symmetry 74
plankton **78**, 79, 105, 195
plants 79
  spores 105
plates 98
playa lakes **164**
point bar **154**–157, *154*, *155*, *156*
pollen 121
polymerization 13
polymorph 54
polyp **99**
pore pair **88**
posterior **81**
potential energy 152
prawn 78
precipitates 184
predation 135
predators **78**, 85, 92, 94, 102, 118, 122, 135
  gastropods 135
preferred orientation 123
pre-fossilization **124**
pressure dissolution **58**, 59
prodelta **179**
producers 131
progradation **143**, 179, *180*
Proterozoic 105
protractor and retractor muscles **83**
proximal **140**
Pseudodiadema 85–90, 129
pseudofossil **66**, *66*
pyrite 13, 67, 121, 126, 127, 174
pyroxenes 12, 13

quartz 12, 13, 55, 59
quartz sandstone 63

radial symmetry 74, 104
radiolarians 212
radula **92**, 94, 103
rates
  of deposition 215–17
  of weathering 14
red beds **12**, 25, 60
reduction of iron 13
reef 193
reef framework 193
reef-building corals 100, 195
reefs 64, 100, 105
regular echinoids **89**, 120, 123
replacement **127**
reptiles 105
Rhizocorallium 113–114, *113*
ribs **93**, 95
rimmed shelf carbonate platform 197, 198
rip current **171**
ripples 33, 34, 40
  current-formed **30**–31, 40, 43, 156
  ladder-back 41
  mud-draped 168, 177
  wave-formed 40, 168, 174
river valley 152
rounding 24, 28
roundness 7, 24, 28
RS 5 16
RS 10 6–8, 25, 28, 59, 60, 65, 148, 163, 186
RS 12 63, 208
RS 15 58
RS 18 63, 65, 186
RS 19 16
RS 20 148
RS 21 125, 126, 142, 147, 148, 186, 201
RS 22 148, 186, 201
RS 24 97, 98, 142, 201
RS 27 148, 174, 208
rudists 196
runnels 172
Rusophycus 112

salt pan 164
saltation **20**, *20*, 21, 23, 30, 44, 151
sand 62
  drift 44
  waves **34**, 44, 171
sandbodies 170, 175
sandstones 59, 61–3
saturation 54
scales 105
scallops 74, 85
scavengers **78**, 92, 94, 102, 120, 121
scours, 192
scroll 92
sea-anemones 99
sea-fans 104
sea-level 141, 192
sea-lily 75

sea-slugs 90
sea-snails 90
seatearth 152
sea-urchin 75
sea-whips 104
seatearth 152
sediment gravity flows **45**–48, *45*, 151, 160, 202, 205–211
sediment
  producers 193
  transport 18
sedimentary
  rocks 58
  structures **7**, 192
segmented body 104
septa **73**, 93, 94, 99
septum 73
sessile **78**
  benthos **195**
sets **29**, 30, 36
sexual
  dimorphism **95**
  reproduction 100
shales 58
shape 18
shell
  bed **124**, 135
  development 94
  fragments 122
shell-crushing predators 196
shelly marine invertebrates 72
shoreface *166*, **167**, 174, 197
shrimps 104
siderite 60, 61
silica 127
silica cement 59
silicate minerals 13
siliceous
  phytoplankton 53, 60
  skeletal debris 213
siliciclastic
  carbonates 186, *pl. 16.1*
  classification 62–63
  sedimentary rocks 28
  sediments 13
silt 23, 26, 62
sinuosity **153**
siphons **83**, 91
siphuncle **94**
skeletal
  elements 73–76
  frameworks 100
  material 61
  remains 123
slugs 90
snails 73, 90
sockets **82**, **96**
soils 17, 152
solitary **73**
sorting 21, 23, 27, 28, 38
South Florida 197–198
sparite 60–61, 64
species 95, **102**
specimens
  with many skeletal elements 74
  with one skeletal element 73
  with two skeletal elements 74
speed 19
sphericity 7, 24
spicules 105
spines **87**, 89
sponges 77, 105, 122
spores 121
sporopollenin 121
spreite **113**, 134
spring and neap tides 166
squids 77, 78, 93, 94
stalk 97
star dune 163
starfish 85, 111
stinging cells 100, 104
stony deserts 163
storm
  activity 120
  wave-base **41**, 167
strandplains **139**, *145*, 172–175, *172*, *175*
strata 29
stratigraphical correlation 105
stratigraphy 67
stratum 29
stromatolites 56, **105**, 188
stunting 134
subduction 202
submarine
  canyons **202**, 205, 209–10
  fan **209**–211
subsidence 163
subtidal *166*, **167**, 191
subvertical tapering desiccation cracks 189
sudden burial 120
sulfate ions 121
supply of sediment 163
supratidal *166*, **167**
  flats 105, *pl. 16.7*
surface creep **21**, *21*, 23, 42
suspension 20–3, 26, 29, 151, 195
  deposits **26**, 27, 29
  feeders **79**, 94, 102, 105 113, 115
  load **20**, *20*, 30, 38, 151, 178
suspension-feeding worms, 136
sutures **73**, 93, 95
swales 41
swaley cross-stratification 41, *41*, **42**
swallowers **79**
swash **172**, 173
symbiosis 100, 195

tailpiece **75**
taphonomic alteration 194
taphonomy **118**, 132
taxa **103**
taxon **103**
taxonomic hierarchy **103**
tectonic setting 140
teeth **82**, **96**, 105, 124
tentacles 79, 96, 99, 104
terrigenous **212**
test **85, 96**
textural
  immaturity 49, 63
  maturity 27, **28**, 46, 56, 58
texture **7**, 14
Thalassinoides 213
tidal 167
  bundles **37**, *37*, *pl. 4.7*
  currents 37
  dunes 171
  flats **37**, 176, 188–191, *190*, *pl. 2.5*
  processes 176
  range **37**, 167, 170
tides 36–37, 169–170, *170*, 179
time 215–218
time-averaged **135**
tool marks **47**–48, *48*
torsion **90**
trace fossils **66**, 106–117, 119, 131–132, 134, 173, 174, 188
  and water depth 114
tracks 106
trails 106
transport 17–26, 38, 45, 46, 132
transverse dune **163**
Triassic 99
tributaries 152
Trilobita 104
trilobites **75**, 100–102, *101*, 111, 133
trophic pyramid **131**
trough cross-stratification **35**, *35*
TS B 65
TS R 125, 126
TS U 97
TS W 6, 65
tube feet **88**, 98, 111
tubercles **87**
turbidites **205**–208, 215, *pl. 17.1*
turbidity currents 46–48, *47*, 109, 110, 204, 205, *207*, 216
turbulent flow **19**, *19*, 46

umbo **81**, 97
umbones **81**
unidirectional 184
uniformitarian comparison 69, 193
upper flow regime **33**, 34, *34*, 173

vagrant **78**
valley profile 152
valves **74**, 81, 96
vegetation 152, 162, 163
ventral **81**
  valve 97
vertebrates 105
visceral mass **90**

viscosities 18, 19, 20, 46
visual system 101

wadi **164**, *pl. 14.2*
Walther's Law **144**
washover fan 177
water
  in the desert 164
  table **58**, 150
  transported grains 24
  vascular system **87**, 98
wave *38*
  bedform 39, 37–42, *39–41*
  processes 167
wave-base **38** *see also* fairweather wave-base and storm wave-base
wave-formed ripples 39, *39*, **40**, *40*, 174
wave-produced bedforms 37
waves 37, 38, 39, 169, 179
weathering 3–16, *17*, 140, 151, 162
well-oxygenated conditions 107
whelks 90
wind
  in the desert 162–163
  pattern 163
winnowing 23, 38, 124, 204
woolly mammoths 66, 69
worms 77, 85, 107, 108

zooplankton 53, 57, 60